Springer
Tokyo
Berlin
Heidelberg
New York
Hong Kong
London
Milan
Paris

Kenji Kashiwaya (Ed.)

Long Continental Records from Lake Baikal

With 172 Figures

Springer

Kenji Kashiwaya, Dr.Sc.
Professor
Institute of Nature and Environmental Technology
Kanazawa University
Kakuma, Kanazawa 920-1192, Japan

Cover design by Asako Demura

ISBN 4-431-00643-5 Springer-Verlag Tokyo Berlin Heidelberg New York

Library of Congress Cataloging-in-Publication Data

Long continental records from Lake Baikal / Kenji Kashiwaya (ed.).
p. cm.
Includes bibliographical references and index.
ISBN 4-431-00643-5 (acid-free paper)
1. Climatic changes--Russia (Federation)--Baikal, Lake. 2. Baikal, Lake (Russia)--Climate. 3. Lake sediments--Russia (Federation)--Baikal, Lake--Analysis. 4. Global temperature changes. 5. Global environmental change. 6. Paleoclimatology--Cenozoic. 7. Paleoclimatology--Pleistocene. I. Kashiwaya, Kenji, 1948-

QC884.2.C5L66 2003
551.69575--dc22

2003058445

Printed on acid-free paper

Typesetting: SNP Best-set Typesetter Ltd., Hong Kong
Printing and binding: Shinano Inc., Japan
SPIN: 10905685

Preface

Terrestrial sediment records, especially lacustrine sediment records, have some advantages compared with deep-sea sediment and ice-core records. Those advantages include high-resolution environmental records, not only physical and chemical records but also biological-information-rich records. With the appropriate approach and methods, terrestrial sediments can furnish clues of detailed biological evolution on Earth in addition to physical and chemical process-based geological history. Thus we may provide more detailed and different stories than the ones based on information obtained from ocean sediments or ice cores.

Lake Baikal in eastern Siberia, located in the middle latitudinal zone, is highly sensitive to solar insolation, as the entire lake remained uncovered by ice sheets throughout the late Cenozoic era. The sediments include abundant long-sequent information in and around the lake because they have accumulated continuously for at least 10 million years. This indicates that observation-based and process-based information in recent periods can be compared to long-term information in a lake-catchment system. The past is a key to the future, through the present. Paleoinformation can be suitably used for considering the future in this field. In addition, Baikal sedimentation is estimated to have begun 30 million years ago; invaluable information on long-term environmental changes not only in eastern Eurasia but globally may be recorded in the sediments.

In October 2001, an international workshop was held in Ulan Bator, Mongolia, jointly organized by the Institute of Geology and Mineral Resources (Mongolian Academy of Sciences), the Institute of Geochemistry (Siberian Branch of the Russian Academy of Sciences), and the Japanese Association for Baikal International Research Program (JABIRP). The paper and poster sessions were followed by a field excursion around Ulan Bator. The main theme of the meeting was paleoclimatic changes and the evolution of life based on long Baikal sediment core samples. Following the stimulating and productive discussions in that meeting, recent valuable findings have been compiled in this one volume after a full refereeing and editorial process.

Long core samples as well as short ones have been obtained since 1993 by the Baikal Drilling Project (BDP) team. The long cores are named for the year in which they were obtained: BDP-93, BDP-96, BDP-97, BDP-98, and BDP-99. The information from the samples has gradually clarified environmental changes in and around Lake Baikal. BDP-98, in particular, includes information covering more than 10 million years. Our data possibly suggest a gradual continental cooling and much greater fluctuation after approximately 3 million years ago coinciding with global climatic change during the past 10 million years.

Some environmental shifts have been found: a shift at about 8–9 million years ago may relate to the uplift of the Himalayas and the Tibetan Plateau, which initiated the Indian monsoon. Some shifts at about 5–6 million years ago coincided with a "diminution of upwelling" in the Indian Ocean and may be compared to the Messinian salinity crisis. The amplitude of fluctuations increased at about 3.6–4.0 million years ago, suggesting an overall environmental shift at that time. The interval 2.5–2.8 million years ago, with even higher amplitude, may reflect the expansion of ice sheets in the Northern Hemisphere. The interval 1.6–1.9 million years ago may be related to the environmental change at the beginning of Quaternary, and the interval 1.0 million years ago may signal the initiation of full glaciation in the mid-Pleistocene. It has also been ascertained that oceanic long data indicate a trend similar to that in the current Baikal data, with some differences in fluctuation. New Milankovitch-related cycles have been found in the core samples; new genera of diatoms and their evolution have been discovered; steryl chlorin esters have been found to be effective indicators of paleolimnological environments — all suggesting that the Baikal sediments include long high-resolution information on climatic changes and environmental changes. It also has been found that the records may be related to tectonic and structural development of the Baikal Rift Zone and Lake Baikal.

Comparatively short records from Lake Baikal and its surrounding area, discussed in this book, reveal glacial and interglacial transitions and environmental changes that are directly linked to present changes. These records are also available for proper understanding of long-term changes concerning lake-catchment systems. Discussions on observational and current records and a model for sedimentation processes here are of great use for process-based data interpretation. All these suggest that Baikal data should be included in the standard datasets for global environmental changes.

Members of the editorial board were K. Kashiwaya, T. Kawai, G.I. Matsumoto, K. Minoura, H. Takahara, and T. Takamatsu. We hope that this book will provide not only a reflection of the results obtained to date, but that it also will establish a foundation for further research in long-term Lake Baikal studies.

The assistance of the following manuscript reviewers is gratefully acknowledged: K. Chikita, D. Demske, H. Haraguchi, N. Hasebe, M. Imamura, Y. Inouchi, T. Iwashita, D. Janussen, R.W. Jordan, T. Kanematsu, K. Kashima, P. Kershaw, H. Kitagawa, M. Kumagai, D. Lal, F. Masuda, T. Masuzawa, G.I. Matsumoto, K. Matsuoka, K. Minoura, N. Miyoshi, M. Murayama, T. Nagao, M. Nakai, T. Nakano, M. Nishimura, J. Safanda, T. Saito, H. Sakai, H. Simazu, M. Soma, Y. Soma, M. Sugiyama, H. Takahara, N. Takamatsu, A. Takeuchi, A. Tanaka, Y. Tanaka, F. Riedel, T. Uchiyama, and D.M. Williams.

During the editorial process, the editor was assisted in a variety of ways by Y.

Masuda, N. Hasebe, R. Ohbayashi, Kanazawa University (Y. Hayashi, President), number of individuals and organizations. He especially wishes to thank T. Kawai, Y. Masuda, N. Hasebe, R. Ohbayashi, Kanazawa University (Y. Hayashi, President), the Kanazawa University 21st-Century COE Program Project Team (K. Hayakawa, Leader) and JABIRP. The patience and assistance of the editorial staff of Springer-Verlag Tokyo were highly appreciated.

KENJI KASHIWAYA
Kanazawa, Japan
February 2003

Editorial Board

Table of Contents

Contributors

Long-Term Climatic Changes and Solar Insolation

Long-Term Late Cenozoic Global Environmental Changes Inferred from Lake Baikal Sediments

KENJI KASHIWAYA[1], SHINYA OCHIAI[2], HIROAKI TSUKAHARA[3], HIDEO SAKAI[4], and TAKAYOSHI KAWAI[5]

Summary. Long, high-resolution records preserved in Lake Baikal sediments (BDP-98) suggest that the climate of the Asian continental interior cooled gradually, although not steadily. Major climatolimnological fluctuations had long periods of about 2000 kyr, 1000 kyr, or 600 kyr, in addition to shorter 400-kyr and 100-kyr periods. These possibly correspond to Milankovitch parameters of eccentricity, implying that changes in solar insolation were printed as long-term environmental variations in the eastern Eurasian continental interior. The amplitude of these fluctuations began to increase at about 4.0 Ma, and most orbit-related periods have fluctuated more distinctly since that time. In particular, the 400-kyr filtered fluctuations in the Lake Baikal datasets correspond well with 400-kyr filtered insolation. The beginning of larger fluctuations seems to have been related to obliquity-related insolation minima.

Key words. Lake Baikal records, Milankovitch periods, Insolation

1 Introduction

Recent studies of the long-term global environment have demonstrated that long continental records, such as those preserved in lake sediments (Kashiwaya et al. 1998, 2001a) and loess (An et al. 2001; Ding et al. 1999), are highly relevant for reconstructing long-term, high-resolution environmental changes. It has also

[1] Institute of Nature and Environmental Technology, Kanazawa University, Kakuma, Kanazawa 920-1192, Japan
[2] Department of Earth Sciences, Kanazawa University, Kakuma, Kanazawa 920-1192, Japan
[3] Department of Environmental Sciences, Shinshu University, Matsumoto 930-8621, Japan
[4] Department of Earth Sciences, Toyama University, Toyama 930-8555, Japan
[5] Department of Earth and Environmental Sciences, Graduate School of Environmental Studies, Nagoya University, Furo-cho, Chikusa-ku, Nagoya 464-8602, Japan

been shown that continental climate change roughly synchronizes with global climate change. Compared with oceanic records, terrestrial records have the advantage that sedimentation rates are relatively rapid, which allows higher-resolution studies. In addition, terrestrial deposits more often record changes in the immediately surrounding environments, instead of preserving spatially averaged records for very large areas. The disadvantages of terrestrial records are that the exposed surfaces of loess deposits have usually been eroded by water, solifluction, and other processes. Similarly, lake sediments have been influenced by bioturbation, turbidity currents, and other processes. As a result, working with terrestrial records requires a greater degree of care than working with oceanic sediments.

Lake Baikal, in eastern Siberia, is located in a crucial area for reconstructing insolation-related, long-term climate changes, because the area is known to be very sensitive to solar insolation (Short et al. 1991). Lake Baikal sediment samples recently obtained mainly from ridges are relatively undisturbed, preserve a continuous sedimentation record, and have been used to elucidate long-term environmental changes (Baikal Drilling Project Members 1995; Baikal Drilling Project Leg II Members 1997; Baikal Drilling Project Group 2000).

Our analyses of Lake Baikal sediments from Academician ridge (Kashiwaya et al. 1998, 1999, 2000, 2001a) show that climatolimnological fluctuations in Lake Baikal sediments are closely related to global climate changes as seen in paleo-oceanographic records. These fluctuations reflect orbital cycles, including 400- and 100-kyr periods of eccentricity, the 40-kyr period of obliquity, and the 20-kyr period of precession.

We have recently shown that additional, longer periods may be related to orbital movement and to the initial influence of insolation on the Lake Baikal climate system (Kashiwaya et al. 2001b). Studying the initial influence of insolation is crucial to understanding the current long-term climate system, since it might reflect the beginning of major Northern Hemisphere glaciation. It is possible that the initial influence of insolation and Northern Hemisphere glaciation correlate closely with each other.

Major glaciations have been widely discussed with reference to understanding current climate systems (Raymo et al. 1992; Lourens et al. 1996), and an abrupt environmental shift around 2.8 Ma has been a salient topic. This abrupt shift occurs both in Lake Baikal records (Kashiwaya et al. 1998) and in many oceanic records, and various models have been proposed to explain it (e.g., Maslin et al. 1995; Berger et al. 1999). Most explanations assume that the main causes of this climate shift were closely related to tectonism, e.g., the opening and closing of isthmuses, and the uplift of the Himalayas–Tibetan plateau and southwestern North America (e.g., Ruddiman and Kutzbach 1991).

Atmospheric CO_2 concentrations and insolation fluctuations have also been cited as triggers for the climate shift (Raymo 1994; Maslin et al. 1998). However, recent long marine and loess records imply that major glaciation may have begun well before 2.8 Ma, within the 3.6- to 4.6-Ma interval (Haug and Tiedemann 1998;

Rea et al. 1998; Guo et al. 2001). It is essential to make this difference clear, so that we can better understand the current climate system with its major Northern Hemisphere glaciation and insolation cycles. This goal requires more information to refine models and clarify the main causes of the climate system. We have previously shown that orbit-related climate signals are imprinted in Lake Baikal records (Kashiwaya et al. 1998, 1999, 2000). Here, we discuss the relationship between solar insolation and long-term environmental change, based on analyzing long BDP-98 cores to understand the beginning of the current climate regime.

2 Data and Age Models

We will mainly discuss 600 m of Lake Baikal sediments in core BDP-98, obtained from Academician ridge (53°44′40″N, 108°24′30″E) in the central part of the lake. Academician ridge is topographically isolated and has a continuous sediment record that is thought to have been little influenced by fluvial input or turbidity flows. The age range of the BDP-98 core is estimated to be approximately 12.0 million years, based on paleomagnetic correlations (Sakai et al. 2003; Cande and Kent 1995). Here, we use a tentative age scale based on geomagnetic polarity reversals, as discussed by the Baikal Drilling Project (BDP) magnetic research group (Kashiwaya et al. 2001b; Sakai et al. 2003).

An alternative age model has recently been proposed (Antipin et al. 2001). We will briefly touch on this alternative age model and explain why it is difficult to accept it at the present stage of the study for two reasons (Kashiwaya et al. 2002). First, the core segment with an abnormally high sedimentation rate (267–370 m; subchrons C3An.2n and C3Bn, or 6.567–6.935 Ma) might not have a similar structure or origin as other core segments (Fig. 1). A core segment with an abnormally high sedimentation rate clearly stands out from others in a curve for the γ-ray intensity of the core (Fig. 2a). This difference is clearly seen in the spectral analysis of each segment, which shows that the dominant frequency of the high-sedimentation segment clearly differs from that of other segments (Fig. 2b), indicating that the high-sedimentation segment differs structurally from other segments, and implying that the core is not structurally continuous.

Second, can the increase in grain size in the lowest part of the core be explained by the present geomagnetic age model? Simple calculations of porosity and grain density show that the sedimentation rate in the lowest core segment, based on the geomagnetic age model, is nearly twice as high as that in the uppermost core segment (Fig. 3a). This means that the grain size in the lowest segment must be nearly 1.5 times greater than that in the upper segment, for similar hydrological conditions, if Stokes' law is valid. In fact, grain-size differences are supported by data on mineral grain-size distribution (Fig. 3b). Consequently, there are no robust data that require a revision of the

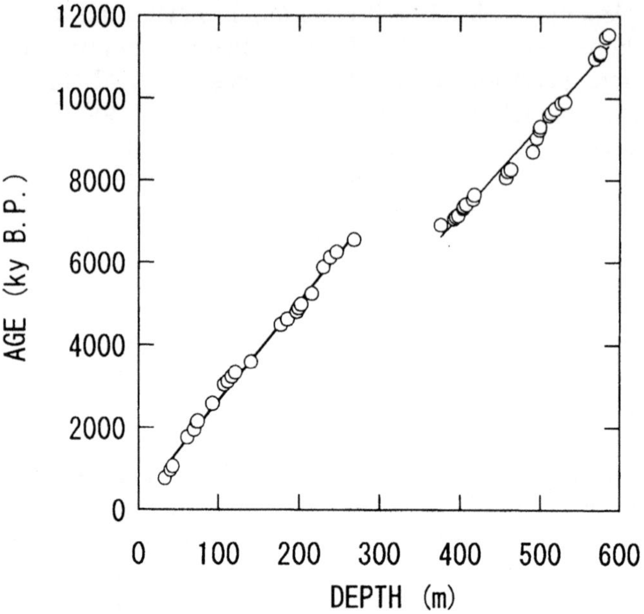

FIG. 1. Age model for BDP-98 based on paleomagnetic records (Kashiwaya et al. 2001b; Sakai et al. 2003). Average sedimentation rate: 4.1 cm/kyr (0–267 m), 29.3 cm/kyr (267–370 m), and 4.6 cm/kyr (370–600 m)

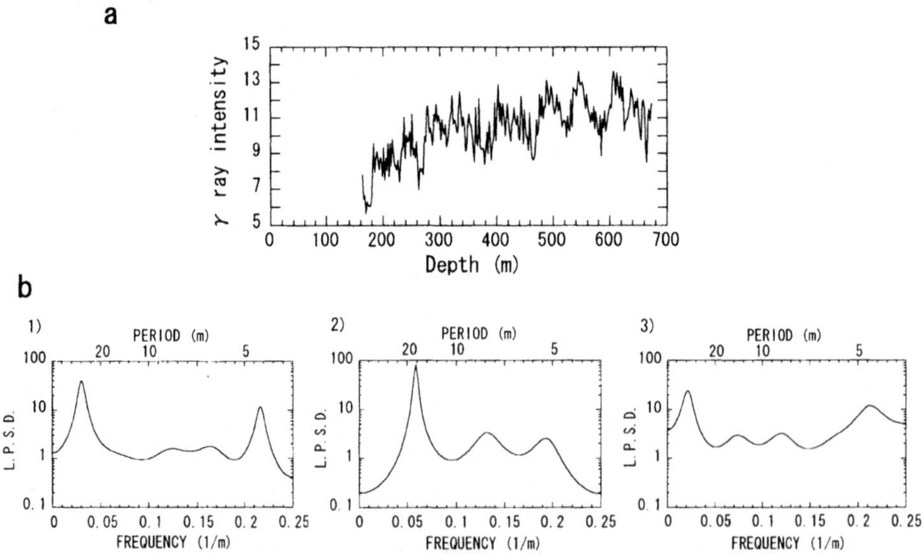

FIG. 2. **a** Longitudinal change in γ-ray intensity for the BDP-98 core (hole 2). **b** Results of spectral analyses for three segments: (1) upper segment (163–261 m), (2) middle (high sedimentation rate) segment (263–361 m), and (3) lower segment (363–673 m) (Kashiwaya et al. 2002, with permission from Elsevier Science)

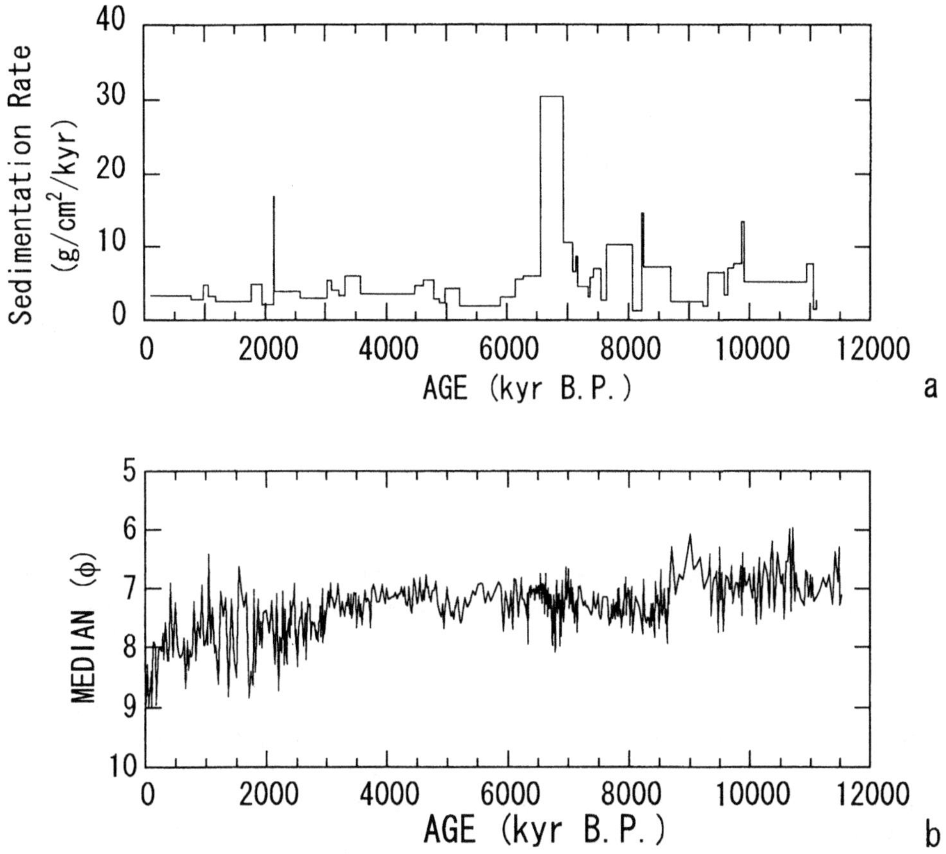

FIG. 3. **a** Changes in sedimentation rate based on the geomagnetic age model for the BDP-98 core. **b** Longitudinal change in the mean mineral grain size of BDP-98 (Kashiwaya et al. 2002, with permission from Elsevier Science)

age model proposed by the geomagnetic research group at the current stage of this study.

There may be some explanations for the data from this core, but we must construct models that incorporate reasonable explanations for given data in order to permit further discussion. At the present time, we recommend excluding this part of the core from statistical analyses in order to avoid data distortion.

3 Cycles Included in the BDP-98 Datasets

We have already reported that the climate of the Asian continental interior cooled gradually, although not steadily, that there were several major shifts in the environment (Fig. 4), and that there were long-term periods related to inso-

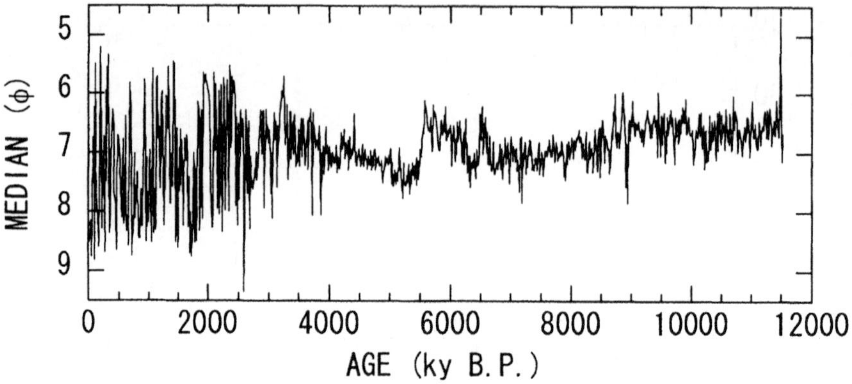

FIG. 4. Temporal change in the mean of whole grain size of BDP-98

lation. Such periods had durations of about 1000 kyr, 600 kyr, or 400 kyr, in addition to 100-kyr, 40-kyr, and 20-kyr periods (Kashiwaya et al. 1998, 2000, 2001a, b). Here, we will introduce another long period and discuss climate shifts which are related to major Northern Hemisphere glaciation.

Figures 5 and 6 show filtered variations in grain size for the last 11.5 myr and 6.5 myr, respectively, indicating that there was little difference in long-term filtered fluctuations between these time intervals. We will limit our discussion here to the 6.5-myr interval in order to avoid statistical distortion, as mentioned above. The numerical filters used here are those around 2.0 myr, around 1.0 myr, and around 0.6 myr. The data clearly indicate that there was a longer period of around 2.0 myr in addition to the 1.0-myr and 0.6-myr periods. The 1.0-myr and 0.6-myr periods may be related to orbital oscillations (Kashiwaya et al. 2001b).

The mineral grain-size datasets also show a similar periodicity (Fig. 7), which implies similar periods of fluctuation. The long period of around 2.0 myr is also seen in marine datasets (Fig. 8), which suggests the existence of such a period in environmental changes globally. Spectral analysis also supports these long periods within long-term fluctuations (Fig. 9).

What is this long period primarily related to? Most long periods known until now are related to orbital cycles. Here, we will examine the fluctuation and periodicity of insolation (Fig. 10, original data by Laskar et al. 1993). The numerical filters used here are the same as were applied to the calculations above. It is certain that there is a period of around 2.0 myr, in addition to other long periods. The results of spectral analysis for insolation also suggest a similarly long period (Fig. 11), implying that this long period is closely related to fluctuations in insolation.

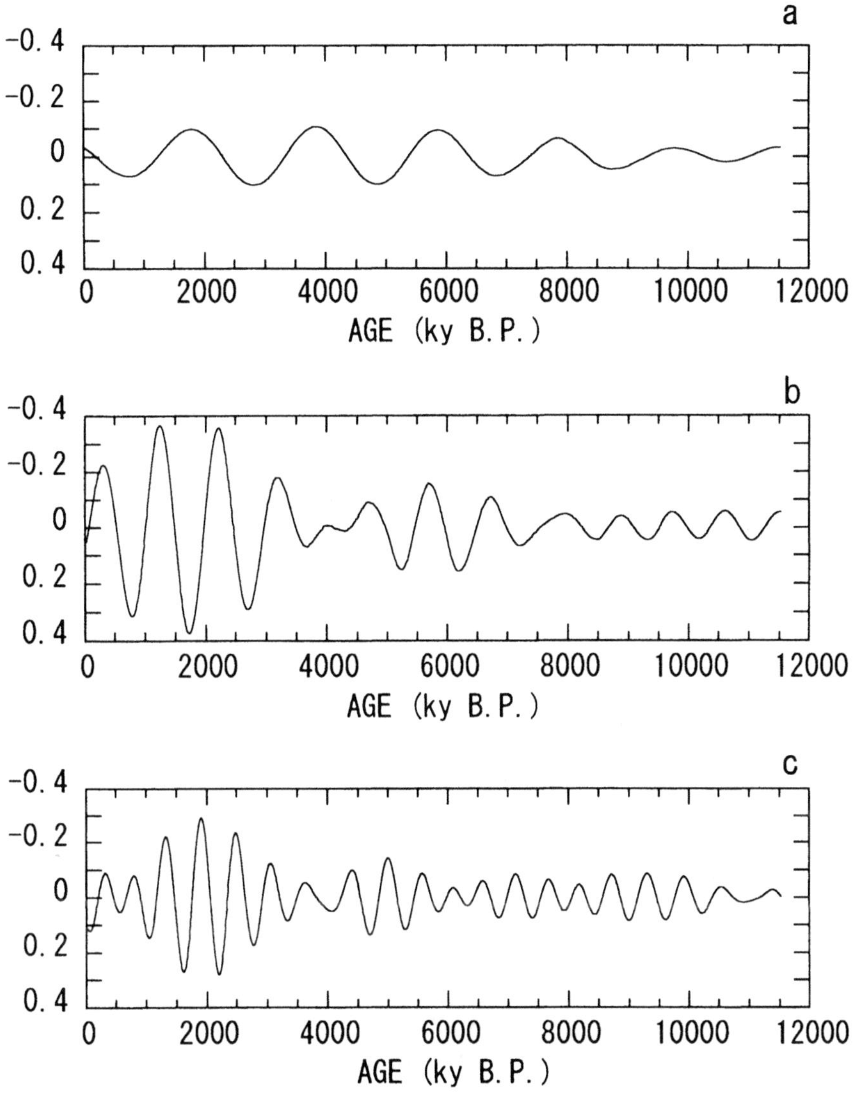

FIG. 5. Filtered curves for the mean of whole grain size during the past 11.5 Ma: **a** 2500–1500 kyr; **b** 1500–750 kyrl; **c** 750–500 kyr band-pass filter

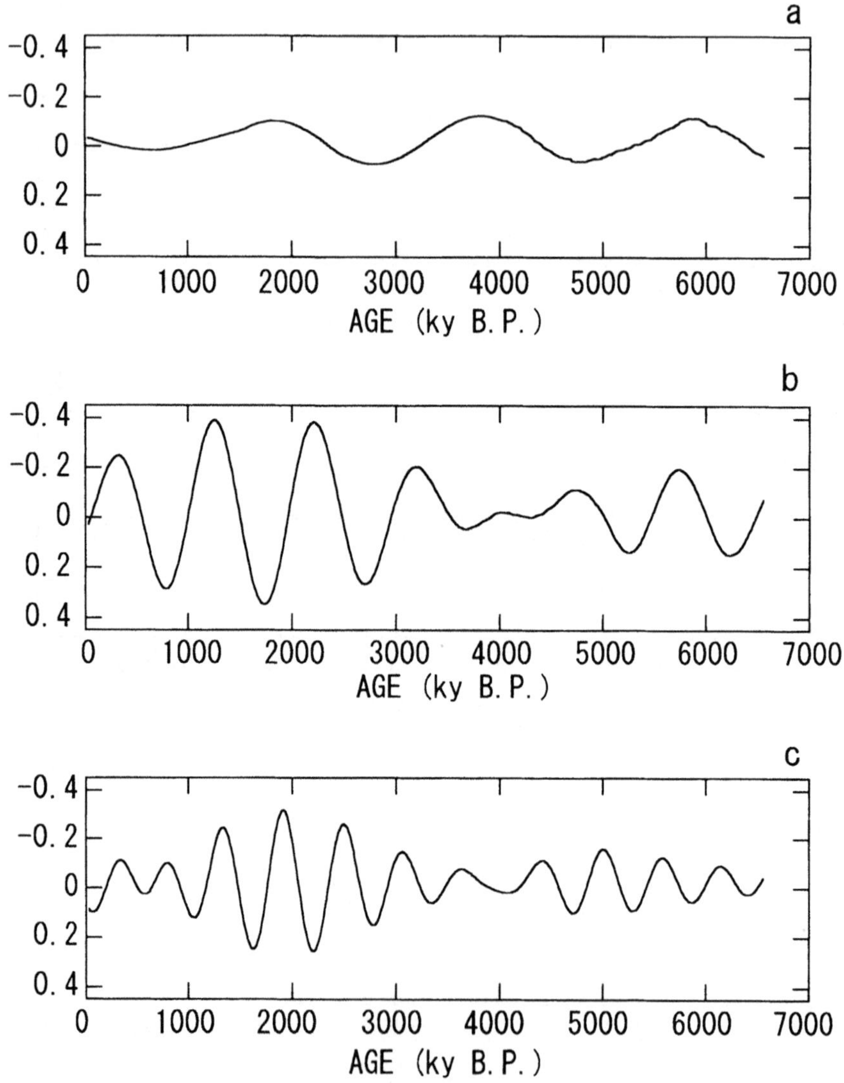

FIG. 6. Filtered curves for the mean of whole grain size during the past 6.5 Ma: **a** 2500–1500 kyr; **b** 1500–750 kyr; **c** 750–500 kyr band-pass filter

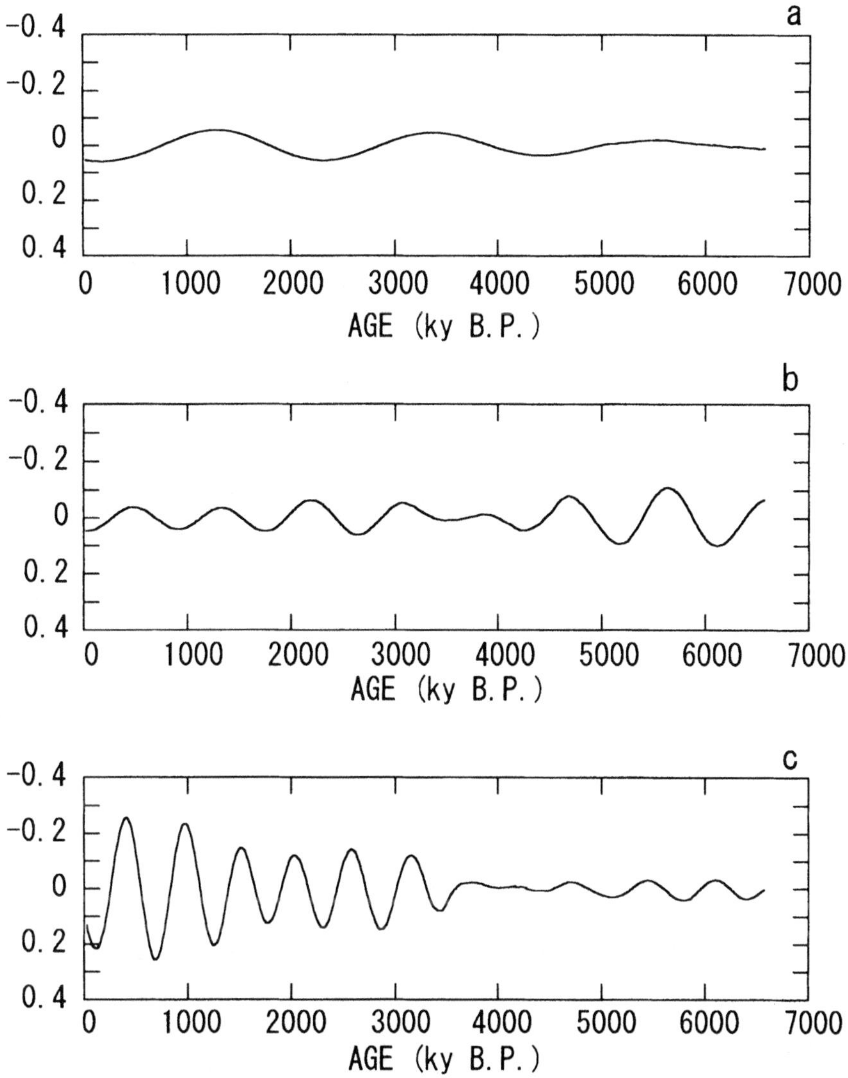

FIG. 7. Filtered curves for the mean of mineral grain size: **a** 2500–1500 kyr; **b** 1500–750 kyr; **c** 750–500 kyr band-pass filter

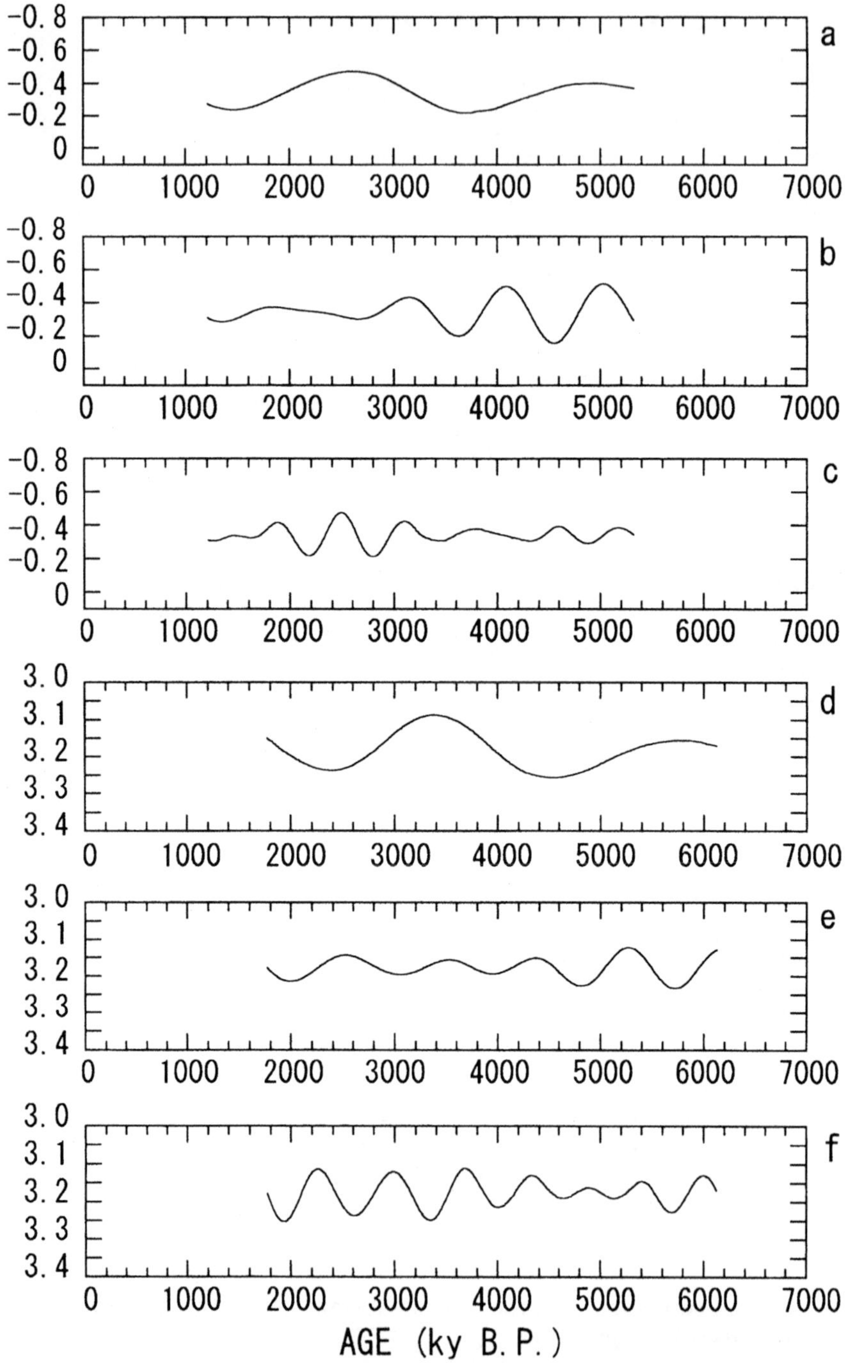

FIG. 8. Filtered curves for oceanic $\delta^{18}O$ records for Mediterranean data (Lourens et al. 1996): **a** 2500–1500 kyr; **b** 1500–750 kyr; **c** 750–500 kyr. Also oceanic $\delta^{18}O$ records from ODP (site 846) (Shackleton et al. 1995); **d** 2500–1500 kyr; **e** 1500–750 kyr; **f** 750–500 kyr band-pass filter, $\delta^{18}O$ (‰)

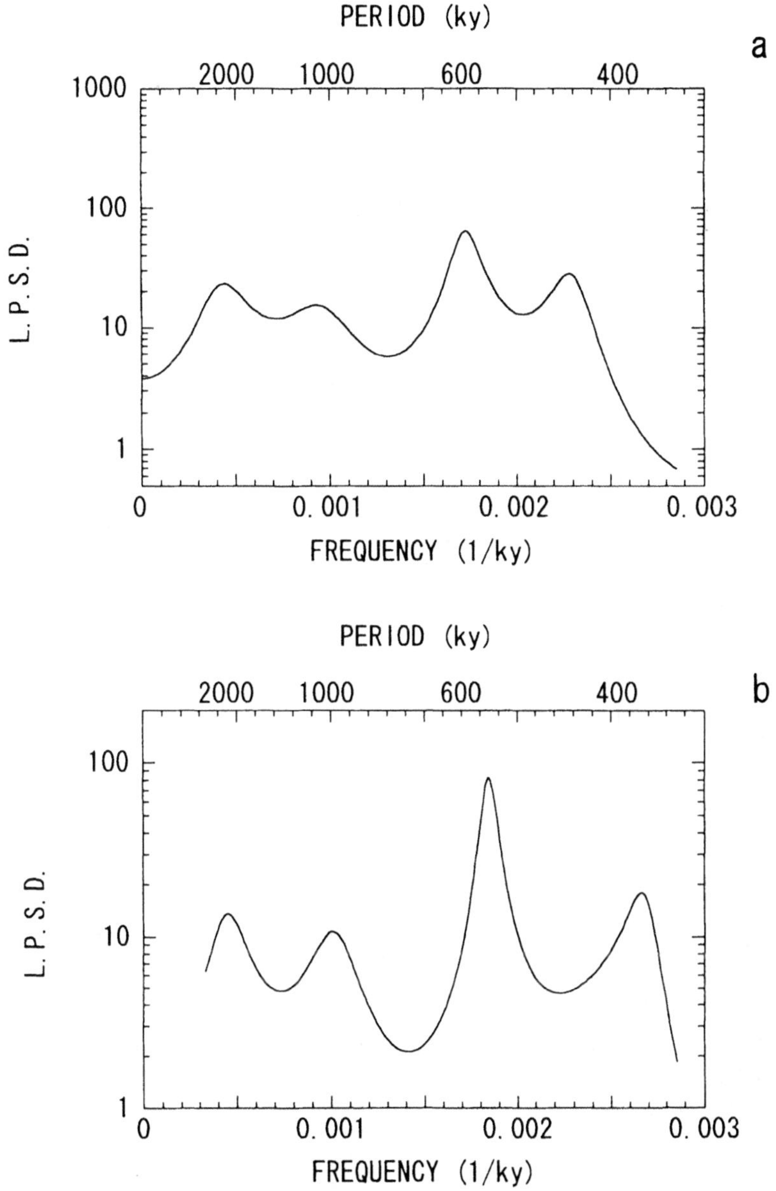

FIG. 9. Results of spectral analyses for the mean **a** whole grain size, and **b** mineral grain size

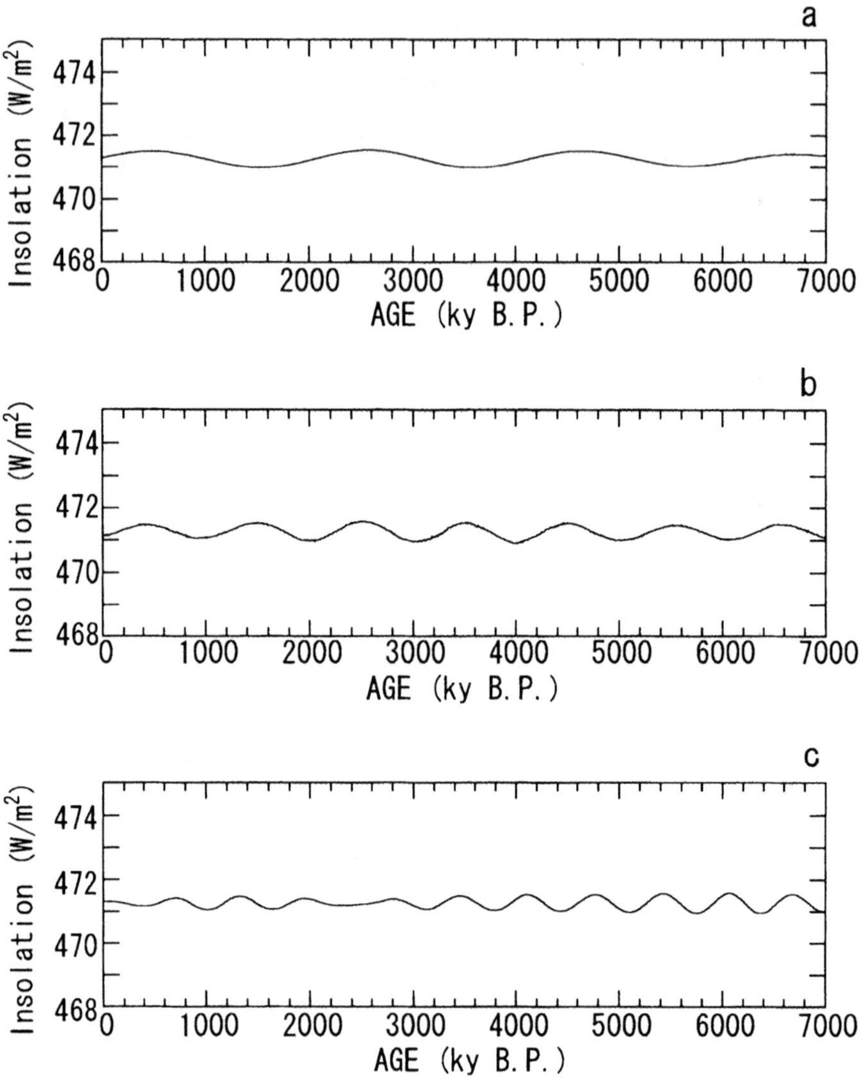

FIG. 10. Filtered curves for 65° N July: **a** 2500–1500 kyr; **b** 1500–750 kyr; **c** 750–500 kyr band-pass filter

4 Major Northern Hemisphere Glaciation

As described above, several climate shifts are recorded in the long BDP-98 core. A shift in about the 2.6- to 2.8-Ma interval has been widely discussed in relation to major Northern Hemisphere glaciation. Shifts at around 3.6–4.6 Ma have

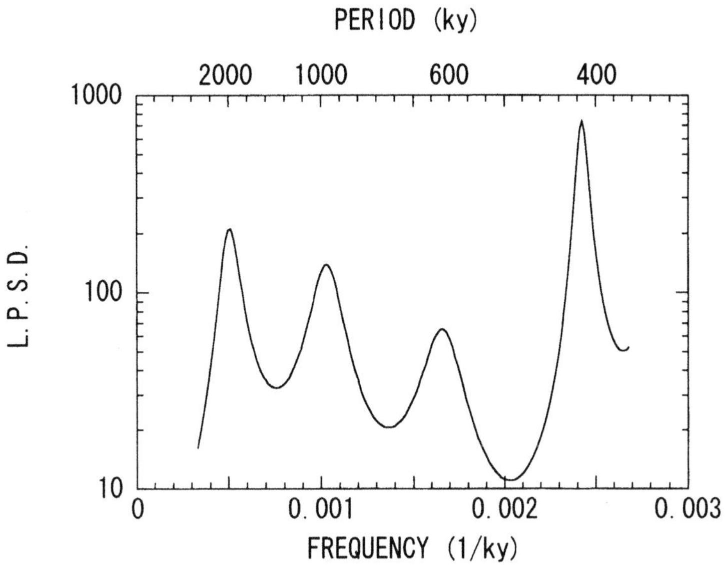

FIG. 11. Results of spectral analysis for insolation 65° N in July

recently been proposed, and have mainly been attributed to tectonic movements such as the formation of isthmuses, the uplift of the Himalayas–Tibetan plateau, or the presence of islands (Haug and Tiedemann 1998; Rea et al. 1998; Cane and Molnar 2001). The Lake Baikal records also indicate that most orbit-related oscillations began to fluctuate more clearly at ca. 4.0 Ma, especially an oscillation at about the 400-kyr period (Kashiwaya et al. in preparation).

To consider these changes in detail, we applied the same method to other BDP-98 datasets (whole grain size and mineral grain size) and to datasets from the equatorial Pacific and Mediterranean (Shackleton et al. 1995; Lourens et al. 1996). An intensification of amplitude at about 4.0 Ma is found in most filtered fluctuations (Fig. 12). The shift at about 2.6–2.8 Ma is also seen in the fluctuations, and is especially clear in the mineral grain-size fluctuation. These relationships suggest that a shift around 4.0 Ma was present in addition to other shifts, which implies that the first shift initiated the current climate regime.

Until now, major shifts in climate oscillation have been attributed to tectonism, as have the shifts at around 3.6–4.6 Ma and 2.6–2.8 Ma, as noted above. Other causes have been proposed more recently, because these rapid climate shifts cannot be explained by tectonism alone. It has been suggested that the first shift at about 4.0 Ma may have been related to solar insolation (Kashiwaya et al. in

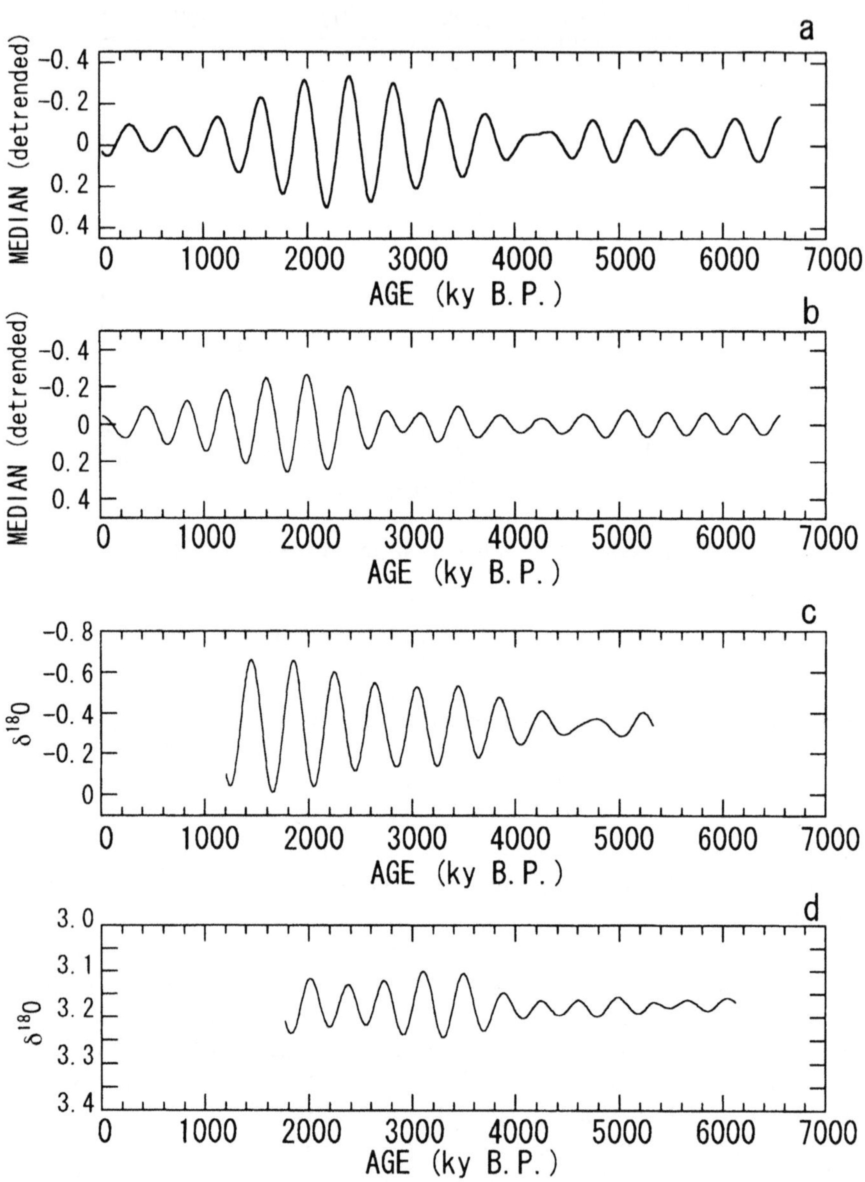

FIG. 12. Band-pass filtered curves (500–350 kyr) for **a** whole grain size; **b** mineral grain size; **c** oceanic $\delta^{18}O$ records from the Mediterranean; **d** oceanic $\delta^{18}O$ records from ODP (site 846)

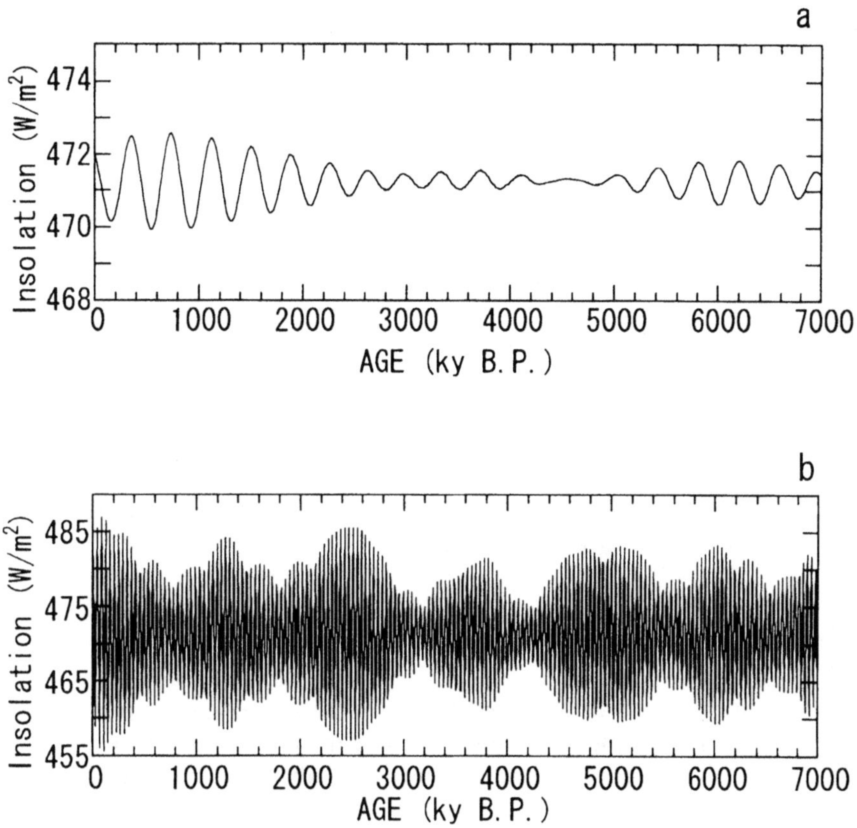

FIG. 13. Band-pass filtered curves **a** 500–350 kyr and **b** 50–35 kyr for insolation 65° N in July

preparation), so we checked some components of insolation using the same filter as discussed above. Figure 13 shows the calculated results, which indicate that an increase in the amplitude of 400-kyr filtered (eccentricity component) insolation clearly began at about 4.0 Ma, and that the amplitude of 40-kyr filtered (obliquity component) insolation increased at about 4.0 Ma as well as at 3.0 Ma, although there was no large fluctuation in other filtered insolations. These data suggest that the shift at about 4.0 Ma was closely related to changes in insolation, especially to obliquity and larger eccentricity components. This idea is supported by the numerical simulations of Li et al. (1998), in which the ice-sheet volume in the Northern Hemisphere began to increase with a pulse after 4.0 Ma, and more frequently after 2.8 Ma in the case of a gradual decrease in atmospheric CO_2 concentration (Fig. 14).

These studies lead to the conclusion that the beginning of the current climate regime (characterized by major Northern Hemisphere glaciation) was at around 4.0 Ma, and that it intensified at about 2.8 Ma, both dates being closely related

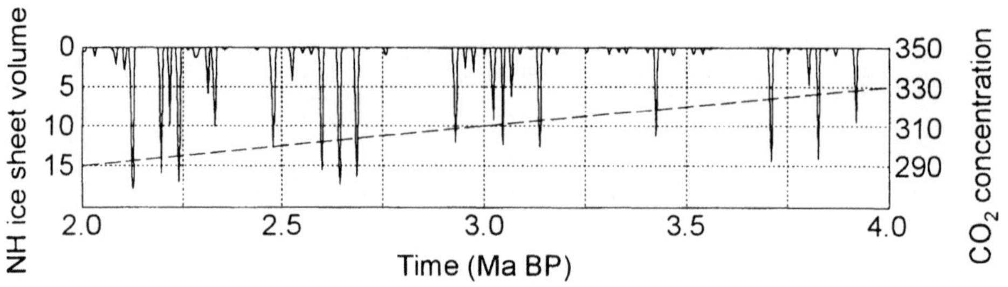

FIG. 14. Simulated result for Northern Hemisphere ice-sheet volume with insolation fluctuations (*solid line*) and CO_2 concentration (*dashed line*) decreasing from 4.0 Ma to 2.0 Ma (Li et al. 1998, with permission from American Geophysical Union)

to changes in insolation. The 400-kyr eccentricity and obliquity parameters are interrelated for the 4.0-Ma shift, but the obliquity parameter alone is related to most of the 2.8-Ma shift. A small shift at about 3.6 Ma may represent a minimum point of eccentricity-related insolation.

5 Concluding Remarks

The research discussed here shows that Lake Baikal sediments provide significant evidence regarding insolation-related climate oscillations, especially for long-term, orbit-related cycles, and provide insights into the beginning of the current climate regime. There is a period of around 2.0 myr, related to insolation, which has fluctuated with almost the same amplitude since at least 12.0 Ma, in addition to other long periods of about 1.0 myr, 0.6 myr, and 0.4 myr, whose amplitudes have increased and decreased over time, particularly at 4.0 Ma. There has been a major global cooling trend for the past 6–7 myr that may be connected to tectonism or to a gradual decrease in CO_2 concentration. Northern Hemisphere ice sheets may have grown during low-insolation summers if winter snow persisted all the year. The global climate system began to satisfy these parameters at about 4.0 Ma, when insolation minima may have acted as triggers for climate change. The intensification of the major glaciation at about 2.8 Ma is probably connected to an additional insolation episode, as reported by Li et al. (1999). This was the beginning of large late Pliocene and Pleistocene fluctuations in the current climate regime.

Acknowledgments. We thank the BDP Leg II members from Russia, America, and Japan who worked to drill the BDP-98 sediment cores. We also thank colleagues in the hydrogeomorphological laboratory at Kanazawa University and in the geomagnetic laboratory at Toyama University for their kind support in this research.

References

An Z, Kutzbach JE, Prell WL, Porter SC (2001) Evolution of Asian monsoons and phased uplift of the Himalaya–Tibetan plateau since late Miocene times. Nature 411:62–66

Antipin V, et al. (2001) The new BDP-98 600-m drill core from Lake Baikal: a key late Cenozoic sedimentary section in continental Asia. Quat Int 80–81:19–36

Baikal Drilling Project Group (2000) Late Cenozoic paleoclimate record in bottom sediments of Lake Baikal. Russ Geol Geophys 41(1):1–29

Baikal Drilling Project Leg II Members (1997) Continuous continental paleoclimate record for the last 4.5 to 5 million years revealed by leg II of Lake Baikal scientific drilling. EOS 78(51):597–604.

Baikal Drilling Project Members (1995) Results of the first drilled borehole at Lake Baikal near the Buguldeika Isthmus. Russ Geol Geophys 36(2):3–32

Berger A, Li XS, Loutre MF (1999) Modelling Northern Hemisphere ice volume over the last 3 Ma. Quat Sci Rev 18:1–11

Cande SC, Kent DV (1995) Revised calibration of the geomagnetic polarity timescale for the late Cretaceous and Cenozoic. J Geophys Res 100:6093–6095

Cane MA, Molnar P (2001) Closing of the Indonesian seaway as a precursor to east African aridification around 3–4 million years ago. Nature 411:157–162

Ding ZL, Xiong SF, Sun JM, Yang SL, Gu ZY, Liu TS (1999) Pedostratigraphy and paleomagnetism of a ~7.0 Ma eolian loess–red clay sequence at Lingtai, Loess plateau, north-central China and the implications for paleomonsoon evolution. Palaeogeogr Palaeoclimatol Palaeoecol 152:49–66

Guo ZT, Peng SZ, Hao QZ, Biscaye PE, Liu TS (2001) Origin of the Miocene–Pliocene red earth formation at Xifeng in northern China and implications for paleoenvironments. Palaeogeogr Palaeoclimatol Palaeoecol 170:11–26

Haug GH, Tiedemann R (1998) Effect of the formation of the Isthmus of Panama on Atlantic Ocean thermohaline circulation. Nature 393:673–676

Kashiwaya K, Ryugo M, Sakai H, Kawai T (1998) Long-term climato-limnological oscillation during the past 2.5 million years printed in Lake Baikal sediments. Geophys Res Lett 25:659–663

Kashiwaya K, Ryugo M, Horii M, Sakai H, Nakamura T, Kawai T (1999) Climato-limnological signals during the past 260 000 years in physical properties of bottom sediments from Lake Baikal. J Paleolimnol 21:143–150

Kashiwaya K, Tanaka A, Sakai H, Kawai T (2000) Paleoclimatic signals printed in Lake Baikal sediments. In: Minoura K (ed) Lake Baikal. Elsevier, Amsterdam, p 53–70

Kashiwaya K, Sakai H, Ryugo M, Horii M, Kawai T (2001a) Long-term climato-limnological cycles found in a 3.5-million-year continental record. J Paleolimnol 25:271–278

Kashiwaya K, Ochiai S, Sakai H, Kawai T (2001b) Orbit related long-term cycles revealed in a 12-Myr continental record from Lake Baikal. Nature 410:71–74

Kashiwaya K, Ochiai S, Tsukahara H, Sakai H, Kawai T (2002) Some issues to be considered in establishing an age model for the long Lake Baikal sediment records. Quat Int 95/96:205–207

Laskar J, Joutel F, Boudin F (1993) Orbital, precessional, and insolation quantities for the Earth from −20 Myr to +10 Myr. Astro Astrophys 270:522–533

Li XS, Berger A, Loutre MF, Maslin MA, Haug GH, Tiedemann R (1998) Simulating late Pliocene Northern Hemisphere climate with the LLN 2-D model. Geophys Res Lett 25:915–918

Lourens LJ, Antonarakou A, Hilgen FJ, Van Hoof AAM, Vergnaud-Grazzini C, Zachariasse WJ (1996) Evaluation of the Plio-Pleistocene astronomical time scale. Paleoceanography 11:391–413

Maslin MA, Haug GH, Sarnthein M, Tiedemann R, Erlenkeuser H, Stax R (1995) Northwest Pacific site 882: the initiation of Northern Hemisphere glaciation. Proc ODP Sci Rep 145:315–329

Maslin MA, Li XS, Loutre MF, Berger A (1998) The contribution of orbital forcing to the progressive intensification of Northern Hemisphere glaciation. Quat Sci Rev 17:411–426

Raymo ME (1994) The Himayalas, organic carbon burial and climate change in the Miocene. Paleoceanography 9:399–404

Raymo ME, Hodell D, Jansen E (1992) Response of deep ocean circulation to the initiation of Northern Hemisphere glaciation (3–2 Ma). Paleoceanography 7:645–672

Rea KD, Snoeckx H, Joseph LH (1998) Late Cenozoic eolian deposition in the North Pacific: Asian drying, Tibetan uplift, and cooling of the Northern Hemisphere. Paleoceanography 13:215–224.

Ruddiman WF, Kutzbach JE (1991) Plateau uplift and climatic change. Sci Am 264:66–75

Sakai H, Nomura S, Horii M, Kashiwaya K, Kawai T, Kravchinsky V, Peck J, King J (2003) Paleomagnetism and paleo-environmental magnetism studied on BDP-98 sedimentary cores from Lake Baikal. In: Kashiwaya K (ed) Long continental records from Lake Baikal. Springer, Berlin, p 233–244

Shackleton NJ, Hall MA, Pate D (1995) Pliocene stable isotope stratigraphy of site 846. Proc ODP Sci Results 138: 337–355

Short DA, Mengel JG, Crowley TJ, Hyde WT, North GR (1991) Filtering of Milankovitch cycles by Earth's geography. Quat Res 35:157–173

Rift Basin History and Geological Environment

Paleoclimate Records from the Lake Baikal Sediments and Lava Formations of the South Baikal Volcanic Area

Mikhail I. Kuzmin[1], Vladimir V. Yarmolyuk[2], Eugene B. Karabanov[1], Takayoshi Kawai[3], Alexander A. Prokopenko[1], Valeryi A. Bychinskyi[1], Marina Y. Khomutova[1], Vladimir F. Geletyi[1], and Eugene V. Kerber[1]

Summary. Five boreholes, drilled in different geological structures of Lake Baikal by the International Baikal Drilling Project, allowed us to obtain unique paleoclimatic records, which can be regarded as being representative of continental Asia. One of the most sensitive paleoclimate proxies in the sedimentary core is the number of diatoms, and correspondingly the amount of biogenic silica. The major and trace elements, the magnetic susceptibility, the clay mineral composition, and a number of other parameters also reflect changes in the environment and the climate.

Age estimates of sedimentary cores are very accurate, and the evidence from biogenic silica in the BDP-98 core shows a regular, gradual cooling episode in Siberia during the last 10 Myr. The Baikal record also indicates climate deterioration from 2.8 to 2.5 Ma. This is evident from glacial clay layers in the BDP-96 and BDP-98 drill cores, which contain almost no diatoms around the Matuyama/Gauss paleomagnetic reversal boundary. The glacial clay layers are interpreted as being linked with the growth of young mountain ranges around Lake Baikal, where glaciers could develop at a height of 2000 m or more. This cooling phase is coincident with the formation of the Chinese loess, which is an indicator of a cooler climate in the region.

Paleoclimatic changes are also recorded in the Baikal region in the morphology of late Cenozoic lavas. Volcanoes formed during glacial periods contain hyaloclastites with inclusions of globular lava, while those in interglacial time consist of massive lavas with distinct slaggy zones. The K–Ar ages of these vol-

[1] Vinogradov Institute of Geochemistry, Siberian Branch of the Russian Academy of Sciences, Favorsky str. 1A. Irkutsk 664033, Russia
[2] IGEM, Moscow, Russia
[3] National Institute for Environmental Studies, 16-2 Onogawa, Tsukuba 305-0053, Japan (present address: Department of Earth and Environmental Sciences, Graduate School of Environmental Studies, Nagoya University, Furo-cho, Chikusa-ku, Nagoya 464-8602, Japan)

canics correlate well with the Lake Baikal paleoclimatic record obtained from sediments.

These results can be used to address questions about the tectonic and structural development of the Baikal rift zone (BRZ) and Lake Baikal, as well as past climatic changes in different time scales.

Key words. Hyaloclastites, Insolation, Paleoclimate, Rift, Volcanism

1 Introduction

An urgent problem at present is to identify the reasons for climatic changes, and to forecast the future climate. In order to predict oscillations of climate in the future, we need to know the regularities and nature of fluctuations in the past. In this respect, the past is the key to the future.

This chapter makes an attempt to reconstruct the paleoclimate and paleoenvironment of the Lake Baikal region from sedimentary sequences and from different types of volcanic eruption. The area studied is ideal for the purpose, with several deep rift depressions (e.g., the history of Lake Baikal for 30 Ma, of Lake Khovsgol for 6 Ma, and of other depressions of the Baikal type) and abundant Cenozoic volcanic occurrences (Yarmolyuk et al. 2000) (Fig. 1).

We have made a comparison of paleoclimatic records from Lake Baikal sediments with the ages of volcanic eruptions in glacial and interglacial periods.

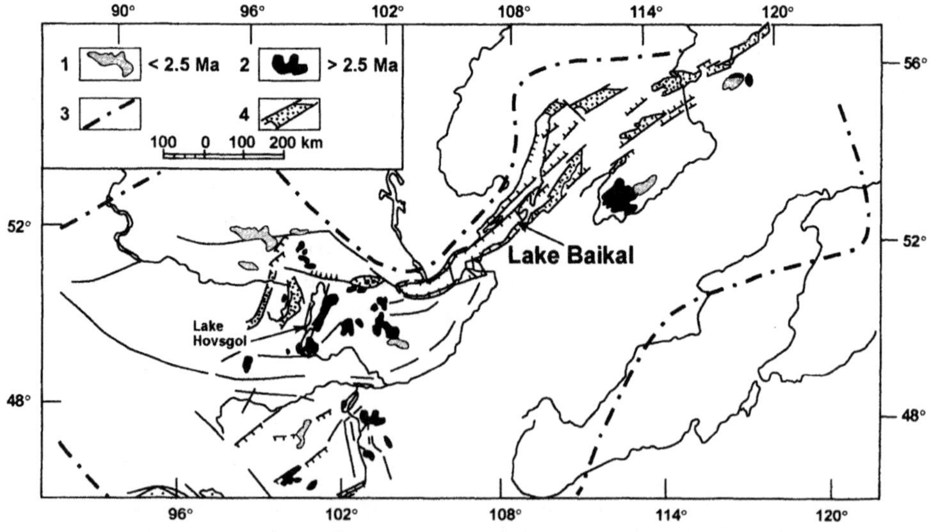

FIG. 1. Distribution of basalts of different ages in the Baikal rift zone. *1*, basalts younger than 2.5 Ma; *2*, basalts older than 2.5 Ma; *3*, distribution of anomalous mantle; *4*, depressions of the Baikal type

These investigations, based on accurate age estimates of Baikal sedimentary sequences and K–Ar age determinations of volcanic eruptions in glacials and interglacials, provided a refined evolution of the southwestern flank of the Baikal rift zone, and identified the influence of climatic and geological factors on the evolution of the Baikal region.

2 Sedimentary Record

2.1 Paleoclimatic Studies from Lake Baikal Deep Sedimentary Cores

Five boreholes were drilled in different geological structures in Lake Baikal by the International Baikal Drilling Project (BDP), which has been in operation for more than 10 years (Kuzmin et al. 2001) (Fig. 2). The BDP-93 core (102 m long)

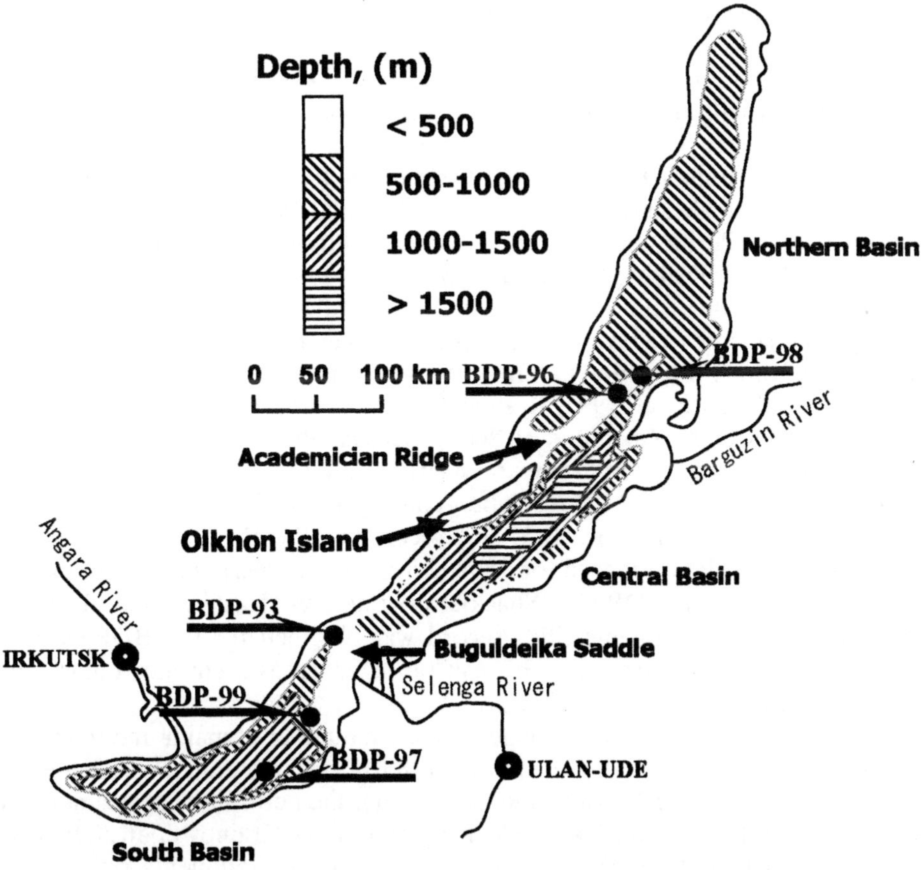

FIG. 2. The bathymetric scheme of Lake Baikal. *Dots*, drill site locations

obtained from the Buguldeika saddle spans from the present to 660 ka, and exhibits an average sedimentation rate of 17 cm/ka (Kuzmin et al. 2000a). High-resolution Holocene and late Pleistocene records were recovered from this core (Karabanov et al. 2000b; Prokopenko et al. 2001b,c). The BDP-96 drill core (200 m long) retrieved from Academician ridge gives a record for 5 Ma. This core allowed the reconstruction of paleoclimatic and paleoenvironmental changes in central Asia for the past 5 Myr (Baikal Drilling Project Members 1998; Bezrukova et al. 1999; Williams et al. 1997; Karabanov et al. 2000a). A detailed history of endemic diatom algae evolution has been recovered from the BDP-96 core; three new genera and ten new species have been described (Khursevich et al. 2000, 2001a,b). The BDP-97 core was retrieved from the south basin. The first freshwater lacustrine gas hydrates were discovered and studied from this core (Kuzmin et al. 2000a). The BDP-98 core (600 m long) from Academician ridge spans ca. 10 Myr. Scientific results recovered from this core provided a better understanding of the major geological stages in the evolution of Lake Baikal (Baikal Drilling Project Members 2000, 2001). Sediments of a new BDP-99 core (350 m long) from the Posolskaya bank are under investigation. They will represent unique information about the tectonic evolution of the region and lake level fluctuations (Baikal Drilling Project Members 2003).

Investigations on these cores revealed that Lake Baikal sediments are composed of rhythmically laminated fine-grained deposits, represented by alternating diatom-rich layers and layers of a dense terrigenous clay barren of diatom valves (Baikal Drilling Project Members 1997, 1998, 2000; Karabanov 1999; Kuzmin et al. 2001). Diatom-rich sediments accumulated during warm interglacials, while the terrigenous clay formed during the glacials (Bezrukova et al. 1991; Baikal Drilling Project Members 1997, 1998, 2000). Spectral analysis of the biogenic silica and diatom distribution revealed cycles of 23, 41, 100, and 400 kyr in the cores recovered from Academician ridge and the Buguldeika saddle, confirming the orbital nature of the sedimentary rhythms of Lake Baikal deposits (Williams et al. 1997; Kashiwaya et al. 1998; Baikal Drilling Project Members 1998, 2000; Kuzmin et al. 2001). Because these cycles have also been found in the marine record, they are clearly related to changes in the Earth's orbital characteristics and the insolation coming to the Earth.

The Lake Baikal climatic (BDP-96) record for the Brunhes chron agrees well with the insolation record for 65°N (Laskar et al. 1993) and the oceanic isotope record of core ODP-677 (Shackleton et al. 1984; Kashiwaya et al. 1998). An orbitally tuned biogenic silica record was obtained for the BDP-96 core (Prokopenko et al. 2001a), and this can be regarded as a chronostratigraphic standard for Siberia and central Asia.

As well as the distribution of biogenic silica, a number of major and trace elements (Goldberg et al. 2001; Kuzmin et al. 2001), clay minerals (Yuretich et al. 1989; Kuzmin et al. 2000b; Solotchina et al. 2001), and petrophysical sedimentary characteristics (Peck et al. 1994, 1996; King et al. 1993; Krainov et al. 2001) also show a cyclical pattern. The chemical composition of sediments reflects the weathering process of rocks in the Baikal catchment basin, and is therefore different in cold and warm periods. Goldberg et al. (2001) distinguished these so-called "cold" and "warm" elements. Cold elements include alkaline (Rb) and rare-alkaline (Ba) elements. These elements are related to the feldspar component of granitoid and granite–gneiss rocks which occur in the Lake Baikal catchment basin. Physical weathering is predominant during cold periods, when glaciers crushed the bedrock and a fine fraction came into Lake Baikal as a suspension and was deposited in the sediment (Kuzmin et al. 2001; Karabanov 1999). This is verified by a comparison of average concentrations of petrogenic oxides in sediments. Sediments formed in cold periods exhibit heightened K, Na, and in particular Al_2O_3 contents, which are the main elements of feldspar and plagioclase. In warm periods, marked mainly by a chemical weathering, feldspars are destroyed first. The elements they contain are brought into the lake and can be deposited on soil geochemical barriers.

Recent investigations have shown that Lake Baikal sediments contain good records of paleoclimatic changes in central Asia.

2.2 Reconstruction of Climatic Changes in Central Asia from Lake Baikal Sedimentary Records

Sedimentary BDP-96 (200m) and BDP-98 (600m) cores recovered from Academician ridge have been studied in order to reconstruct environmental and climatic changes, because the sedimentation has been continuous since the late Miocene, and the lake was not covered by ice throughout the Pleistocene. According to the decrease in diatom valves and biogenic silica concentrations in the lake sediments, a regular cooling in the region started in the Miocene (Baikal Drilling Project Members 1998, 2000; Karabanov et al. 2000a). A sharp cooling in the Northern Hemisphere started 2.8–2.5 Ma (Fig. 3) (Karabanov et al. 2000b; Williams et al. 2001). Dense glacial clay began to be accumulated in the lake at that time (Karabanov et al. 2000c; Williams et al. 2001).

It should be noted that the above time span marks the origination of loess in China (Kukla et al. 1988; Bloemendal et al. 1995) as well as an accumulation of iceberg rafted (IRD) material in the north Atlantic and north Pacific Oceans (Shackleton et al. 1984; Maslin et al. 1995), both of which are indicators of a cooler climate in those regions. The growth of Tibet and the Himalayas, with glaciers on their tops, would have changed the air circulation in the Northern Hemisphere and led to a significant cooling (Kuhle 2000; Ruddiman and Kutzbach 1989, 1991).

In the Lake Baikal region, we believe that the young mountain ranges were high enough for glaciers to form on their tops 2.5–2.8 Ma. The melting of these glaciers led to the accumulation of the first glacial lacustrine sediments in Lake Baikal, which were found in the BDP-96 core at a depth of 110–115 m as a dense clay (Karabanov 1999; Baikal Drilling Project Members 2000; Karabanov et al. 2000b; Kuzmin et al. 2001). This clay contains coarse detritus, indicating significant ice and iceberg rafting and intensive glaciation in the region

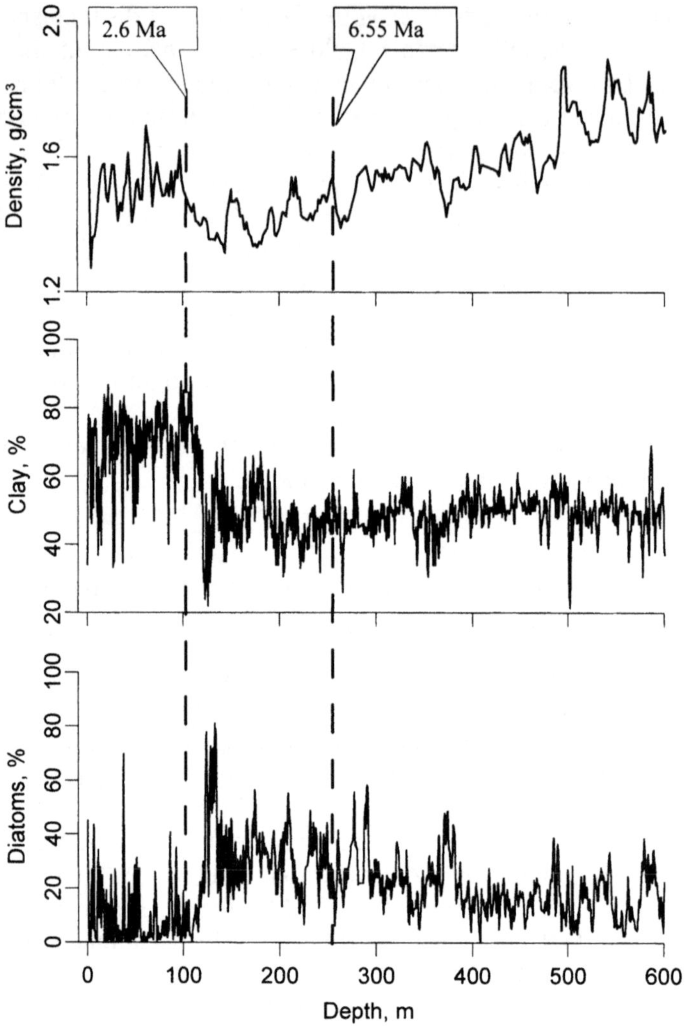

FIG. 3. Distribution of diatoms and clays, and the density of sediments in the BDP-98 core. A sharp change in the values is seen at a depth of 100 m

(Karabanov 1999; Karabanov et al. 2000c). The glacial clay indicates that iceberg rafting accumulated with cycles of 23, 41, and 100 kyr. It suggests a cyclic formation of glaciers in the mountains driven by rhythmic climatic oscillations (Karabanov 1999; Baikal Drilling Project Members 2000). The assumption of mountain growth is confirmed by sharp changes in element concentrations or element ratios found at a depth of 110–115 m in the BDP-96 core. Figure 4 shows that the K_2O/Al_2O_3 ratio critically increases below this depth, which indicates an increase in the weathering of the bedrock, and a supply of weathering products

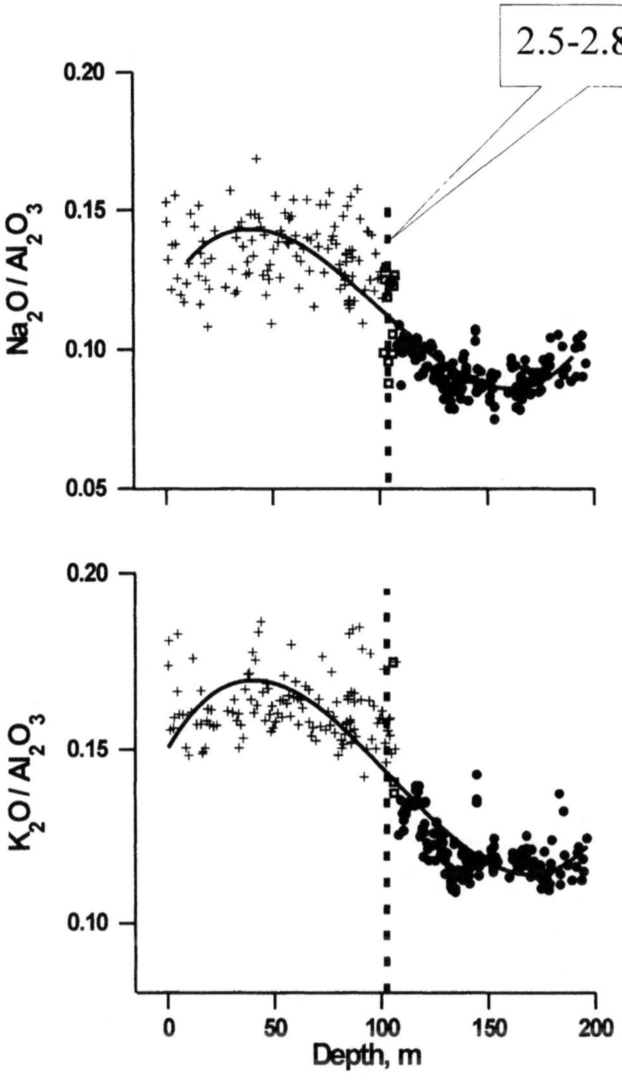

FIG. 4. Distribution of K_2O/Al_2O_3 (weight %) in the BDP-96 core. A sharp change in the values was found at a depth of 100–105 m

into the lake. Most weathering would result from the rugged terrain formed by the growth of mountains and the consequent exposure of the bedrock subaerially. Granitoids are predominant in the bedrock in the region. Prior to these events, the terrain was weakly rugged, with few bedrock outcrops (Logatchev et al. 1974).

Age models have been obtained for Lake Baikal sediments based on detailed paleomagnetic data from the BDP-96 and BDP-98 cores (0–270 m section with a constant average sedimentation rate) (Baikal Drilling Project Members 1998, 2000). These models provide exact dates for all glacial and interglacial episodes recorded in the Lake Baikal sequence during the late Cenozoic (6 Myr). A more precise, orbitally tuned biogenic silica record of Lake Baikal has been provided for the entire Brunhes (0–780 ka) (Prokopenko et al. 2001a). Using these records, it should be possible to determine the age of any climatic events in central Asia in the late Cenozoic accurately.

3 Volcanic Records

3.1 Climatic Conditions (Glacial–Interglacial) Determined from the Morphology of Lava Sequences

The morphology of volcanic eruptions and the types of volcanic structure are controlled by the climate during an eruption. Cenozoic basalts are abundant in the south Baikal volcanic area (Fig. 5). The formation of this volcanic area started in the late Oligocene (30–35 Ma) and is still continuing (Yarmolyuk et al. 2000).

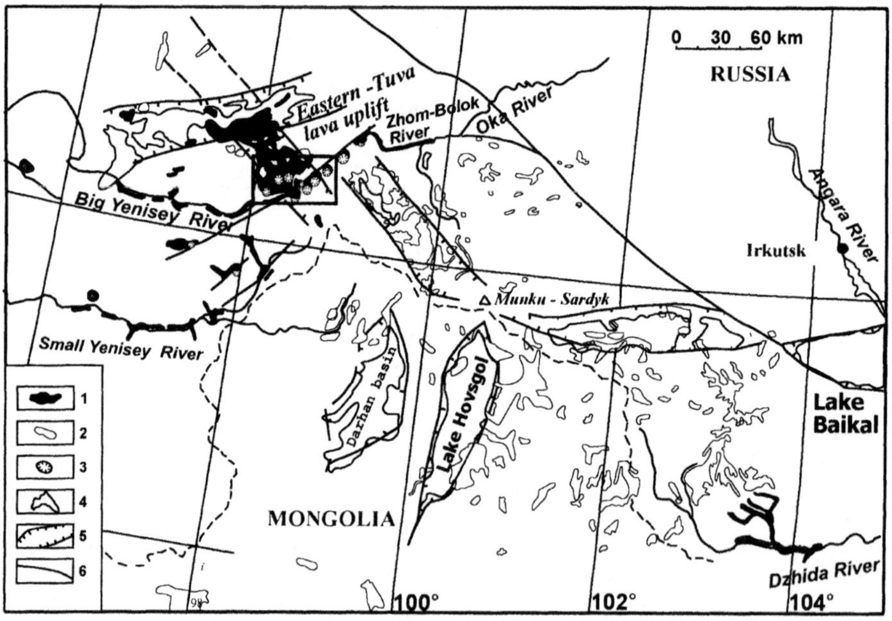

FIG. 5. Location of late Cenozoic volcanic formations within the south Baikal volcanic area. *1*, Pliocene–Holocene lava fields; *2*, late Oligocene–Pliocene lava fields; *3*, Pliocene–Holocene volcanoes of the eastern Tuva volcanic uplift; *4*, late Cenozoic terrigenous formations; *5*, grabens; *6*, faults. The area enclosed by a rectangle shows the eastern Tuva volcanic uplift (see Fig. 6)

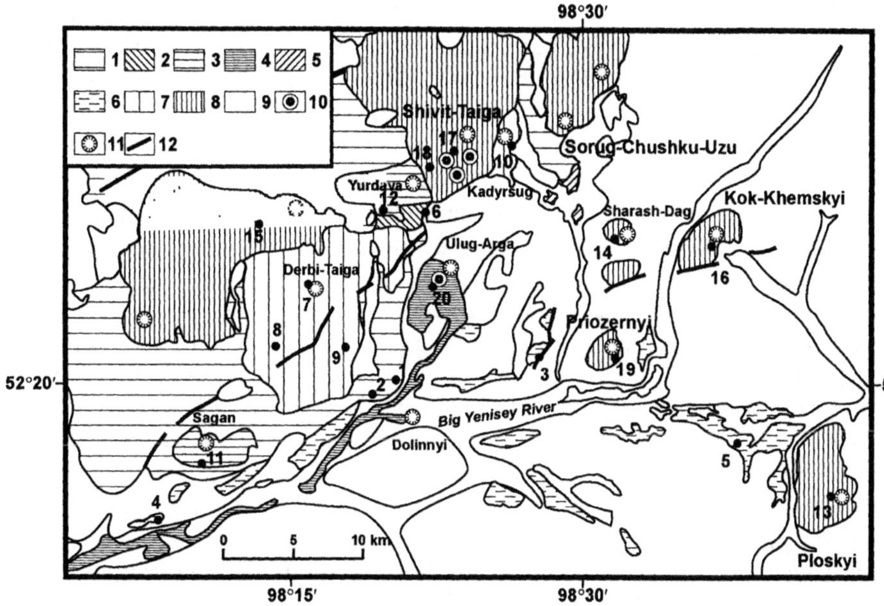

FIG. 6. Diagram of the eastern Tuva lava upland, showing volcanic eruptions of different types and age (Ma): *1–4*, interglacial lavas: *1*, basalt lava of plateau and ancient valley eruptions (2.07–2.14); *2*, lava of hawaiites (1.21); *3*, basanite lavas of "ancient" volcanoes (0.29–0.35); *4*, basanite lava of "young" volcanoes (≤0.05). *5*, middle Miocene lavas (14.7). *6–8*, glacial lavas: *6*, lava and hyaloclastites of ancient valley eruptions (1.65–1.75); *7*, hyaloclastites of "ancient" volcanoes (0.565–0.76); *8*, hyaloclastites of "young" volcanoes (0.06–0.225). *9*, loose sediments. *10, 11*, centers of volcanic eruptions: *10*, volcanic funnel; *11*, slag cones. *12*, faults. *Solid circles*, sampling sites of volcanic are as for K–Ar dating. Numbers near circles correspond to the numbers in Table 1

Recent volcanism (younger than 2.5 Ma) is abundant here, as shown in black in Fig. 5. Volcanic eruptions occurred at times of frequently changing climate. The eastern Tuva lava uplift is of particular interest for studying the influence of climatic conditions on volcanic eruptions (Fig. 6). Volcanic activity has been taking place during the past 2.2 Myr. The major portion of the volcanic upland is a lava plateau composed of a group of shield volcanoes of different ages, and dissected by river valleys. The volcanic material forming the lava plateau and the shield volcanoes can be divided into two types in terms of its morphologic features: (1) normal massive lava with clear upper slaggy zones, which indicate eruptions in subareal conditions, and (2) lava clastic formations, including hyaloclastites, globular lavas, lahars, and products of the rewashing of these rocks. Hyaloclastites are rocks formed by nonsorted fragments of glassy lavas and glass. Small fragments are often of black basalt; larger ones always have a glass-like outer zone. They have the shape of zonal lava balls and pillows, which are typical of globular lavas. The matrix consists of fine volcanic glass. Variously sized fragments are randomly mixed in layers and lenses 10–15 m thick. Rocks of volcaniclastic complexes make large, generally isolated, dome-like volcanic structures. Walls of rock show that the volcaniclastic sequences have a coarse, layered texture. All volcaniclastic sequences recovered in the walls show the succession of lava flows (Fig. 7). Volcaniclastic sequences are found throughout the whole volcano, and their thickness can reach several hundred meters. Volcanoclastic sequences contain abundant interbeds and lenses of well-sorted volcanomictic gravelites, sand, and silt. The occurrence of these rocks indicates lacustrine formation conditions.

For the origin of the volcaniclastic complex, it should be remembered that hyaloclastites, and globular and pillow lavas are formed when they erupt into much colder conditions (water or ice), as is observed in Iceland (Bullard 1980). Eruption into a large deep lake is unlikely, because the area is located at high altitudes in the eastern Sayan Ranges. Because large glaciers covered it repeatedly during the Pleistocene (Grossval'd 1965), only glacial eruptions can be suggested for the eastern Tuva uplift. Therefore, hyaloclastites and globular lavas indicate eruptions into large glaciers and confirm the glaciation on the top of the eastern Sayan Ranges at the time of eruption. Absolute age determinations of these rocks allow exact dating of glacial periods.

3.2 K–Ar Dating of Lavas

K–Ar chronology was applied for volcanic sequences in the south Baikal area. The determinations were carried out at Institute of Geology of Ore Deposits, Petrography, mineralogy and Geochemistry (IGEM), Moscow, using the mass spectrometric complex with an MI-1201 IG device (Goldberg et al. 2001). The main characteristics of this device (sensitivity from Ar 5×10^3 A/torr and level of blank experiments from ^{40}Ar 9×10^{-7} ng) mean that extremely small amounts of radiogenic ^{40}Ar (0.001–0.01 ng/g) can be determined. We obtained dates for the middle– upper Pleistocene rocks using samples of not more than 100 mg. The accuracy was checked by systematic analyses of biotite (Bern 4) and muscovite (P-207) as reference samples. The radiogenic argon was measured using ^{38}Ar as a tracer. The age was calculated using the constants $\lambda_\kappa = 0.581 \times 10^{-10}$ per year, $\lambda_\beta = 4.962 \times 10^{-10}$ per year, and $^{40}\kappa$ = 0.01167 (at.%). The geochronology and lava morphology characteristics obtained showed that volcanism occurred in times of systematic climate changes, marked by alternating glacial and interglacial periods (Table 1). Massive lava flows and hyaloclastites containing globular and pillow lava fragments are of different ages. The chilled glass-like lava and hyaloclastites are found in volcanic sequences of age (ka): (1) 1650–1750; (2) 725–760; (3) 565–600; (4) 110–225. Volcanic sequences containing rocks without signs of being chilled in water or a glacial environment are of age (ka): (1) ~1210; (2) 290–350; (3) 50; (4) younger than 50. The first sequences are considered to be glacial, while the second ones are believed to be interglacial. The latter also include eruptions along the Big Yenisey river valley, which occurred during the Kargapolovo–Lashamp excursion (42–44 ka) (Yarmolyuk et al. 2001), and eruptions along the Zhom–Bolok river valley (late Holocene).

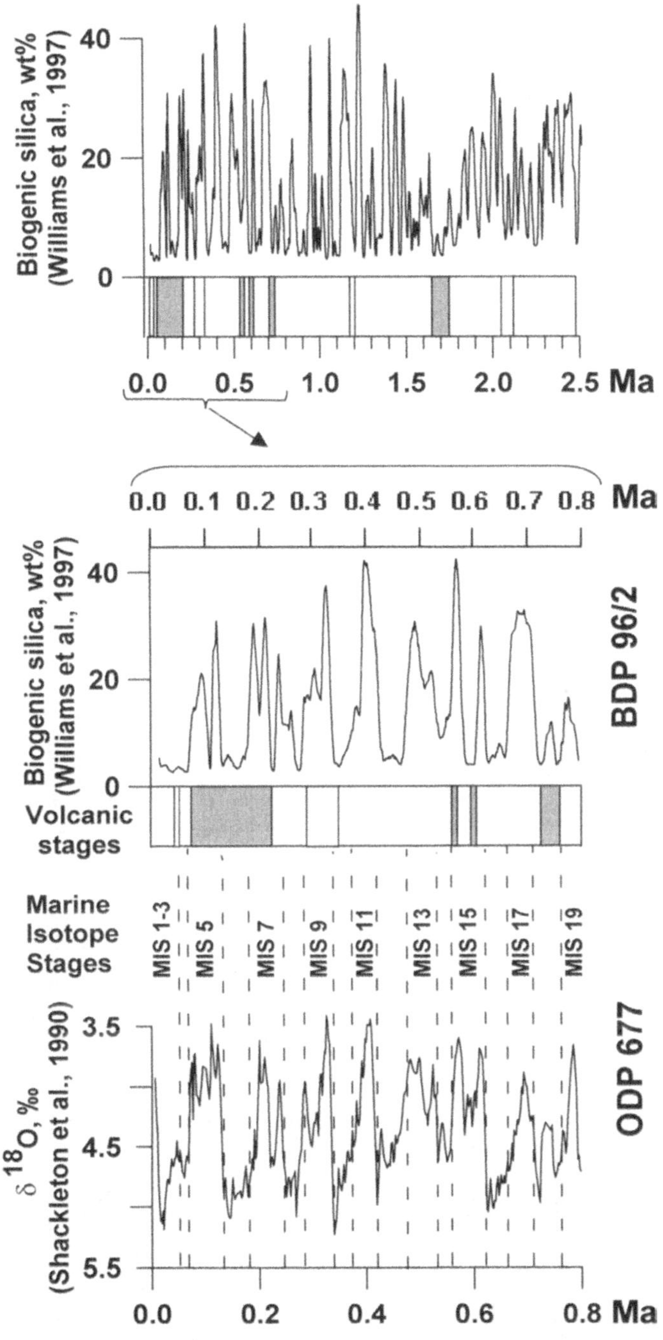

FIG. 7. Comparison of basalts erupted in the glacial (*black*) and interglacial (*gray*) periods, compiled from the distribution of biogenic silica in the BDP-96 core for the past 2.5 Myr (*upper plot*). *Lower plot*, the same for the Brunhes chron

TABLE 1. K–Ar dating of recent volcanic occurrences in the eastern Tuva

No.	Sample	Volcanic structure	Rock associations	Type of eruptions	K (%) (±σ)	^{38}Ar $g \cdot 10^{-9}$	$^{40}Ar_{rad}$, $g \cdot 10^{-9}/g$ (±σ)	Atmospheric ^{40}Ar (%)	Age (±1.6σ) (ky)
Lava of plateau and ancient valley eruptions									
1	Be-1/9	Lava plateau	Basalt lava	Subaerial, Interglacial	1.35 ± 0.03	0.09	0.200 ± 0.01	68.0	2140 ± 200
2	Be-1/4	Lava plateau	Basalt lava	Subaerial, Interglacial	1.10 ± 0.025	0.109	0.158 ± 0.007	90.0	2070 ± 150
3	Pr-4/2	Lava river of Paleo-Yenisey	Lava and hyaloclastites of basanites	Glacial, valley glaciers <100 m	1.68 ± 0.02	0.108	0.204 ± 0.005	63.1	1750 ± 100
4	Be-1/18	Lava river of Paleo-Yenisey	Lava and hyaloclastites of basalts	Glacial, valley glaciers <100 m	0.80 ± 0.02	0.108	0.096 ± 0.004	77.5	1730 ± 150
5	Pl-2/5	Lava river of Paleo-Yenisey	Lava and hyaloclastites of basalts	Glacial, valley glaciers <100 m	1.05 ± 0.02	0.106	0.120 ± 0.004	80.2	1650 ± 130
6	Sct-2/6	Shield volcano on the lava plateau	Lava of hawaiites	Subaerial, Interglacial	1.12 ± 0.03	0.13	0.095 ± 0.03	72.5	1210 ± 80
Lava of "ancient" volcanoes									
7	De-1/9	Derbi–Taiga volcano	Hyaloclastites of hawaiites	Glacial	1.48 ± 0.03	0.134	0.078 ± 0.002	80.2	760 ± 50
8	De-1/2	Derbi–Taiga volcano	Hyaloclastites of basanites	Cover glaciers >700 m	2.16 ± 0.04	0.137	0.109 ± 0.003	74.1	725 ± 50
9	Be-1/12	Bezymyannyi volcano	Lava and hyaloclastites of basanites	Glacial, valley glaciers <100 m	1.75 ± 0.03	0.089	0.073 ± 0.006	86.9	600 ± 80
10	Kd-1/3	Kadyrsuk volcano	Lava and hyaloclastites of basanites	Glacial, valley glaciers <100 m	1.55 ± 0.02	0.064	0.061 ± 0.001	88.4	565 ± 80

11	Be-1/2	Sagan volcano	Lava of basanites	Subaerial, Interglacial	1.68 ± 0.015	0.038	0.041 ± 0.003	91.5	350 ± 40
12	Sht-2/8	Yurdava volcano	Lava of basanites	Subaerial, Interglacial	1.91 ± 0.015	0.038	0.039 ± 0.003	90.0	290 ± 40
Lava of "young" volcanoes									
13	Pl-1/5	Ploskyi volcano	Hyaloclastites of basanites	Glacial, cover glaciers >700 m	1.78 ± 0.02	0.065	0.028 ± 0.005	93.1	225 ± 50
14	Pr-2/4	Charash–Dag volcano	Hyaloclastites and lavaclasts of hawaiites	Glacial, cover glaciers >700 m	1.30 ± 0.02	0.048	0.019 ± 0.004	94.4	210 ± 50
15	De-2/3	Al'bine–Boldok volcano	Hyaloclatites and lavaclasts of basanites	Glacial, cover glaciers >700 m	1.66 ± 0.02	0.047	0.022 ± 0.004	96.8	195 ± 50
16	Pr-3/3	Kok–Khem volcano	Hyaloclastites and lavaclasts of basanites	Glacial, cover glaciers >700 m	1.82 ± 0.02	0.058	0.019 ± 0.004	96.4	150 ± 50
17	Sht-1/8	Shivit volcano	Hyaloclastites and lavaclasts of basanites	Glacial, cover glaciers >700 m	1.14 ± 0.03	0.112	0.0098 ± 0.0015	96.6	130 ± 40
18	Sht-1/1	Shivit colcano	Hyaloclastites and lavaclasts of basanites	Glacial, cover glaciers >700 m	1.40 ± 0.03	0.111	0.0104 ± 0.0021	98.0	110 ± 40
19	Pr-1/7	Priozernyi volcano	Hyaloclastites and lavaclasts of basanites	Glacial, cover glaciers >700 m	1.68 ± 0.03	0.063	0.0081 ± 0.0025	98.6	75 ± 40
20	UA-1/5	Ulug–Arga volcano	Lava of basanites	Subaerial, Interglacial	1.78 ± 0.02	0.062	0.0056 ± 0.0006	98.8	48 ± 20

4 Comparison of Paleoclimatic Reconstructions Obtained from Volcanic Formations and Lake Baikal Sedimentary Records

If our idea that volcanic rocks composed of massive lava formations are from interglacial periods and those containing abundant hyaloclastites originated in glacial periods is true, the age of glacial and interglacial periods identified from volcanic rocks should coincide with warm and cold periods found in the Lake Baikal sedimentary record. Such a comparison is shown in Fig. 7, which shows the paleoclimatic record for the past 2.5 Myr and for the Brunhes chron using an orbitally tuned age model (Prokopenko et al. 2001a). This figure shows the good correlation of the Lake Baikal paleoclimatic record with reconstructions obtained from the morphology of volcanic rock.

The lava plateau (samples 1 and 2, 1.95–2.2 Ma, Table 1) and a small shield volcano (sample 6, 1.2 Ma) are believed to be from an eruption during an interglacial period, based on observations of rock facies. This eruption took place in epochs with abundant diatoms, corresponding to a warm climate. The latter eruptions of the Yurdava (sample 12, 290 ± 40 ka), Sagan (sample 11, 350 ± 40 ka), and Ulug-Arga (sample 21, 48 ± 20 ka) volcanoes, as well as valley eruptions along the Big Yenisey river (50 ka), correlate with sediments from interglacial stages in the Lake Baikal archive (MIS 9 and MIS 3, respectively).

The fragmented glass-like lavas and hyaloclastites were formed at nearly 1.65–1.75 Ma in the valley of the Paleo–Yenisey (samples 3–5, see Fig. 6). Fragments of lava sequences were observed at sites in the paleovalley with a constant slope of 7–10°. The absence of sedimentary rocks indicates that large lake basins had not been available. A considered time span coincided with the inception of rapid cooling lasting from 1.75 to 1.45 Ma (see Fig. 7). Taking into account the small thickness of the chilled lava (to 50 m), we can assume the existence of thin glaciers along the main valley during these eruptions, which would have melted afterward.

The next stage of glacial eruptions is recorded in volcaniclastite rocks of the Derbi–Taiga volcano (samples 7 and 8, 725–760 ka) (Fig. 8). The rocks are as thick as 550 m. The volcano dominates the lava plateau by 700 m. Large lenses of volcanomictic sandstone and siltstone are abundant in the volcaniclastic complex. The rocks origin in a lacustrine environment attest to the lake basin being comparable in size to the area of volcanic eruptions. The lake overlapped the upper part of the volcanic structure. The volcano did not have a crater, and thus a lake basin over 500 m above the base of the volcano could be formed in a glacial setting. Thus, we consider that the thickness of the glacial cover within the lava uplift was not less than the volcano's height at that time. In the Lake Baikal record, this time interval corresponds to a glacial period of age 714–764 ka, and thus belonging to the cold MIS-18 and the Manciyckiy glacial period (Arkhipov et al. 1997; Karabanov et al. 2001).

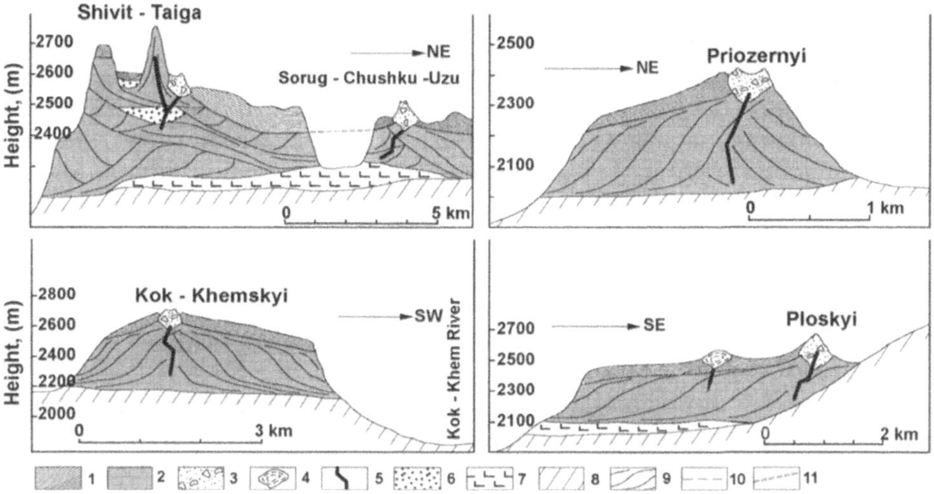

FIG. 8. Schemes of volcanoes arising from glacial eruptions between 225 and 100 ka. *1*, lava sequences; *2*, volcanoclastic sequences; *3*, slag cones; *4*, lava dome; *5*, feeding dykes; *6*, sandstone, gravelites, siltstone of lacustrine sediments; *7*, sequences of lava plateau; *8*, prevolcanic basement; *9*, boundaries of layered pockets of volcaniclastic sequences; *10*, a supposed upper boundary of the cover glaciation during the Shivit volcano formation; *11*, supposed location of the bottom of the glacial lake

The chilled lava of the Kadyrsug volcano (sample 10) and the Bezymyannyi volcano (sample 9) formed between 565 and 600 ka (see Table 1) along the foot of the Derbi–Taiga volcano. Comparison with the sedimentary record shows that those volcanic eruptions took place during MIS 15 (600 ka) or the beginning of the cold MIS 14 (565 ka). If the Derbi–Taiga and Kadyrsug volcanoes had been generated at different times, the links with the cooling episodes would have been more evident, which could have indicated the existence of thin glaciers. On the other hand, if both volcanoes had formed simultaneously (taking into account any error in age determinations), they could have arisen either during a cooling episode of MIS 15 or during MIS 14.

Abundant glacial eruptions took place in the middle–late Neopleistocene (samples 13–20, 225–100 ka). A large group of volcanoes originated at that time. They include Shivit (samples 17 and 18, 130–110 ± 40 ka), Kok-Khem (sample 16, 150 ± 50 ka), Priozernyi (sample 20, 75 ± 40 ka), and Ploskyi (sample 13, 225 ± 50 ka) (see Figs. 6 and 8, and Table 1). If we look for temperature minima in the Lake Baikal climatic chronostratigraphy, the above time spans correspond to MIC 7d (220–233 ka), stage MIS 6 (127–185 ka), and MIS 5d (105–117 ka). It should be noted that the above volcanoes, which are composed of hyaloclastites, form a common field and can be regarded as eruptions occurring during a common cold episode. Within the error of K–Ar age determinations, they correspond to MIS 6 (127 ± 50 to 185 ± 50 ka). Further radiometric age determinations of volcanic rocks and geological investigations of the volcanoes should provide a more exact correlation of volcanic rocks with cooling episodes recorded in the Lake Baikal sediments. At this time, the volcanoes included separate plateau mountains with a thick (400–600 m) hyaloclastitic lower part and a lava summit (to 100 m). Lenses of fine-layered volcanogenic sandstone and siltstone are abundant among hyaloclastites from these volcanoes. Their hypsometric location indicates their formation within glacial lakes, uplifted above the volcano's base by 600 m (see Fig. 5). With the above data, we suggest that volcanoes formed at a time of complete glaciation, which covered the whole lava uplift. The thickness of the glacier can be estimated from the thickness of volcanoclastic complexes over 600 m.

Thus, there is a good correlation between climatic changes obtained from the Lake Baikal paleoclimatic record and the lava morphology of eruptions in glacial and interglacial periods. This suggests that each cooling epoch revealed in the sediment composition was accompanied by glacial periods of varied strength in the mountains framing Lake Baikal.

The time resolution of the Lake Baikal sedimentary archives and the volcanic records is different, and the number of glacial and interglacial episodes recorded in the volcanic record of the eastern Tuva uplift is significantly lower than that in the archive from Lake Baikal. This is probably the result of insufficient radiometric age determinations of volcanic rocks. The impulses of volcanic activity were not common to all glacial and interglacial stages recorded in the Lake Baikal sedimentary cores.

5 Conclusions

1. Good agreement was found between the age of climatic events recorded in the Lake Baikal paleoclimatic records and geochronological determinations of glacial and interglacial stages reconstructed from lava formations in the south Baikal volcanic area.
2. The good correlation between these two records demonstrates that each cold stage recorded in the Lake Baikal sediments is associated with glacial episodes in the south Baikal area, at least in mountainous regions. A rapid cooling episode occurring 2.8–2.5 Ma ago was a sign of the first glacial period in the region.
3. Data on hyaloclastite sequences indicate that the thickness of the ice sheets varied significantly. However, during long cooling episodes, in particular in MIS 18 and MIS 6, glaciers could reach a hundred meters and could possibly exceed 1 km.

Acknowledgments. The researches were supported by Russian Foundation for Basic Research (grant number 01-05-97223; 00-15-98528; 02-05-64196; 01-05-65155)

Russian Ministry of Industry Science and Technology, Science and Technology Agency of Japan and by U. S. NSF grant EAR-96-14770. The authors are thankful to Dr. A. Polozov and Mrs. S. Kotomanova for the preparation of the manuscript.

References

Arkhipov SA, Zykina VS, Krukover AA (1997) Stratigraphy and paleomagnetism of glacial and loess solid deposits of the west Siberian valley. Russ Geol Geophys 38:1027–1048

Baikal Drilling Project-93 Members (1997) Preliminary results of the first drilling on Lake Baikal, Buguldeika site, south-eastern Siberia. Quat Int 37:3–17

Baikal Drilling Project Members (1998) Continuous record of climatic changes in Baikal sediments for the last 5 MA. Russ Geol Geophys 39:139–156

Baikal Drilling Project Members (2000) The new BDP-98 600-m drill core from Lake Baikal: a key late Cenozoic sedimentary section in continental Asia. Quat Int 80–81: 19–36

Baikal Drilling Project Members (2001) Late Cenozoic paleoclimatic record in Baikal sediments (from studies of the 600-m core). Russ Geol Geophys 41:3–32

Baikal Drilling Project Members (2003) High-resolution sedimentary record of the new BDP-99 deep drill core from the Posolskaya Bank of Lake Baikal (in Russian). Russ Geol Geophys, in press

Bezrukova EV, Bogdanov YA, Williams DF, Granina LZ, Ignatova HV, Karabanov EB, Kuptzov VM, Kurulev AV, Letunova PP, Lichoshvai EV, Chernaeva GP, Shimaraeva MK, Yakushin AO (1991) Deep changes of the ecosystem of the north Baikal in the Holocene. Dokl AN SSSR 321:1032–1037

Bezrukova EV, Kulagina NV, Letunova PP, Shestakova ON (1999) Trend of vegetation and climate change in the Lake Baikal region for the last 5 Ma from palynological data on Lake Baikal sediments. Russ Geol Geophys 40:735–745

Bloemendal J, Lice XM, Rolph TC (1995) Correlation of the magnetic susceptibility stratigraphy of Chinese loess and the marine oxygen isotope record: chronological and paleoclimatic implications. Earth Planet Sci Lett 131:371–380

Bullard FM (1980) Volcanoes of the Earth. University of Texas Press, Austin, p 580

Goldberg EL, Phedorin MA, Grachev MA, Zolotarev KV, Khlystov OM (2001) Geochemical indicators of paleoclimate changes in sediments of Lake Baikal. Russ Geol Geophys 42:76–87

Grossval'd MG (1965) Development of the relief of the Sayan–Tuva uplift. Nauka, Moscow, p 165

Karabanov EB (1999) Geological structure of sedimentary strata of Lake Baikal and reconstruction of climatic changes of central Asia during the late Cenozoic (in Russian). Thesis, Institute of Lithosphere of Russian Academy of Sciences, Moscow, p 72

Karabanov E, Kuzmin MI, Williams DF, Bezrukova EV, Prokopenko AA, Kerber EV, Gvozdkov AN, Geletiy VF, Weil D, Schwab M (2000a) The global cooling of central Asia according to the Lake Baikal sedimentary record (in Russian). Dokl Acad Sci Russia 370:61–66

Karabanov EB, Prokopenko AA, Wiliams DF, Khursevich GK (2000b) A new record of Holocene climate change from bottom sediments of Lake Baikal. Palaeogeogr Palaeoclimatol Palaeoecol 156:1211–1224

Karabanov EB, Kuzmin MI, Prokopenko AA, Williams DF, Khursevich GK, Bezrukova EV, Kerber EV, Gvozdkov AN, Geletiy VF, Weil D, Schwab M (2000c) Glaciations of central Asia in the late Cenozoic according to the sedimentary record from Lake Baikal. In: Minoura K (ed) Lake Baikal: a mirror in time and space for understanding global change processes. Elsevier, Amsterdam, p 71–84

Karabanov EB, Prokopenko AA, Kuzmin MI, Williams DF, Gvozdkov AN, Kerber EV (2001) Glacial and interglacial periods of Siberia: paleoclimate records of Lake Baikal and correlation with west Siberia stratigraphic schemes (the Brunhes Chron) (in Russian). Russ Geol Geophys 42:48–63

Kashiwaya K, Ryugo M, Sakai H, Kawai T (1998) Long-term climato-limnological oscillation during the past 2.5 million years printed in Lake Baikal sediments. Geophys Res Lett 25:659–663

Khursevich GK, Karabanov EB, Williams DF, Kuzmin MI, Prokopenko AA (2000) Evolution of freshwater centric diatoms within the Baikal rift zone during the late Cenozoic. In: Minoura K (ed) Lake Baikal: a mirror in time and space for understanding global change processes. Elsevier, Amsterdam, p 146–154

Khursevich GK, Karabanov EB, Prokopenko AA, Williams DF, Kuzmin MI, Fedenya SA, Gvozdkov AN, Kerber EV (2001a) Detailed diatom biostratigraphy of Baikal sediment during the Brunhes Chron and climatic factors of species formation (in Russian). Russ Geol Geophys 42:108–129

Khursevich GK, Karabanov EB, Prokopenko AA, Williams DF, Kuzmin MI, Fedenya SA (2001b) Biostratigraphic significance of new fossil species of the diatom genera Stephanodiscus and Cyclotella from upper Cenozoic deposits of Lake Baikal, Siberia. Micropaleontology 47:47–71

King J, Peck J, Gangemi P, Kravchinsky VA (1993) Paleomagnetic and rock–magnetic investigations of Lake Baikal sediments. Russ Geol Geophys 34:174–191

Krainov MA, Kravchinskyi VA, Peck JA, Sakai H, King JW, Kuzmin MI (2001) Paleoclimate record in bottom sediments of Lake Baikal from magnetic susceptibility data. Russ Geol Geophys 42:87–98

Kuhle M (2000) The uplift of Tibet above the snow line and its complete glaciation as a trigger of the Quaternary ice. A. Hypothesis for ice age development. IPPCCE Newsl 13:36–39

Kukla G, Heller F, Ming LX, Chun XT, Lin TS, Sheng AZ (1988) Pleistocene climates in China dated by magnetic susceptibility. Geology 16:811–814

Kuzmin MI, Karabanov EB, Prokopenko AA, Gelety VF, Antipin VS, Williams DF, Gvozdkov AN (2000a) Sedimentation processes and new age constraints on rifting stages in Lake Baikal: results of deep-water drilling. Int J Earth Sci 88:183–192

Kuzmin MI, Solotchina EP, Vasilevskyi AN, Stolpovskaya VN, Karabanov EB, Geletiy VF, Buchinskyi VA, Anoshin GN, Shulgenko SG (2000b) Clay minerals of Baikal bottom sediments as a paleoclimate proxy. Russ Geol Geophys 41:1347–1349

Kuzmin MI, Karabanov EB, Kawai T, Williams D, Bychinskyi VA, Kerber EV, Kravchinskyi VA, Bezrukova EV, Prokopenko A, Geletii VF, Kalmychkov GV, Goreglyad AV, Antipin VS, Khomoutova MY, Soshina NM, Ivanov EV, Khursevich GK, Tkachenko LL, Solotchina EP, Ioshida N, Gvozdkov AN (2001) Deep drilling on Lake Baikal: main results. Russ Geol Geophys 42:8–35

Laskar J, Joutel F, Boudin F (1993) Orbital, precessional and insolation quantities for the Earth from –20 Myr to +10 Myr. Astron Astrophys 270:522–533

Logatchev NA, Antoschenko-Olenev IV, Bazarov DB, Galkin VI, Goldurev GS, Endrihinskiy AS, Zolotarev AG, Sizikov AI, Ufimtzev GF (1974) Uplifts of Pribaikalie and Transbaikalia. Nauka, Moscow, p 358

Maslin MA, Haug GH, Sarnthein M, Tiedemann R, Erlenkeuser H, Stax R (1995) Northwest Pacific site 882: the initiation of Northern Hemisphere glaciation. In: Rea DK, Basov IA, Scholl DW, Allan JF (eds) Proc ODP Sci Results 125, Ocean Drilling Program, College Station, Texas, p 315–329

Peck J, King JW, Colman SM, Kravchinskyi VA (1994) A rock–magnetic record from lake Baikal, Siberia: evidence for the late Quaternary climate change. Earth Planet Sci Lett 122:221–238

Peck JA, King JW, Colman SM, Kravchinskyi VA (1996) An 84 kyr paleomagnetic record from the sediments of Lake Baikal, Siberia. J Geophys Res 101:11356–11385

Prokopenko AA, Karabanov EB, Williams DF, Kuzmin MI, Shackleton NJ, Crowhurst SJ, Peck JA, Gvozdkov AN, King JW (2001a) Biogenic silica record of the Lake Baikal response to climatic forcing during the Brunhes. Quat Res 55:123–132

Prokopenko AA, Karabanov EB, Williams DF, Kuzmin MI, Khursevich GK, Gvozdkov AN (2001b) The detailed record of climatic events during the past 75 000 yrs BP from the Lake Baikal drilling core BDP-93-2. Quat Int 80–81:59–68

Prokopenko AA, Kuzmin MI, Karabanov EB, Williams DF (2001c) Response of Lake Baikal archive to the catastrophic Henrich event in the northern Atlantic for the last 80 000 years. Dokl RAS 379:391–397

Ruddiman WF, Kutzbach JE (1989) Forcing of the late Cenozoic hemisphere climate by plateau uplift in southern Asia and the American west. J Geophys Res 94:18409–18427

Ruddiman WF, Kutzbach JE (1991) Plateau uplift and climatic change. Sci Am 246:66–77

Shackleton NJ, Backman J, Zimmerman H, Kent DV, Hall MA, Roberts DG, Schitker D, Baldauf JG, Desparairies A, Homrighausen R, Huddlestun P, Keene JB, Kaltenback AJ, Krumsiek KAO, Morton AC, Murray JW, Westberg-Smith J (1984) Oxygen isotope calibration of the onset of ice-rafting and history of glaciation in the North Atlantic region. Nature 307:620–623

Shackleton NJ, Berger A, Peltier WR (1990) An alternative astronomical calibration of the Lower Pleistocene timescale based on ODP site 677. Trans Roy Soc Edinburg Earth Sciences 81:251–261

Solotchina EP, Prokopenko AA, Kuzmin MI, Vasilevskyi AN, Shulzhenko SG (2001) Differences in glacial and interglacial clay mineral associations of Baikal sediments from BDP-93-2 and BDP-96 cores. Russ Geol Geophys 42:146–157

Williams DF, Peck J, Karabanov EB, Prokopenko AA, Kravchinsky V, King J, Kuzmin MI (1997) Lake Baikal record of continental climatic response to orbital insolation during the past 5 million years. Science 278:1114–1117

Williams DF, Karabanov EB, Prokopenko AA, Kuzmin MI, Khursevich GK, Gvozdkov AN, Bezrukova EV, Solotchina EP (2001) Late Miocene–Pleistocene records from Baikal sediments: markers for paleoclimatic and biostratigraphic reconstructions. Russ Geol Geophys 42:35–48

Yarmolyuk VV, Lebedev VI, Sugorakova AM, Bragin VYu, Litasov YuD, Prudnikov ST, Arakelyan MM, Lebedev VA, Ivanov VG, Kozlovskyi AM (2000) Eastern Tuva area of recent volcanism in central Asia: stages, products, and pattern of volcanic activity. Volcanol Seismol 28(3):3–32

Yuretich R, Melles M, Sarata B, Grobe H (1989) Clay minerals in the sediments of Lake Baikal: a useful climate proxy. J Sediment Res 69:588–596

Geothermal Studies of Underwater Boreholes in Lake Baikal

R.P. Dorofeeva and M.V. Shapova

Summary. As part of the Baikal Drilling Project (BDP), the temperature and thermal conductivity (λ) of the bottom sediments were measured in the BDP-93, -96, and -98 boreholes for a subsequent estimation of heat flow. The values of these parameters at great depths were obtained for the first time.

The thermal conductivity values, measured by a comparator of thermal conductivity of the bottom sediments, were 1.76, 1.24, and 1.19 W/mK for BDP-93, -96, and -98, respectively.

The relationships among the thermal conductivity of the bottom sediments, water content, conductivity, resistivity, and natural γ-rays are traced.

Key words. Underwater borehole, Temperature, Thermal conductivity

1 Introduction

The geothermal regime in eastern Asia has been studied in the southern part of the Siberian platform, the Baikal rift zone, the Trans-Baikal folded area, and Mongolia, four provinces which differ in their geology and tectonics (Fotiadi 1987).

In most territory of eastern Asia, the heat flow varies within the range 30–50 mW/m^2. Low flux (less than 20 mW/m^2) is typical of some localities in the Siberian platform. Thermal highs (60–90 mW/m^2 or higher) have been observed in the Baikal rift zone (Duchkov 1985; Duchkov et al. 1999). However, the heat flow at Lake Baikal was measured by a conventional temperature probe in the bottom sediments at shallow depths (Golubev 1982). Deep-water drilling in the ice of Lake Baikal allowed us to measure the true temperature distribution, and also to measure the thermal conductivity of bottom sediments at greater depths.

Institute of the Earth's Crust, Siberian Branch of the Russian Academy of Sciences, Lermontov str. 128, Irkutsk 664033, Russia

In order to study the paleoclimate in the framework of the International Program of the Baikal Drilling Project (BDP), three underwater boreholes, 102, 300, and 673 m deep, were drilled and studied from a barge frozen in the ice from 1993 to 1998 (Fig. 1, Table 1). The geothermal work carried out in the project included temperature measurements and measurements of the thermal conductivity of the bottom sediments, which were required for the determination of heat flow values. The thermal conductivity of the sediments was determined in

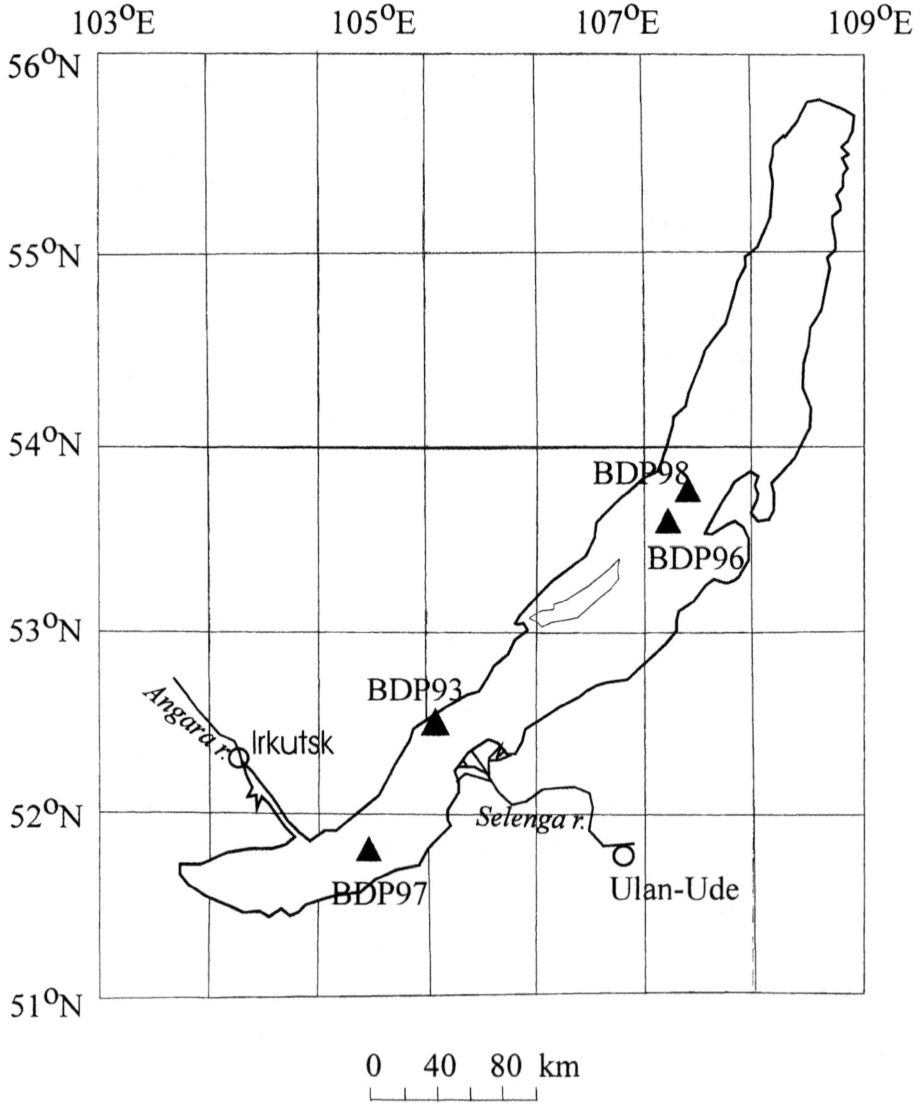

FIG. 1. Map of the location of underwater boreholes in Lake Baikal

TABLE 1. Location of underwater boreholes in Lake Baikal

Boreholes	Site	Location		Depth of water (m)	Depth of borehole (m)	Notes
		Latitude N	Longitude E			
BDP-93	Buguldeyka saddle	52°31′05″	106°09′11″	354	102	–
BDP-96	Academician ridge	53°41′48″	108°21′06″	321	302	–
BDP-97	South basin	51°47′50″	105°29′13″	1436	225	Thermal conductivity was not measured
BDP-98	Academician ridge	53°44′46″	108°24′32″	337	674	–

the laboratory by the comparator of thermal conductivity method, which is described later (Kalinin 1981). A total of 553 thermal conductivity measurements were made in the BDP-93 borehole, 512 in the BDP-96/1 borehole, 360 in the BDP-96/2 borehole, and 1391 in the BDP-98 borehole. The vertical distribution of the thermal conductivity of the sediments was compared with the results of logging, including gamma log, conductivity, and resistivity measurements.

2 Techniques and Methods of Measurement

2.1 Temperature

Temperature measurements in the boreholes were made by the Institute of the Earth's Crust (V.A. Golubev), the Institute of Geophysics SB RAS (A.D. Duchkov and S.A. Kazantsev), and Irkutskgeofizika and Nedra Enterprises. The temperatures were measured using a 500-m-long cable equipped with three 12-kOhm thermistors, 30 m apart, at the lower end of the cable (Duchkov and Kazantsev 1996). Measurements were taken automatically at preset sampling intervals and the data were saved and stored on a hard disk. Further details are given by Fotiadi (1987).

The drilling operation in BDP-93 lasted for about 10 days (250–270 h) to reach a steady-state regime, including hole flushing and tear-down. The cable was installed in the borehole just after the drilling. Three thermistors were set at 26, 56, and 86 m beneath the lake floor in hole N2, and measurements were made for the period from 0 to 50 h. Thermistors at 28, 58, and 88 m in borehole N1 recorded temperatures for the period from 50 to 300 h. The combined measurements, with a small statistical adjustment for depth differences in N2 and N1, are shown in Fig. 2. The geothermal gradient varied from 36 to 48 mK/m, with a mean value of 41 mK/m (Duchkov and Kazantsev 1996).

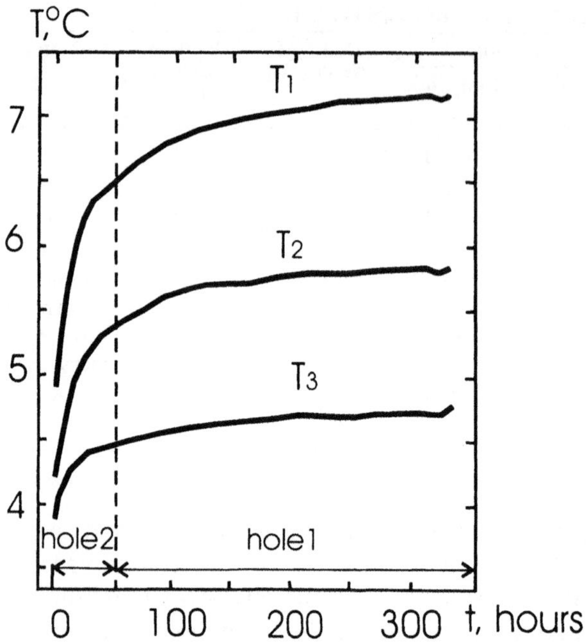

FIG. 2. Variation in the temperature of bottom sediments at three depths in the BDP-93-1 and BDP-93-2 boreholes after drilling (Duchkov and Kazantsev 1996). Temperatures are at depths: T_3, 28m; T_2, 58m; T_1, 88m

There was a disturbance in the temperature regime during the drilling when surface water, at a temperature of about +1°C, flushed a borehole to depths lower than 100m, where the temperature of the water was about 3.5°C. This resulted in a general lowering of the temperature.

Temperature recovery in the BDP-96 and BDP-98 boreholes took 72 and 34h, respectively. Some specific features of the drilling did not allow steady-state conditions in the boreholes to be maintained. The temperature was measured with a TR-5 (borehole resistance thermometer for logging; the rate of measurement is about 600m/h (Antipin et al. 2000). The temperatures in the wellhead and at the face of the borehole were 6 and 21°C, respectively, according to the Nedra Enterprise data. The calculations of stationary values of the geothermal gradient (γ) for the BDP-98 borehole were made within 30–500m; the interval at 0–30m and deeper than 500m were excluded because they were distorted by the effects of the lake bottom, and on abrupt temperature rise from the change in the drilling rate. The measured γ is 20–30mK/m. We estimated a steady-state regime using Kutasov's program (Kutasov et al. 2000). This program allows a thermogram of the borehole, obtained from the thermal log of disturbed drilling, to be transformed to the steady state. This program requires the time of drilling, and the depth and diameter of the borehole. The calculations showed that there is a steady geothermal gradient in the upper sedimentary layer up to 500m of 60 ± 10mK/m on average. With such a gradient, the temperature at the BDP-98 borehole face should be 43–44°C, i.e., 12–13°C higher than the measured one.

TABLE 2. Standards of thermal conductivity used for the calibration of the comparator of thermal conductivity device

##	Name of standard	Temperature (K)	Thermal conductivity λ (W/mK)	Standard error (%)	Total error (%)	
					With lubrication	Without lubrication
1	Polystyrene foam PS-100	295	0.042	2.5	3.5	5.8
2	Organic glass	295	0.195	0.7	1.1	1.7
3	Glass TF-1	295	0.696	0.7	1.2	1.2
4	Glass K-8	295	1.081	0.7	1.3	1.4
5	Glass KV	295	1.335	0.7	1.3	2.5
6	Alloy VT-6	295	6.622	1.1	2.1	2.7
7	Stainless steel 12X18H10T	295	14.740	0.8	3.7	3.9

2.2 Thermal Conductivity

The lake bottom sediments were measured by a comparator of thermal conductivity (Fig. 3). This instrument uses one of the steady-state methods for relative measurements of thermal conductivity, and is precalibrated against standard conductivities in the range 0.042–14.74 W/mK (Table 2), which provide a constant, high instrumental precision within 3%–5%. The sensor of the instrument is installed on an axially split surface of the standard (of Baikal bottom sediments), and a measurement of the thermal conductivity of the sediments is made without disturbing their integrity (Kalinin 1981).

The basic advantage of this device is a high measurement rate (about 1–2 min) and the possibility of measuring samples of an arbitrary shape and size. The operation of the heat comparator is based on a two-point thermal sounding of the surface of the sample by two thermal probes, between the upper ends of which the temperature difference (about 40 K) is created and held constant with the help of a semiconducting battery. The heat conductivity of the sample is indicated by the thermoelectric power of the differential thermal couple which is regulating the temperature difference between the lower ends of the thermal probes where they contact the sample surface in the existing heat mode.

We investigated 100 m of the sediments in the BDP-93 borehole, 200 m of those in the BDP-96/1–2 borehole, and from 350 to 600 m in the BDP-98 borehole. We also measured the thermal conductivity by the needle probe method (Golubev 1982).

3 Results

3.1 BDP-93

A total of 662 thermal conductivity values (λ) in borehole BDP-93 were determined by two different methods at different times. First, 109 measurements on

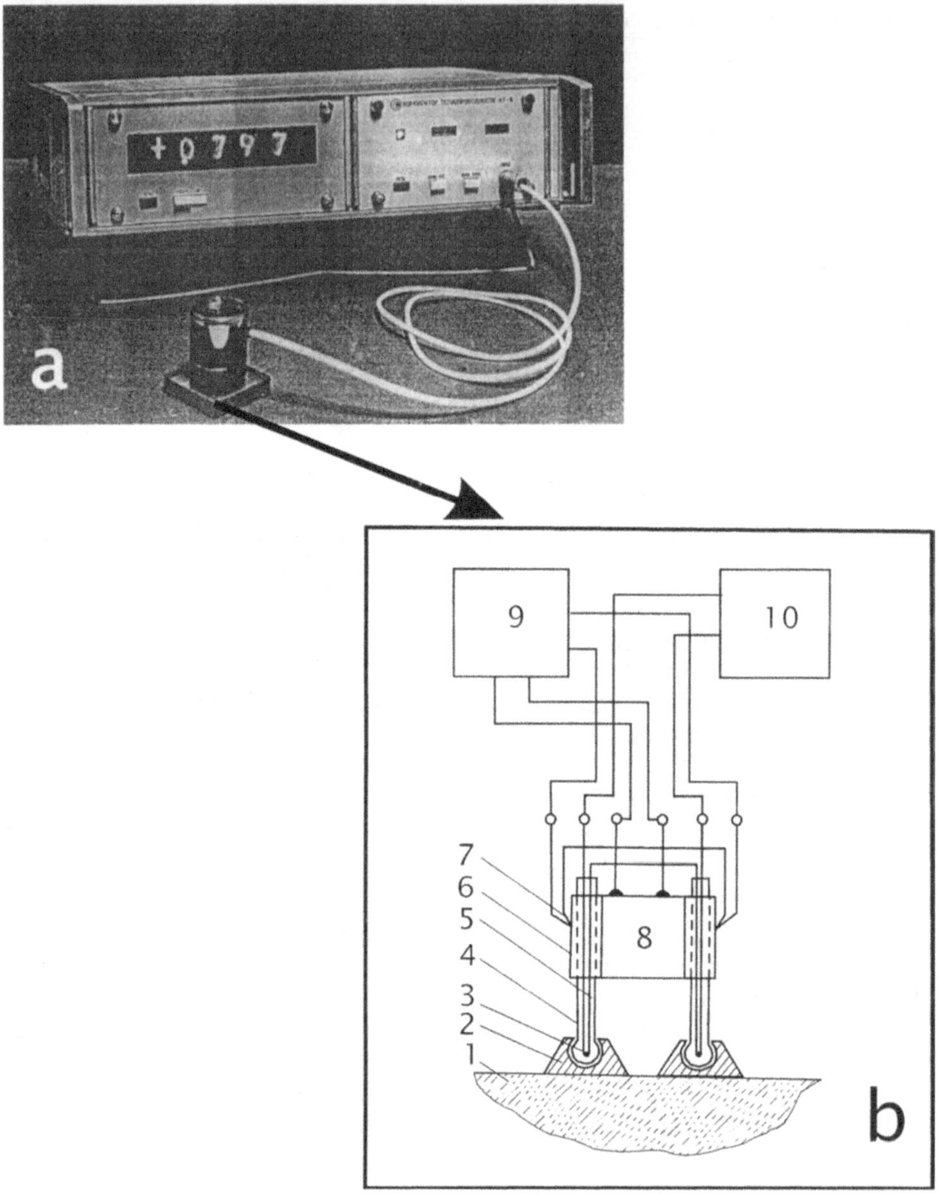

FIG. 3. **a** A general view of a comparator of thermal conductivity for thermal conductivity measurements. **b**. A diagram of the comparator of thermal conductivity. *1*, sample; *2*, copper tips; *3*, spherical ends of thermal probes; *4*, thermal probes, made of brass; *5*, measuring thermocouple; *6*, copper plates; *7*, regulating thermocouple; *8*, thermoelectric battery; *9*, automatic regulator; *10*, voltmeter

core N2 were made on the barge immediately after the core recovery by pressing the probe into its butt-ends. The sensing device was semiautomatic, and was designed following the concept of the constant power needle probe (a pulse–source method) (Golubev 1982; Fotiadi 1987). The measurements yielded a spread of values from 0.73 to 1.61 W/mK. A relatively low conductivity of 0.8–0.9 W/mK was obtained for the shallowest sediments within a depth range of 5–18 m. Below 18 m, the conductivity increased steeply due to the decrease in the water content (1.2 ± 0.2 W/mK).

Several weeks later, the thermal conductivity was measured by a comparator of thermal conductivity in the laboratory. The core samples were split axially into two halves, and their surfaces were sensed (553 determinations on 70 samples in the BDP-93 borehole). The results of thermal conductivity measurements in all boreholes are shown in Fig. 4. An increase from 1.4 to 1.8 W/mK was observed down to a subbottom depth of 50 m, which was followed by a decrease down to 1.6 W/mK between 50 and 80 m depth, and then by another increase starting from below 80 m. The arithmetic mean of the conductivities examined in the laboratory was 1.76 ± 0.04 W/mK, which is 30% higher than that measured on the barge on nonsplit samples by the needle probe. This may be due to the fact that the lower conductivity obtained immediately after the core recovery was caused by a dilution of sediments with water as they were lifted. Gases were observed during coring as expansion pockmarks of 3–10 cm of core liner length. At that time, gas was not sampled or analyzed, and its chemistry and exact amount remain unknown.

Lithologically, the recovered sediments are diatomaceous and pelitic mud with some sand admixture. The portion of pelitic material decreases downward, and the portions of silt and sand increase accordingly. The section showed a sharp demarcation line between the fine- and coarse-grained sediments at 50 m depth, with fine-grained in the upper and coarse-grained in the lower section.

The behavior of the thermal conductivity coefficient for the BDP-93 borehole was considered at various depths. The frequency of occurrence of all the λ values is shown at 20-m intervals (on a centimeter scale) in the form of a spreadsheet (Fig. 5) showing the thermal conductivity increase with depth.

An attempt was made to observe the behavior of the thermal conductivity of the sediments as a function of their lithology. Thermal conductivity diagrams were plotted separately for six lithological groups, divided according to the percentages of silt and sand they contained. The number of measurements varied from 38 on clay silt (50%–90% clay and silt) to 151 on clays. Individual anomalously low values belong to sediments containing organic cells, and the above-normal peaks correspond to consolidated sandy and silty lenses (Fig. 6).

The diagrams with λ values averaged over 2 and 5 m (see Fig. 4) show that the averaging does not influence the mean values of this parameter. As a whole, an abnormally high value of λ (1.76 ± 0.04) was typical of the bottom sediments in the BDP-93 borehole. Repeated measurements of the core in 1996 (BDP-93/2 borehole, core 8, sect. 1, D = 1224–1381 cm) showed a 10% decrease in the λ

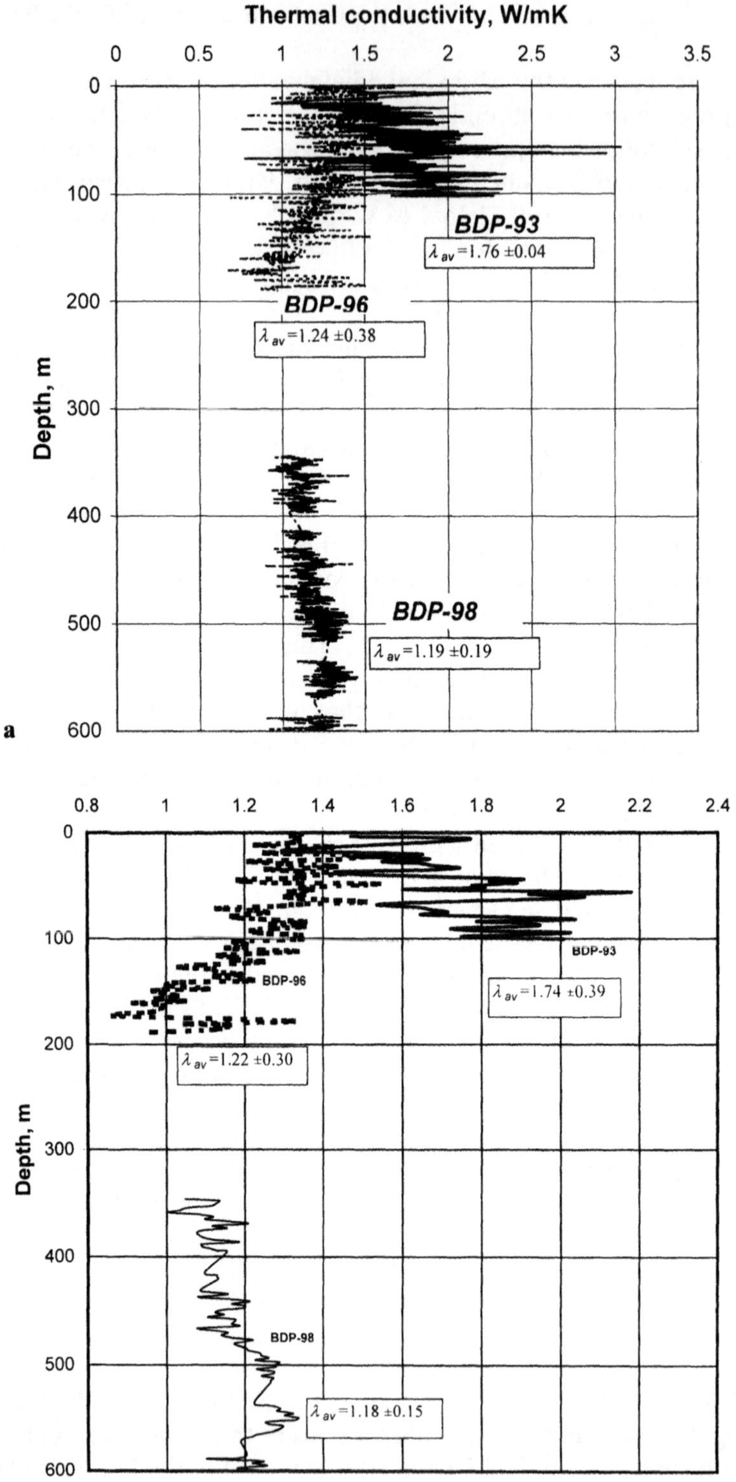

FIG. 4. Thermal conductivity of sediments in BDP-93, BDP-96, and BDP-98 boreholes: **a** total; **b** averaged in 2-m sections

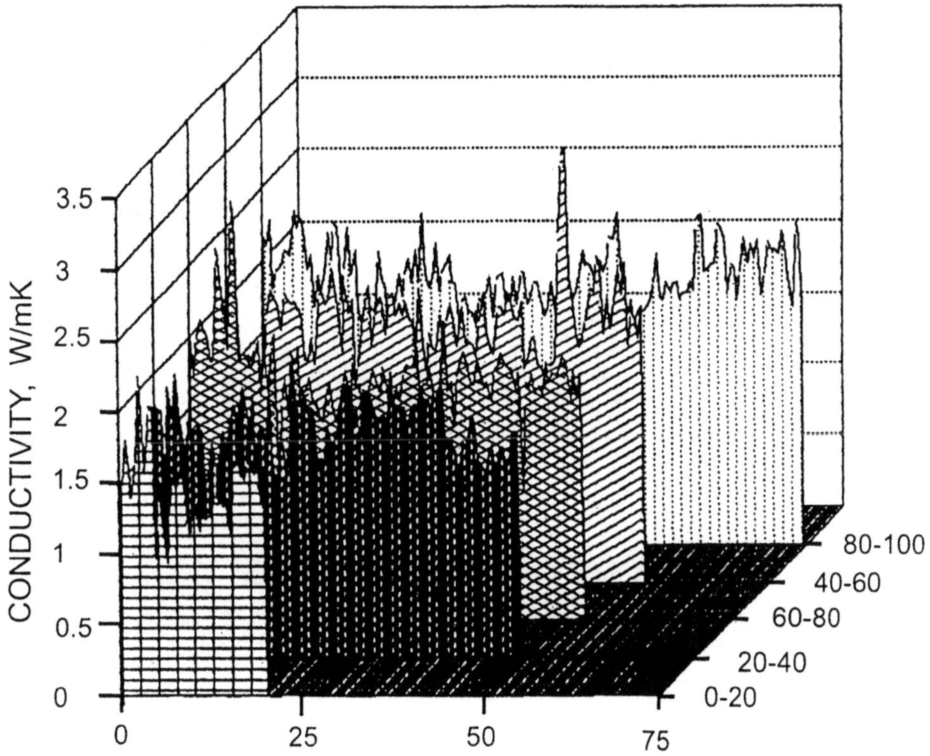

FIG. 5. Distribution of the thermal conductivity of sediments in 20-m successive segments down the column. The x-axis shows the number of measurements in each segment

parameter (1.29 against 1.44, mean over eight measurements), which was obviously due to water loss in the core (Fig. 7). The measurements in the first 5 m showed worse reproducibility (Dorofeeva 1994; Dorofeeva et al. 1994).

3.2 BDP-96

In total, 512 measurements were made in the BDP-96/1 borehole and 360 measurements in the BDP-96/2 borehole. The mean thermal conductivity of bottom sediments (λ) in the BDP-96/1 borehole was 1.24 ± 0.19 W/mK, and the parameter varied from 0.68 to 1.68 W/mK (see Fig. 4). With a large number of measurements averaged over every 2 and 5 m of the bottom sediments, the λ value remained almost constant (1.22 ± 0.30). Statistically processed data present a slight negative correlation between thermal conductivity and depth ($r = -0.55$). Thermal conductivity of the bottom sediments began to decrease with a change in drilling diameter from 78 to 56 mm (Fig. 8) at a depth of about 119.65 m. It was evident that the core had not survived so well in pipes with a large diameter; it was dehydrated by the drilling conditions (partly) or by lithological compaction, or both.

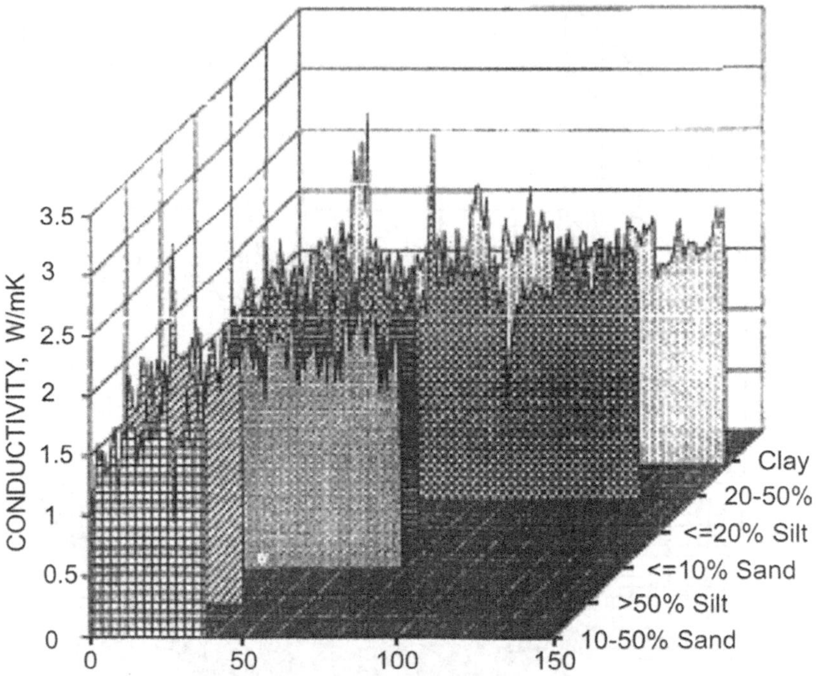

FIG. 6. Distribution of thermal conductivity of sediments in BDP-93/1 core according to their lithology. All data were divided into six sets: clays (clay (>80%), silt+sand (<20%)), <10% and (clay (60%–70%), silt (20%–30%)), 10%–50% sand (clay+silt (50%–90%)), <20% silt (clay (>80%)), 20%–50% silt (clay (50%–80%)), >50% silt (clay (15%–50%), sand (15–50%))

The thermal conductivity coefficient decreased sharply in diatomaceous mud. The value of the parameters investigated in the BDP-96/2 borehole showed good agreement with the BDP-96/1 core (1.31 ± 0.16) (Fig. 8). However, thermal conductivity measured with a needle probe is about 1.5 times lower (1.30/0.88 W/mK). The same tendency was observed in the BDP-93/1 borehole (1.76/1.16).

A similar discrepancy between thermal conductivity values measured by different methods was obtained by Japanese scientists during an investigation of the thermal conductivity of the bottom sediments in a 1000-m borehole drilled in Lake Biwa (central Japan) (1.28 × 0.92). The Japanese quick thermal conductivity meter (QTM) also gives constant power supply probe measurements by the tangent method, as in Golubev's investigations (1982), but the QTM is made in the form of a plate, not a needle (Fujisawa et al. 1985; Horie 1987). The Japanese authors do not propose any straightforward explanation of their results, but they use the QTM data in a computation of the heat flow value (Fig. 9). There is probably some influence from interstitial water evaporated through the surface, and also from inhomogeneities in the bottom sediments of Lake Biwa (Uyeda et al. 1973).

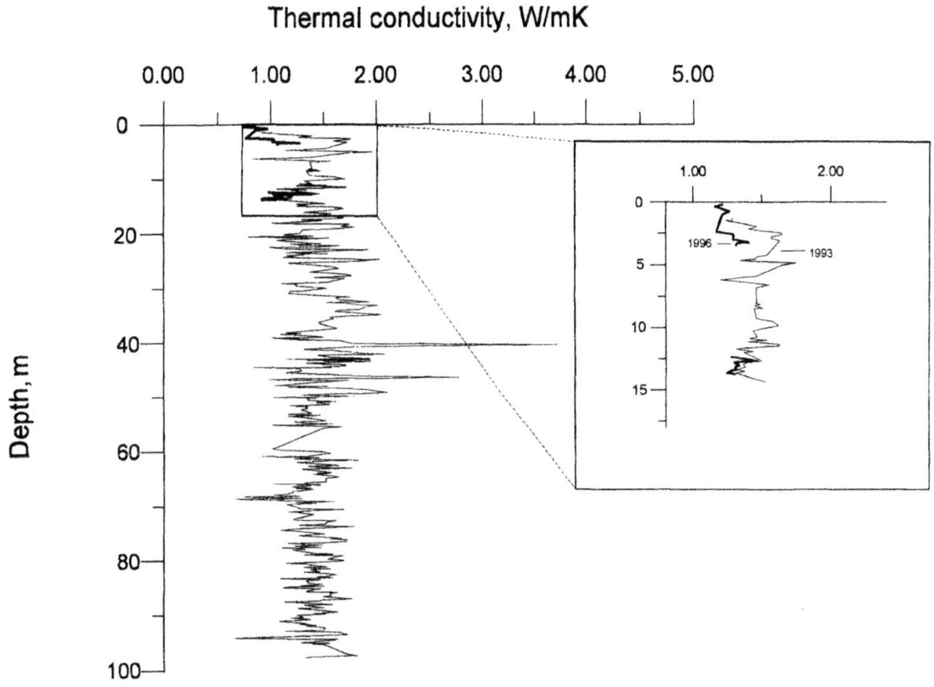

FIG. 7. Thermal conductivity distribution in the BDP-93/2 borehole (*thin line*) and its repeated measurement in 1996 (*bold line*)

3.3 BDP-98

In the BDP-98 core, 1391 measurements of λ were made at a depth range of 345–600 m. Statistical processing was carried out on 1382 measurements of the thermal conductivity of bottom sediments, but excluded nine values which were beyond 2σ. The results of these measurements are shown in Fig. 4.

The mean value of the thermal conductivity ($\lambda \pm 2\sigma$) for 1382 measurements was 1.19 ± 0.19 W/mK. The parameter varied from 0.88 to 1.44 W/mK, but these are individual minimum and maximum values. Generally, in the depth range 345–440 m, the thermal conductivity λ was 1.1 W/mK. It increased (1.25–1.30 W/mK) to a depth of 555 m, and then slightly decreased to 1.20–1.25 W/mK at a depth of 600 m.

The cause of the discrepancy between λ values in the BDP-96 and BDP-93 boreholes is unknown. No discrepancy in λ was obtained in the BDP-98 core at 345–600 m by different thermal conductivity measurement techniques (V.A. Golubev 1998, personal communication).

The mean geothermal gradient in the BDP-98 borehole in the 500-m thick upper sedimentary layer was 60 ± 10 mK/m, and the heat flow was 72 mW/m^2, which is in agreement with earlier published values (Antipin et al. 2000).

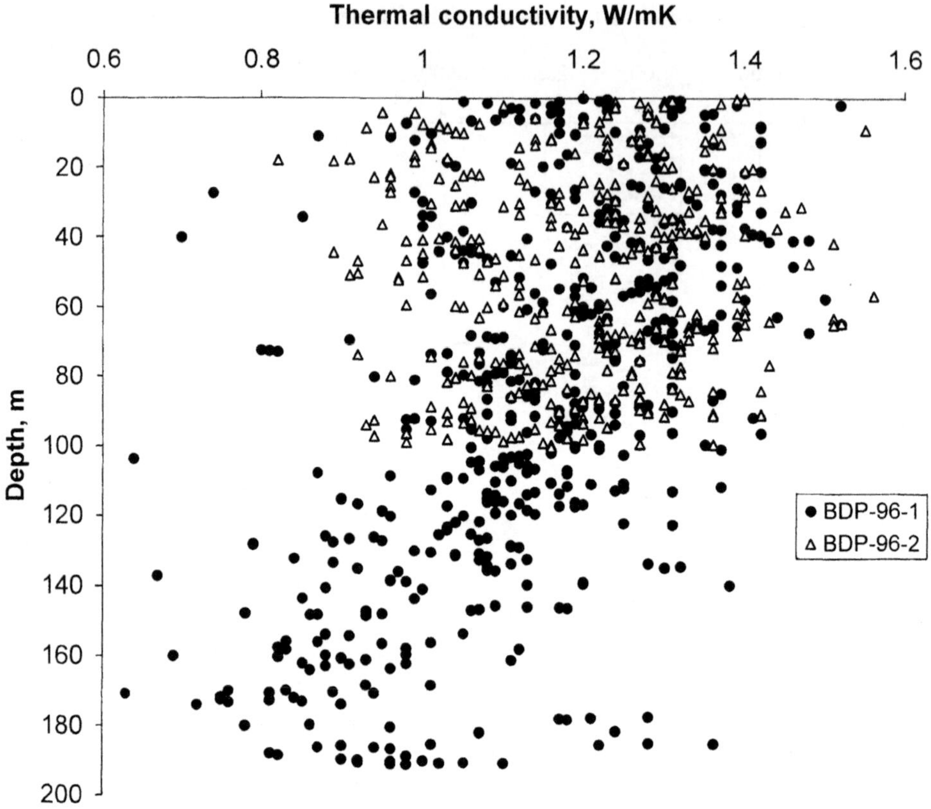

FIG. 8. Thermal conductivity of the bottom sediments in the BDP-96/1 and 96/2 boreholes measured by a comparator of thermal conductivity (n = 512 and n = 360, respectively)

3.4 Relationship between Thermal Conductivity and Other Physical Parameters

The relationships between the coefficient of thermal conductivity of the Lake Baikal bottom sediments and the water content, conductivity, resistivity, and natural γ-rays were only investigated for the BDP-98 borehole. The relationship between these parameters was considered in the range of thermal conductivity measurements from 345 to 600 m.

The water content of these bottom sediments was 40% up to 470 m, and no specific features in thermal conductivity variation were traced in this depth range, but an inverse relationship was partially observed below these depths: the decrease in the water content increased thermal conductivity (depths of 500 and 550 m) (Fig. 10).

The relationship between the resistivity of the bottom sediments and the thermal conductivity is evident (Fig. 11). The negatively correlated behavior of

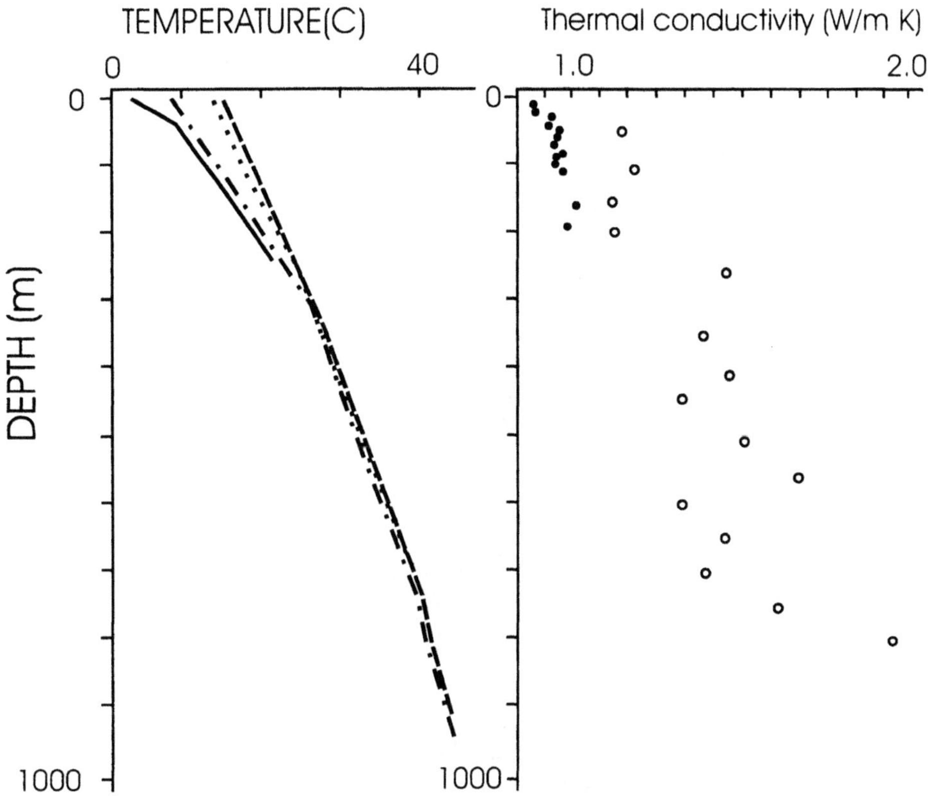

FIG. 9. Temperature gradient and thermal conductivity measured by Japanese scientists (Uyeda et al. 1973; Fujisawa et al. 1985). The *continuous line* in the left-hand corner is the results of Uyeda et al. (1973). The *dot–dash line* shows estimated original temperature profiles. *Dotted lines* and *dashed lines* show measurements on May 8 and May 2, respectively. *Open circles* are thermal conductivities measured by QTM, and *solid circles* are those by the needle probe method of Uyeda et al. (1973)

these parameters is more obvious if the values are averaged over every 10 m (Fig. 11a).

Conversely, a positive correlation is generally observed between thermal and electrical conductivity (electrical log) for depths of 350 and 500 m (Fig. 12). The range of electrical conductivity variation is fairly stable after 400 m, and is 230–250 mS/m.

It is known that the natural radioactivity of the upper part of the Earth is conditioned by a large number of unstable (radioactive) isotopes of various light and heavy elements (long-living, short-living, and single radioactive long-living isotopes) in the rocks and minerals that compose this cover. The behavior of thermal conductivity/natural radioactivity curves is in very good agreement at depths below 450 m. They are in antiphase at depth ranges of 365–385 and

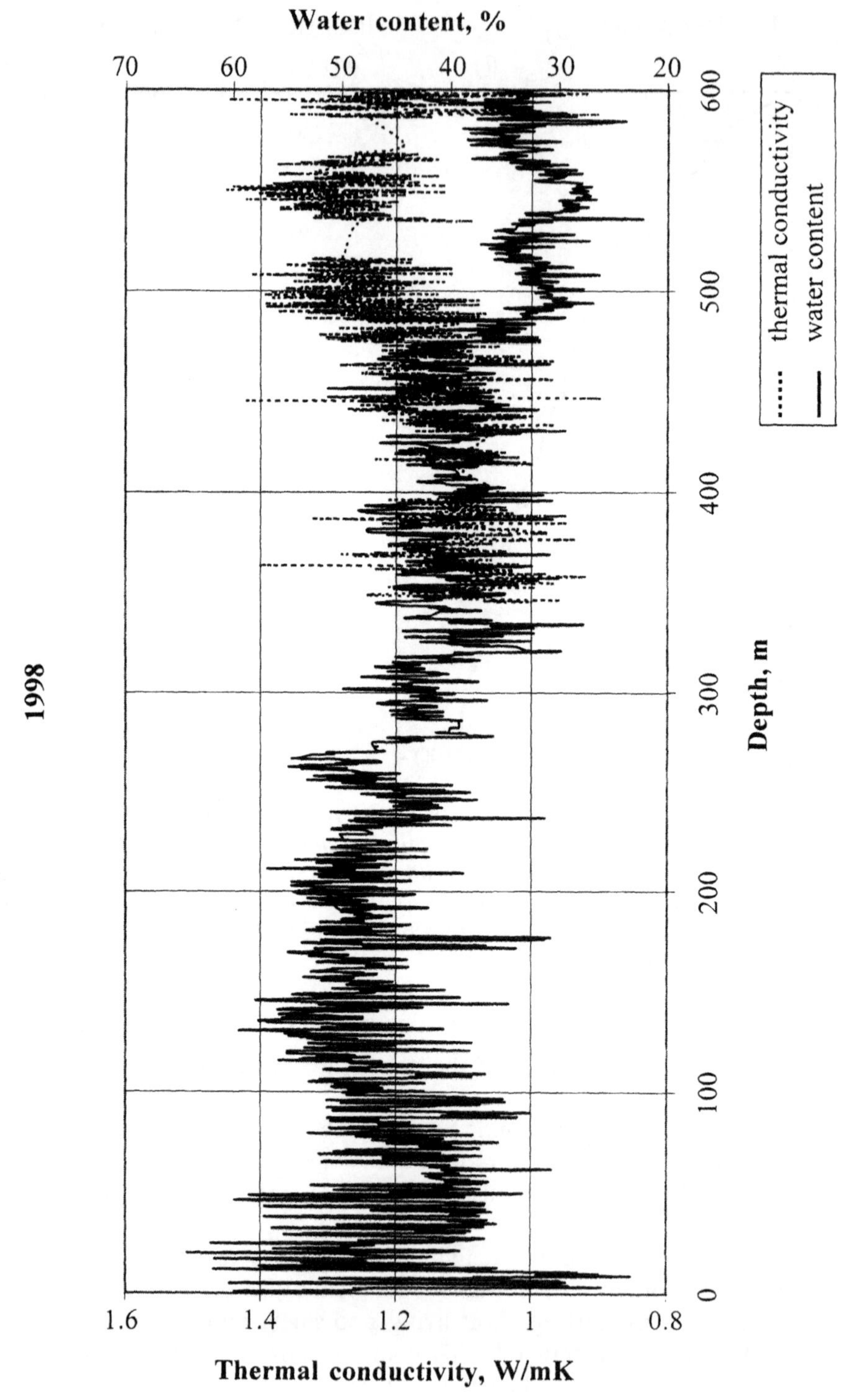

FIG. 10. Thermal conductivity and water content in the BDP-98 borehole

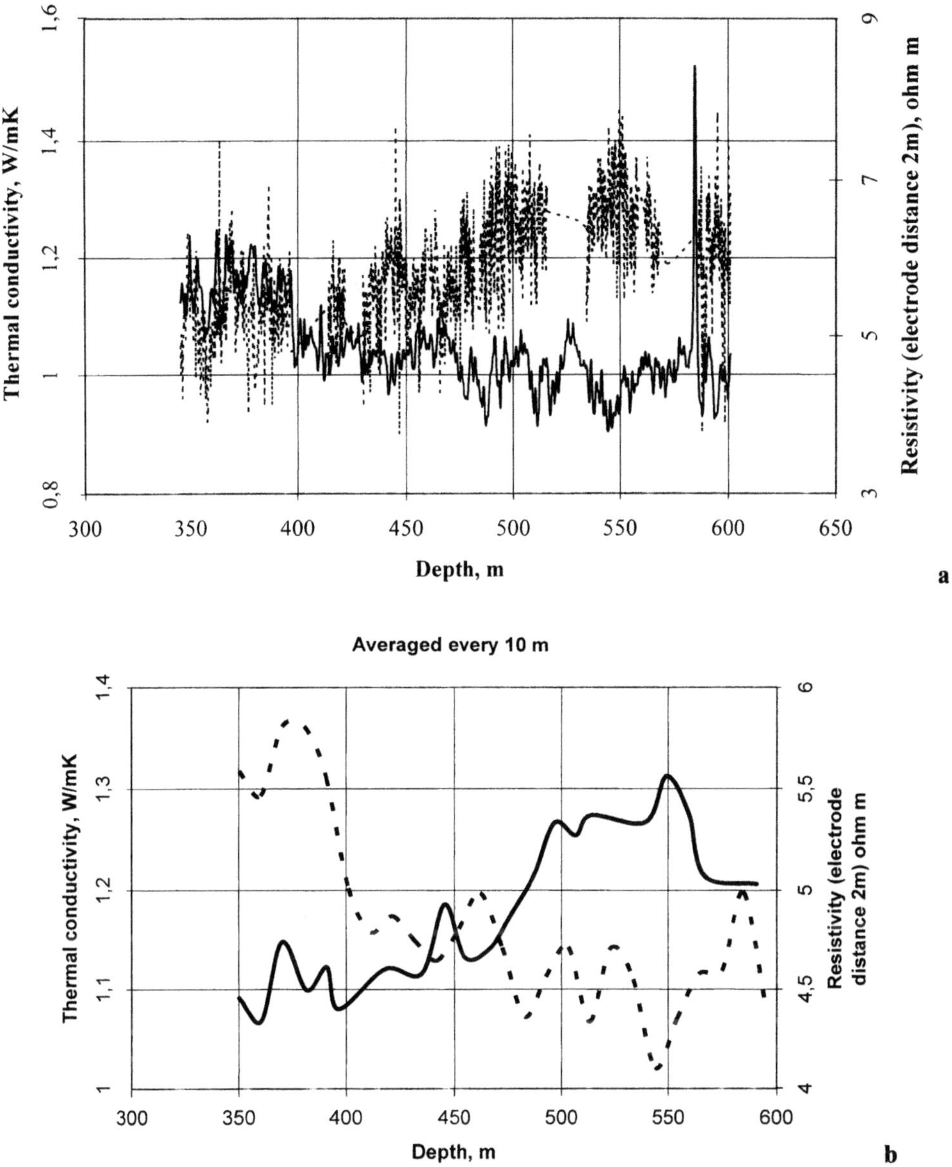

FIG. 11. **a** Thermal conductivity and resistivity distribution in the BDP-98 borehole. *Broken line*, thermal conductivity; *solid line*, resistivity. **b** Values averaged every 10 m. *Solid line*, thermal conductivity; *broken line*, resistivity

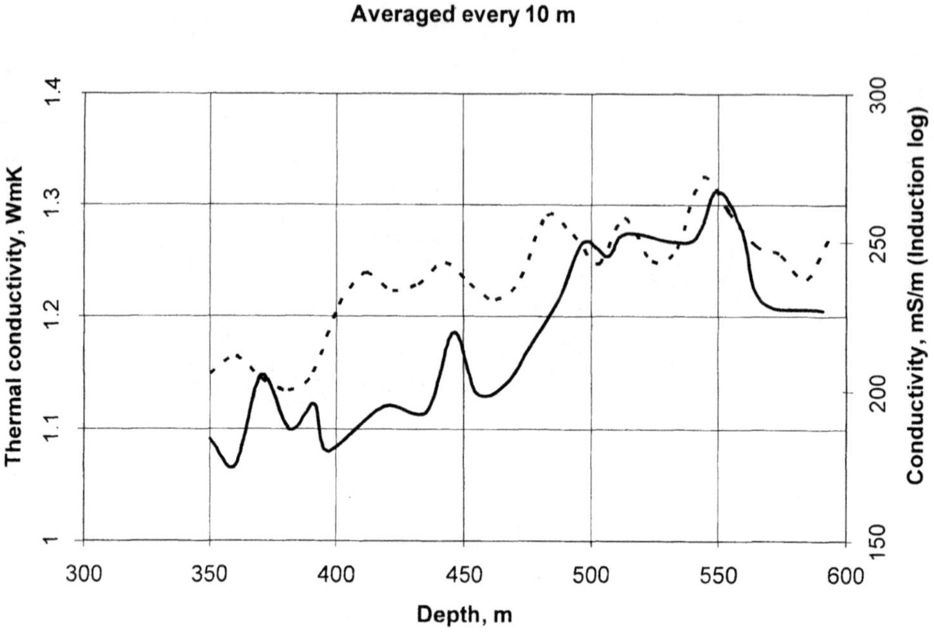

FIG. 12. Thermal conductivity and conductivity distribution in the BDP-98 borehole, averaged every 10 m. *Solid line*, thermal conductivity; *broken line*, electrical conductivity

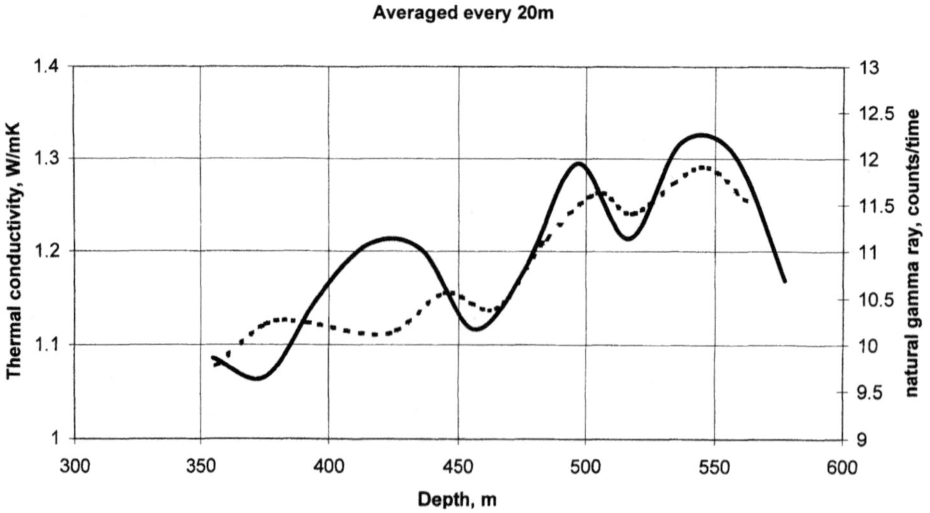

FIG. 13. Thermal conductivity and natural γ-ray distribution in the BDP-98 borehole, averaged every 20 m. *Broken line*, thermal conductivity; *solid line*, natural γ-rays

385–450 m. An inverse relationship in the behavior of the two parameters considered might be explained by an increasing percentage of large silt and sand seams in the Lake Baikal bottom sediments at these depths (Fig. 13).

4 Conclusion

Geothermal studies of BDP-93, BDP-96, and BDP-98 boreholes determined the thermal conductivity of the Lake Baikal bottom sediments to depths of 600 m. Earlier measurements to determine heat flow values using a special probe were only used for shallow (up to 1 m) sediments.

Thermal conductivity values measured in the BDP-93 and BDP-96 boreholes by a comparator of thermal conductivity and a needle probe showed clear discrepencies (1.5 times), but this was not found in the BDP-98 borehole.

The relationships among the thermal conductivity of the bottom sediments, water content, electrical conductivity, resistivity (electrical log), and natural radioactivity (γ-log) were investigated.

Acknowledgments. This work had the financial support of the Russian Fund of Fundamental Investigations (Grant No. 00-05-81154 Бel-2000_a) and the Integration Project SB RAS. We thank T. Leshkevich for the translation. Thanks are also extended to Prof. K. Kashiwaya, T. Nagao, H. Tsukahara, and Dr. V. Gelety who provided us with geophysical data and reports of the heat flow studies in Lake Biwa.

References

Antipin V, Afonina T, Badalov O, Bezrukova E, Bychinsky V, Bukhariov A, Williams D, Gelety V, Gvozdkov A, Golubev V, Goreglyad A, Gorokhov I, Dmitriev A, Dorofeeva R, Duchkov A, et al. (2000) Late Cenozoic paleoclimatic record in the sediments of Lake Baikal, from the investigation data of a 660-m core from deep drilling (in Russian). Russ Geol Geophys 41:3–32

Dorofeeva RP (1994) Thermal conductivity measurements of the bottom sediments in the core from a sub-water borehole in Lake Baikal. Proceedings of an International Conference on Baikal as a Natural Laboratory to Study Environmental and Climatic Changes. Lisna, Irkutsk, vol 4, p 9

Dorofeeva RP, Duchkov AD, Kazantsev SA (1994) Preliminary geothermal results from underwater boreholes in Lake Baikal. IPPCCE Newsl 8:71–76

Duchkov AD (ed) (1985) Catalogue of heat flow data in Siberia (1966–1984) (in Russian). Novosibirsk, Institute of Geology and Geophysics SBRAS, p 82

Duchkov AD, Kazantsev SA (1996) Temperature measurements in the first sub-water boreholes in Lake Baikal (in Russian). Russ Geol Geophys 37:95–103

Duchkov AD, Lysak SV, Golubev VA, Dorofeeva RP, Sokolova LS (1999) Heat flow and geothermal fields of the Baikal region (in Russian). Russ Geol Geophys 40:289–304

Fotiadi EE (ed) (1987) Thermal field of the interior of Siberia (in Russian). Nauka, Novosibirsk, p 196

Fujisawa H, Nagao T, Uyeda S, Honda S, Tsunoda T (1985) Heat flow measurements in Lake Biwa, Central Japan. A preliminary study. Proc Jpn Acad 61:145–148
Golubev VA (1982) Geothermics of Lake Baikal (in Russian). Nauka, Novosibirsk, p 150
Horie S (1987) Die Geschichte des Biwa-Sees in Japan. Wagner, Innsbruck, p 280
Kalinin AN (1981) An instrument for direct measurements of thermal conductivity on solid samples of an arbitrary shape (in Russian). Prom Teplotek 3:48–53
Kutasov IM, Eppelbaum LV, Dorofeyeva RP (2000) Physical–mathematical problem of the recent climate reconstruction from subsurface temperatures logs. Sci Israel Technol Adv 2:79–83
Uyeda S, Watanabe T, Mizushima N, Yasui M, Horie S (1973) Terrestrial heat flow in Lake Biwa, Central Japan. Preliminary Report. Proc Jpn Acad 49:341–346

Geomorphological Development of the Tunka Depression in the Baikal Rift Zone in Siberia, Russia

YOSHITAKA HASE[1], SERGEY K. KRIVONOGOV[2], and AKIKO IWAUCHI[3]

Summary. There is no large lake, such as Lake Baikal or Lake Hovsgol, in the Tunka depression now, but it is one of the depressions of the Baikal rift zone. The bottom of the depression is formed of Miocene to Holocene rocks set on Precambrian and Paleozoic rocks, which mainly crop out in mountains north and south of the depression. In Miocene times, lacustrine sediments were formed and basaltic lava flows erupted. However, in the Pliocene and Pleistocene, fluvial or fluvioglacial sediments were formed. Therefore, it is assumed that the development of the Tunka depression is different from that of Lake Baikal and Lake Hovsgol. The difference has to do with the outflow of water via the Irkut river during the Pliocene, Pleistocene, and Holocene. In the Pliocene, any large lake apparently disappeared, because the sediments are composed of coarse sand and pebbles produced by the rivers. In the Pleistocene, many gravel beds were formed by fluvial and fluvioglacial processes, and, especially in the middle Pleistocene, water probably flowed out of Lake Baikal to the Irkut river via the Kultuk valley (Kononov and Mats 1986; Mats et al. 2000).

Key words. Baikal rift zone, Tunka valley, Irkut river, Drainage system, Sedimentary environment

1 Introduction

The Tunka basin, which is the main part of the Tunka valley, is located about 70 km west of the western border of Lake Baikal (Fig. 1). The valley, with flat land in the bottom, is 150 km from east to west and 30 km from north to south

[1] Department of Earth Sciences, Faculty of Science, Kumamoto University, 2-39-1 Kurokami, Kumamoto 860-8555, Japan
[2] United Institute of Geology, Geophysics and Mineralogy, Siberian Branch of the Russian Academy of Sciences, Academician Koptyug Avenue, 3630090 Novosibirsk, Russia
[3] Avance Co. Ltd., 1-3-39 Shinhoka, Kumamoto 862-0921, Japan

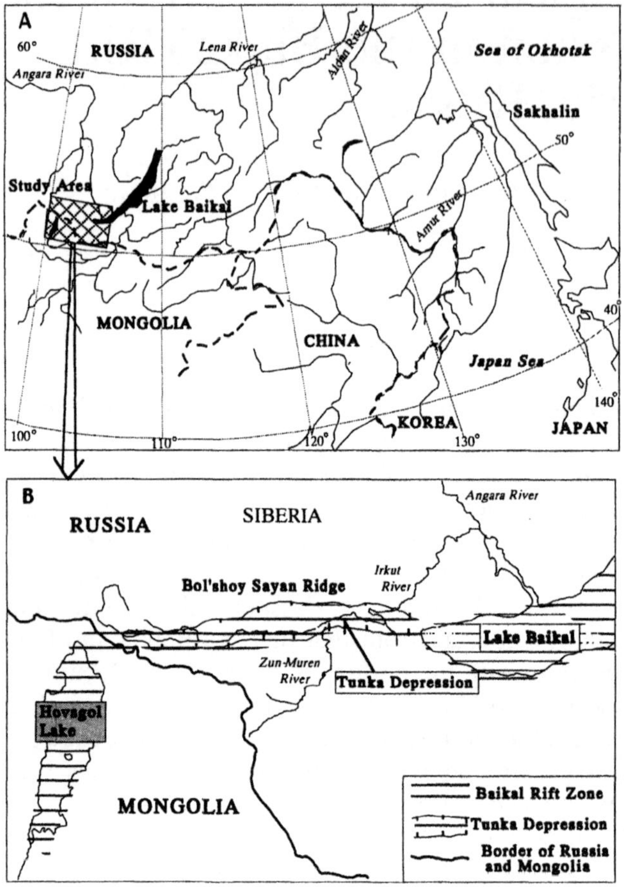

FIG. 1. Map showing the northern area of East Asia (**A**) and the Tunka depression in the Baikal rift zone (**B**)

at its widest part. The valley between the highland on the north and the mountains on the south, the Tunka depression, belongs to the Baikal rift zone. Its geological structure is quite complicated. Many reports on specific aspects of its structural geology and stratigraphy, and descriptions of various localities have been published.

The tectonics, structure, and development of the Tunka area were investigated in detail by Florensov (1960), Sherman et al. (1973), and Logatchev (1974). Mats (1992) and Mats et al. (2000) summarized newer data on the morphostructure and Cenozoic history of the Baikal rift zone, and Ufimtsev (1995) and Ufimtsev and Shchetnikov (2001) reported on the Tunka subarea. The distribution of basaltic lava flows and the age of volcanism in the Tunka were discussed by Logatchev (1954), Logatchev and Kravchenko (1955), Rasskazov (1990, 1993), and Rasskazov et al. (1990). Dmitriev (1968) reported on the results of brown coal exploration, and described lake stages in the Tunka during the Neogene. The late Cenozoic, and particularly the late Pleistocene, sediments and events were also thoroughly studied (Golubeva and Ravsky 1962; Adamenko et al. 1975;

Ravsky 1972). Most of these scientists assumed that lake conditions existed in the Tunka in the Pleistocene. Contrary to their opinion, Krivonogov (1995) briefly discussed nonlake sedimentation in the Tunka valley during the late Pleistocene, and compared it with conditions in the Chara basin, one of the northern territories of the Baikal rift zone, where an ice-dammed lake existed.

Based on our research expeditions of 1995 and 1996, it was shown that the late Cenozoic sedimentary strata in the Tunka valley, as mentioned by Logatchev (1974) and Ravsky (1972), were mainly formed out of sand and gravel. Most of the strata from the Pliocene to the present interglacial were formed in fluvial, fluvioglacial, and eolian environments. There are not many fine lacustrine deposits, except for the Miocene.

The reason that there were not so many lacustrine sediments from the Pliocene to present, although such sediments were deposited in a subsiding area, was clarified by our investigation of the geomorphological development of the Tunka area. We would like to consider the development of the Tunka depression according to its morphological and geological features.

2 Geomorphic and Geological Features of the Tunka Valley

Figure 2 shows the topography of the area to the west of Lake Baikal. The Tunka valley is located in the depressions between Lake Baikal, in Siberia, and Lake Hovsgol, in Mongolia. According to geomorphic and geological characteristics, the area is divided into four parts, as shown in Fig. 2: northern up-lifted highlands, southern mountains, low mountains in the northeast of the area, and basins and hills in the depression.

2.1 Highlands to the North of the Valley

To the north of the Tunka valley, there are highlands which are named Tunkinskiye Gol'tsy and Kitoyskiye Gol'tsy, both in the Bol'shoy Sayan ridge (see Fig. 1). The summit levels are shown in Fig. 2. Some of the mountain peaks are over 3000 m, and many peaks are 2500–3000 m in height. The mountains are deeply dissected by rivers that generally flow eastward.

Two faults define the ridge from the northwest area of Lake Baikal and the Tunka depression (Fig. 2). The main Sayan fault goes through the eastern part of the area in a northwest–southeast direction from the western end of Lake Baikal. The other makes steep cliffs bordering the basin and ridge. This ridge at the north of the depression is mainly composed of Precambrian and Paleozoic rocks (Florensov 1960).

2.2 Southern Mountain Area

The mountain area south of the Tunka depression has peaks of 2000–2500 m, with a few peaks of more than 2500 m (Fig. 2). The watershed of the mountains

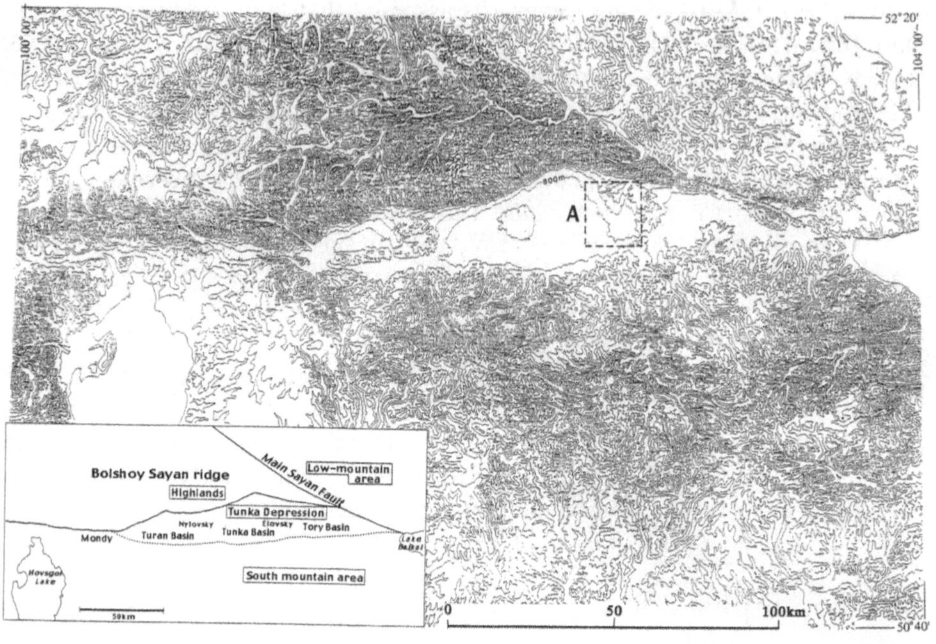

FIG. 2. Contour map showing the topography of the Tunka depression and the surrounding area. The contour interval is 200 m. The minimum contour is a 600-m line next to the shore of Lake Baikal, and the maximum is a 3400-m line at the peak north of Lake Hovsgol. The square, *A*, is the area shown in Fig. 4

runs roughly east and west, and partly north and south. The peaks surrounding the area descend slowly, and the cliffs along the rivers are steeper than around the peaks. The altitude of the peaks gradually decreases from south to north in the northern area of the watershed. These mountains are composed of Precambrian and Paleozoic rocks (Florensov 1960).

2.3 Low Mountain Area

The low mountains (2000–1500 m) spread from southeast to northwest on the northwest of the western margin of Lake Baikal (Fig. 2). The southern border of the low mountains is the main Sayan fault. The low mountains consist of Archean or early Proterozoic granite and gneiss (Logatchev 1974).

2.4 Basins and Hills in the Depression

2.4.1 Basins

The depression is divided into five basins by hills. The basins are Tory, Tunka, Turan, and two other basins, which lie in a straight line from east to west in the depression (Fig. 2). Their bottoms are at an elevation of about 750 m. The main river, the Irkut, flows from west to east, with meanders (Fig. 3). In the main part of its valley, the riverbeds are sandy, but in the upper parts they are characterized by pebble and boulder deposits. The river terraces are very restricted in extent, and the side slopes of the valley are mainly characterized by fan. In many places a thick layer of eolian sands, formed in the periglacial age, covers these sediments. Some cinder cones exist on the bottom of the Tunka basin. The Chersky volcano erupted in the Pleistocene (Table 1).

2.4.2 Hills Separating the Depressed Areas

There are hills between the basins with peaks 1500–1400 m in elevation (Fig. 2). The hills between the Tunka and Turan basins are composed of granitic and metamorphic rocks (Florensov 1960; Logatchev 1974) and they are called the Nylovsky border. The hills between the Tory and Tunka basins, the Elovsky

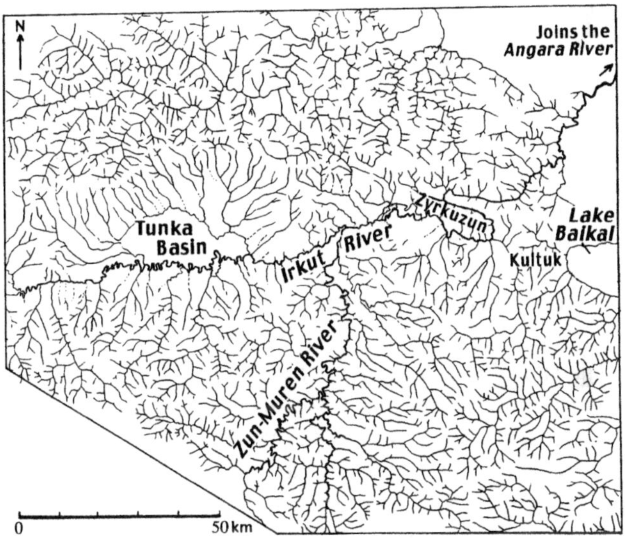

FIG. 3. Map showing the drainage system in the Tunka area. The Irkut river is extremely meandering, and the Zun-Muren river also meanders. Other rivers have less sinuosity. The Irkut river flows into the high northern mountains, but nevertheless does not flow through the low border into Lake Baikal

TABLE 1. K-Ar ages of the basalts situated in the Tunka depression

Sample number	Analyzed material	Isotopic age (Ma)	40Ar[a] (scc/g × 10^{-5})	% 40Ar[a]	% K
KA96-4412 1995916	Whole rock	4.4 ± 0.2	0.020	82.2	1.13
Belyijar intrusive rock			0.19	79.2	1.13
KA96-4413 1995919	Whole rock	0.7 ± 0.4	0.003 (7)	6.0	1.25
Chersky volcano			0.003 (5)	5.4	1.24
KA96-4414 1995920	Whole rock	4.1 ± 0.2	0.030	80.6	1.81
Hobok			0.028	83.7	1.81

[a] Analyzed by Teledyne Isotopes

border, are composed of Proterozoic granite, metamorphosed rocks (green schist), and Cenozoic volcanic and sedimentary rocks (Florensov 1960; Logatchev 1974). Miocene and early Pliocene basalts (Table 1 except for the Chersky volcano) are exposed on the upper half of the hills.

3 Drainage Patterns in the Tunka Area

Figure 3 shows the drainage system in the area to the west of Lake Baikal to the Tunka depression. The patterns of drainage in the northern and southern mountains are dendritic and rectangular, and in the plain, in the depression, the pattern is parallel except for the Irkut river.

There are three characteristics of the drainage system in this area. Firstly, the Irkut river from Mondy to Irkutsk meanders very extensively in the Tory basin. As the result of the extremely curved course, the Irkut river goes through to the Angara river, not into Lake Baikal. Secondly, the channel of the Irkut river in the Zyrkuzun curvature area shows a straight pattern. The Irkut river has a very strange course, because the land dividing the Tory basin and Lake Baikal is lower than the northern mountains; nevertheless, the Irkut river does not flow into Lake Baikal. The Irkut river channels in the mountainous area are deepened along the course of the river. Thirdly, there is only one meandering river, the Zun-Muren river; other streams in the southern mountains do not meander.

4 Radiometric Ages

The samples of basaltic rocks and carbonic matter in the sediments were dated by radiometric methods (Table 1 and Fig. 4). The basalt outcrop on the right side of the Irkut river is 4.4 ± 0.2 Ma, and one of the basalt flows making up the hills in the eastern part of Galbay village is 4.1 ± 0.2 Ma, both by the K–Ar method. These basalt flows demonstrate Pliocene volcanic activity in the Tunka depression. Another activity in the depression is shown by the K–Ar dating of Chersky volcanic rocks. These show 0.7 ± 0.4 Ma. Therefore, early or middle Pleistocene volcanic activity in the depression was demonstrated by our studies. Rasskazov (1993) and Rasskazov et al. (1990) presented basalt dates indicating an origin of volcanism from the Oligocene to the middle Pleistocene.

5 Sedimentary Environments of the Tunka Depression

Basalt and tuff exposed at the cliffs of the Irkut river (Fig. 4) have been assumed to date from the Oligocene to the Pliocene. Our new data involve only the youngest lava flows of the Pliocene age (Fig. 4). The Miocene basalt and sedi-

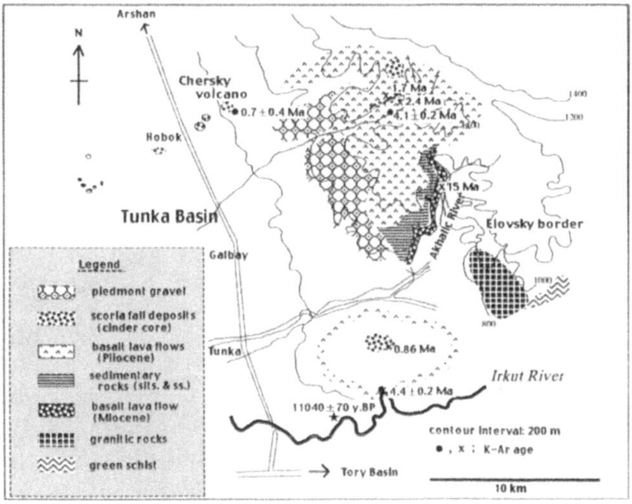

FIG. 4. Route map showing a succession of basaltic lava flows, and sediments of Miocene, Pliocene, and Pleistocene in the Elovsky area of the east Tunka basin. *Solid circles*, K–Ar ages (new data); *star*, Rasskazov (1993)

mentary rocks are situated close to the Elovsky border in the valley of the Akhalic river. The Miocene siltstone includes some plant macrofossils, but the fossils are very poorly preserved except for the fruits of the *Trapa* sp. This indicates that the siltstone was deposited in a lacustrine condition. The early Pleistocene sediments of this area are composed of coarse material such as sand and gravel, and are intercalated partly by fine-grained sand, and partly by silt and clay with carbonaceous matter. It is known that these sedimentary facies were produced in fluvial processes. Krivonogov (1995) proposed that the late Pleistocene to Holocene sediments mainly originated as diluvium accompanied by soil production, solifluction, and eolian deposition.

6 Consideration of Tectonic Movement

The fault system is thought by Sherman and Levi (1977) to be a transform fault concerned with the formation and spread of Lake Baikal. The current movement of the main Sayan fault is concluded as a left-lateral slip, which was shown by our study to have caused the displacement of the flow course of the Irkut river. Figure 5A shows the route of the ancient Irkut river during the Pliocene, and Fig. 5B shows the present Irkut river.

The northern mountainous area has been up-lifted from the Oligocene to recent times, and the southern mountainous area has been up-lifted and tilted from south to north at the same time (Fig. 6). Figure 6B shows the projected profile of the area. An imaginary line of the land surface descends gently down toward the north. Ufimtsev (1995) showed the transverse profile across the strike of the Tunka depression. It was tilted, with south up and

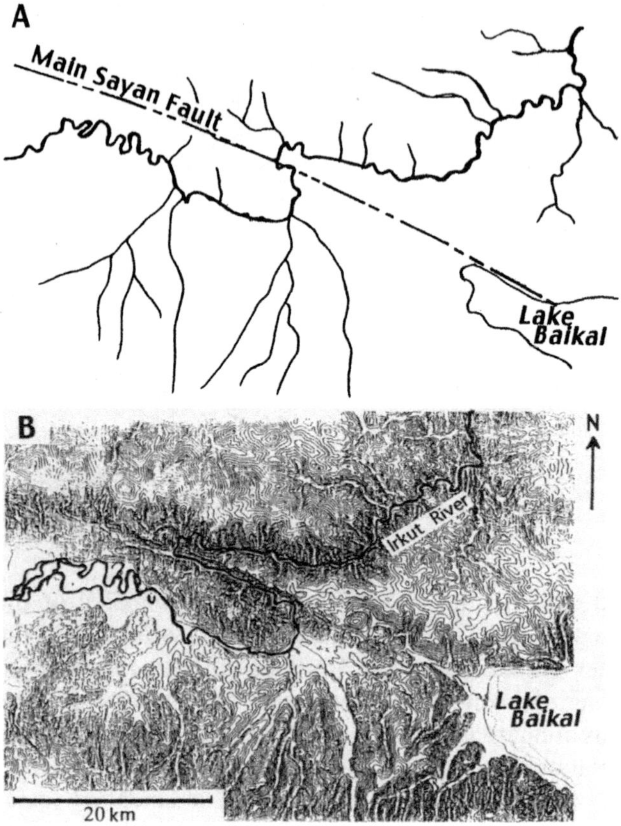

FIG. 5. Maps showing the changed route of the Irkut river from the Pliocene to the present. The contour interval is 50 m. The maps show an estimate of the lateral movement along the main Sayan fault

north down in the northern mountains, and was a bending arch in the southern mountains.

7 Geological Processes Affecting the Geomorphological Development of the Tunka Area

Movements of the Baikal depression started in the Oligocene (Mats 1992). K–Ar age-dating of basalts, evidencing volcanism, suggests that movements of the Tunka depression occurred in the late Cenozoic era. In the Miocene epoch, there were some lakes in this area, and the streams from the lakes flowed into the plain. The plain spread to the northern and southern part of the area. The river kept along the route from the lake as the low mountains rose up in the late Cenozoic era. Fluvial depositional environments continued during that time. Sometimes, eolian deposits were laid down in the depression. Strata intercalated in sediments in the early Pleistocene may suggest a lacustrine environment, but we did not see evidence of any big lakes in the depression.

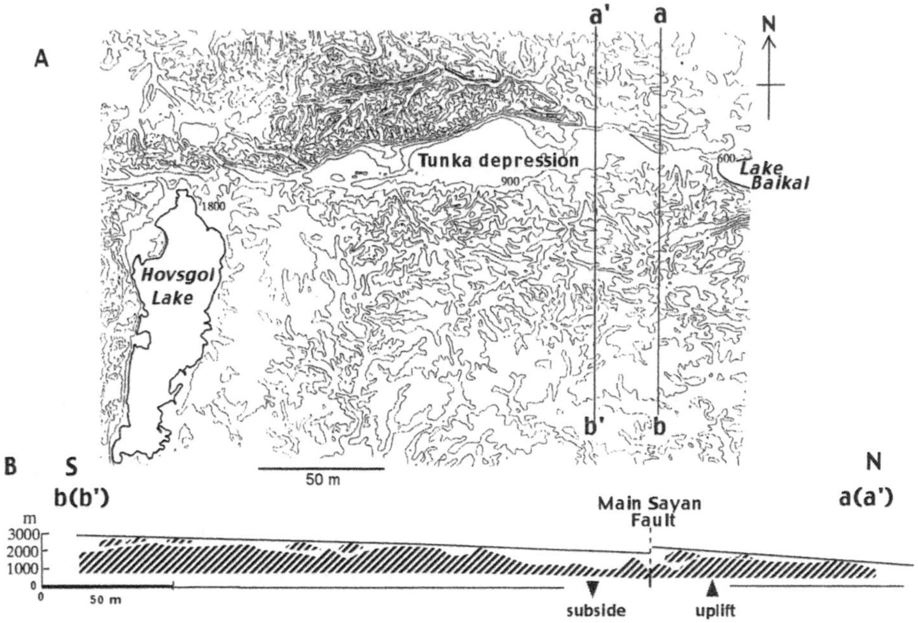

FIG. 6. The topography in **B** is a projection of two cross sections together along the lines a–b and a′–b′ in **A**. **B** Tilting northern and southern blocks at the border of the main Sayan fault

8 Discussion and Conclusion

The route of the Irkut river is aberrant, as mentioned above. Did the ancient Irkut river flow to Lake Baikal, or was it as it is at present? If it had taken a course to Lake Baikal, the stream was eventually cut off and changed course. The reason might be an orogenic rising between the Tory basin and Lake Baikal. Another possibility might be the piracy of a stream from another river that was a tributary of the Angara river. Yet another reason might be as a result of a fault cutting off the stream, or a landslide covering the stream. As a result of any number of possible events, the stream changed its route to the present one.

The course of the ancient Irkut river was likely similar to that of the present one except for the route along the main Sayan fault. When the area became a peneplain in the Oligocene or early Miocene period, the ancient Irkut river took the course from Tory to Irkutsk, as at present. Although the area rose up gradually, the river kept its course and incised a narrow, deep valley throughout the Pliocene and Pleistocene to the present (Fig. 5). In the middle Pleistocene, the water from Lake Baikal temporarily flowed out via the ancient Kultuk valley (Kononov and Mats 1986; Mats et al. 2000).

In conclusion, we have assumed that the Tunka area was a peneplain with meandering rivers. Its uplift in the north and tilt in the south was initiated in the

Oligocene. The main Sayan fault, which was a left-lateral slip fault, defined the north of the area. The south side, being tilted down northward, was accompanied by a depression defined mainly by two gravity fault lines. The ancient Irkut river meandered on the peneplain and became progressively deeper during that time. In the Miocene, it is suggested that some lakes formed when the depressed areas subsided more rapidly than the bed of the ancient Irkut river. The lakes formed lacustrine sediments, but the lakes did not persist over a long time-span. After the Miocene, the sediments in the depression deposited coarse-grained materials, and no big lacustrine condition existed, except in small lakes.

Acknowledgments. We are indebted to Shoji Fujii, Emeritus Professor of Toyama University, for his valuable suggestions on Baikal geology. We would like to thank Dr. Kimiyasu Kawamuro, a chief researcher of the Tsukuba Branch, the Forestry and Forest Products Research Institute, for his role of organizer of the project. We also thank Dr. Mikhail A. Grachev, executive director of the Limnological Institute, Siberian Division, Russian Academy of Science, for his help with our research in Siberia. We are grateful to Dr. Hikaru Takahara, Associate Professor of Kyoto Prefectural University, for his advice on our research expeditions. We thank Emeritus Professor Alfred Traverse of Pennsylvania State University for his comments, which greatly improved the manuscript. This work was supported by the fund of the International Cooperative Investigation Program of STA Japan, the US NFS, and the Russian Academy of Sciences.

References

Adamenko OM, Belova VA, Popova SM, Adamenko RS, Ponomareva EA, Efimova LI (1975) Biostratigraphy of upper pleistocene sediments of the Tunka depression (in Russion). Russ Geol Geophys 6:78–85

Dmitriev GA (1968) Tertiary ancient lakes of the Tunka depression (Transbaikalye) Mesozoic and Cenozoic lakes of Siberia (in Russian). Nauka, Moscow, p 49–59

Florensov NA (1960) Mesozoic and Cenozoic basins of the Baikal region (in Russian). Edition of the USSR Academy of Sciences, Moscow, Leningrad, p 268

Golubeva LV, Ravsky EI (1962) Anthropogene of the Tunka depressions (in Russian, with English abstract). Transactions of the Commission on the Investigation of Quaternary Period 19, Publishing House of the USSR Academy of Sciences, Moscow, p 240–259

Kononov EE, Mats VD (1986) The history of Baikal drainage (in Russian). Izv Vuzov Geol Razved 6:91–98

Krivonogov SK (1995) Late Pleistocene sedimentation in Tunka, south Baikal, Russia, compared with the northern territories. XIV Congress INQUA, Berlin, Abstracts, p 149

Logatchev NA (1954) Cenozoic volcanism of the Tunka depression. Materials on study of the productive forces of Buryat–Mongolian USSR (in Russian). Ulan-Ude, Buryat–Mongolian Publishing House, p 139–145

Logatchev NA (1974) The Sayan–Baikal Stanovoy upland. Uplands of Prebaikalye and Transbaikalye (in Russian). Nauka, Moscow, p 72–162

Logatchev NA, Kravchenko EV (1955) Distribution of basalts in Tunka depression (in Russian, with English abstract). Proc USSR Acad Sci 104:597–600
Mats VD (1992) The structure and development of the Baikal rift depression. Baikal International Center for Ecological Research, Preprint No. 1, Irkutsk, p 1–70
Mats VD, Fujii S, Mashiko K, Osipov EYu, Yefrimova IM, Klimansky AV (2000) Changes in Lake Baikal water levels and runoff direction in the Quarternary period. In: Minoura K (ed) Lake Baikal. Elsevier, Amsterdam, p 15–34
Rasskazov SV (1990) Basalts and structural development of the western end of the Baikal rift zone (in Russian). Geochemistry 4:614–619
Rasskazov SV (1993) Magmatism of the Baikal rift system (in Russian). Nauka, Novosibirsk, p 1–288
Rasskazov SV, Batyrmurzaev AS, Magomedov ShA (1990) Potassium–argon dating of basalts in the Tunka valley (in Russian). Sov Geol 9:62–67
Ravsky EI (1972) Sedimentation and climates of internal Asia during the Anthropogene (in Russian). Nauka, Moscow, p 1–335
Sherman SI, Levi KG (1977) Transform faults of the Baikal rift zone. Proc USSR Acad Sci 233:461–464
Sherman SI, Medvedev ME, Ruzhich VV (1973) Tectonics and volcanism of the southwestern part of the Baikal rift zone (in Russian). Nauka, Moscow, p 1–136
Ufimtsev GF (1995) Rifting and its expression in the near-surface lithosphere of East Siberia (in Russian and in English). Russ Geol Geophys 36:80–88
Ufimtsev GF, Shchetnikov AA (2001) Newest structure of the Tunka rift (in Russian). Geomorphology (Russia) 1:67–87

Late Cenozoic Paleoenvironmental Changes Inferred from Long Baikal Sediment Cores

Paleoenvironmental Changes in the Eurasian Continental Interior during the Last 12 Million Years Derived from Organic Components in Sediment Cores (BDP-96 and BDP-98) from Lake Baikal

Genki Inoue Matsumoto[1], Chizu Fujimura[2], Koji Minoura[2], Nobuki Takamatsu[3], Tetsuo Takemura[4], Susumu Hayashi[5], Koji Shichi[6], and Takayoshi Kawai[7]

Summary. Organic components in combined sediment cores (BDP-96 and BDP-98) of BDP-96-1 and -2 (200 m) and BDP-98-1 and -2 (600 m) from Academician Ridge in Lake Baikal, in south-central Siberia, were studied to clarify the biological production, sources of organic matter, and paleoenvironmental changes during the last 12 million years (Myr) in the Eurasian continental interior as well as in the world. Total organic carbon (TOC) contents ranging from 0.13% to 2.8%, with an average of 0.93% ($n = 7710$), were fairly low, and were similar to those found in some oligotrophic lakes. TOC and total nitrogen (TN) contents had decreased by about half from 12 Myr ago to the present as a result of global climate cooling. Higher biological production periods are mainly attributed to the contribution of vascular plants.

The abundance of long-chain n-alkanes ($>C_{20}$), n-C_{27}, n-C_{29}, and/or n-C_{31}, and long-chain n-alkanoic acids, n-C_{26} and n-C_{28}, reflects the fact that the dominant plants of the Lake Baikal basin were *Pinus* spp., *Betula* spp., *Salix* spp.,

[1] Department of Environmental Information Science, School of Social Information Studies, Otsuma Women's University, Tama, Tokyo 206-8540, Japan
[2] Institute of Geology and Paleontology, School of Science, Tohoku University, Aoba, Sendai 980-8578, Japan
[3] Department of Chemistry, Faculty of Science, Toho University, Funabashi, Chiba 274-8510, Japan
[4] Department of Chemistry, Faculty of Science, Science University of Tokyo, Kagurazaka 1-3, Shinjuku-ku, Tokyo 162-8601, Japan
[5] Faculty of Agriculture, Gifu University, Yanagido 1-1, Gifu 501-1112, Japan
[6] Forest Environment Division, Forestry and Forest Products Research Institute, Matsunosato 1, Kukizaki-cho, Inashiki-gun, Ibaraki 305-0903, Japan
[7] National Institute for Environmental Studies, 16-2 Onogawa, Tsukuba 305-8506, Japan (present address: Department of Earth and Environmental Sciences, Graduate School of Environmental Studies, Nagoya University, Furo-cho, Chikusa-ku, Nagoya 464-8602, Japan)

etc. Normal-C_{29} alkane is mainly attributed to conifers and herbaceous plants, and has increased from 1 Myr ago to the present. C_{27} stanol/sterol ratios suggest that the lake bottom was oxic for 12–10 Myr BP, but anoxic for 8–6 Myr BP.

Key words. Paleoenvironmental change, Lake Baikal, Geochemistry, Organic components

1 Introduction

Studies on long-term paleoenvironmenal changes are important in order to estimate the influence of future global warming induced by human activity. Information about long-term paleoenvironmental changes has mainly been obtained by analyses of marine sediment cores (e.g., Shackleton et al. 1990; Masuda 1991; Raymo and Ruddiman 1992). Very little is known about paleoenvironmental changes in continental interiors. Lake Baikal is a very appropriate place for the study of long-term paleoenviromental changes, since the lake is one of the oldest (30 million years (Myr) old) and largest freshwater lakes in the world, and is located in south-central Siberia in the Eurasian continent. The sediment layer of the lake is over 8000 m deep at its thickest point, and can be expected to be a record of paleoenvironmental changes in the Eurasian continental interior for the last 30 Myr (e.g., Nikolaev et al. 1985; Hutchinson et al. 1992; Mats 1993; Baikal Drilling Project BDP-96 (Leg II) Members 1997). The Baikal Drilling Project (BDP) was carried out by scientists from Japan, Russia, the USA, and Germany to drill holes in the bottom sediments of Lake Baikal, and to elucidate the long-term paleoenvironmental changes in the Eurasian continental interior and in the world. BDP-96-1 and -2, and BDP-98-1, -2, and -3 sediment cores were taken in 1996 and 1998, respectively, at Academician Ridge in Lake Baikal. Analyses of BDP-96-1 and -2 sediment cores revealed the paleoenvironmental changes in the past 5 Myr (e.g., Karabanov et al. 2000; Kawamuro et al. 2000; Matsumoto et al. 2000; Takamatsu et al. 2000).

Organic components in lake sediments are supplied by living and dead organisms, and are the biomarkers of biological production, autochthonous and allochthonous sources, and paleoenvironmental changes in the drainage basin (e.g., Matsumoto and Takamatsu 1998). Here, we report the total organic carbon (TOC), total nitrogen (TN), hydrocarbons, fatty acids, and sterols in the BDP-96 and BDP-98 sediment cores (combined BDP-96-1 and -2, and BDP-98-1 and -2) to elucidate the biological production, autochthonous and allochthonous sources of organic matter, and paleoenvironmental changes during the last 12 Myr in the Eurasian continental interior and in the world.

2 Materials and Methods

2.1 Samples

BDP-96-1 and -2 sediment cores (200 m) were taken from Academician Ridge (53°41′48″N, 108°21′06″E, water depth 321 m) in Lake Baikal in south-central Siberia (Fig. 1). The details of the drilling methods for the BDP-96-1 and -2 sediment cores are given elsewhere (Baikal Drilling Project BDP-96 (Leg II) Members 1997). Academician Ridge is best site for the study of long-term paleoclimatic changes because it is topographically isolated from the influence of fluvial input and turbidity flows. During January and March in 1998, BDP-98-1 and -2 (600 m) sediment cores were taken from Academician Ridge near the BDP-96-1 and -2 sediment core drilling site (53°44′48″N, 108°24′32″E, water depth 333 m, Fig. 1) with a core recovery of 95% (Baikal Drilling Project Group 2000). A gravity core (VER96-2ST3GC, 4.7 m) from near the BDP-96-1 and -2 site was used to cover the missing top samples of the BDP-96 and BDP-98 sediment cores at drilling. Plant leaf and soil samples were obtained from the surroundings of Lake Baikal.

2.2 Analytical Methods

The total organic carbon (TOC) and total nitrogen (TN) in the BDP-96 and BDP-98 sediment cores were analyzed at 5- to 10-cm intervals with a Fisons NCS automatic elemental analyzer (Milan, Italy), after treatment with hydrochloric acid to remove the carbonate-carbon. The analytical uncertainty of TOC and TN was within ±5%. Hydrocarbons, fatty acids, and sterols in the sediment, and plant leaf and soil samples were analyzed by the methods of Matsumoto et al. (1979, 1982) and Matsumoto and Watanuki (1992). The organic components in the sediment cores were analyzed at 1- to 2-m intervals. Briefly, organic components in the samples were extracted with ethyl acetate after saponification with 0.5 M potassium hydroxide/methanol (80°C, 2 h). The ethyl acetate extracts were chromatographed on a silica gel column (160 mm × 6 mm i.d., 100 mesh, 5% water). Hydrocarbon and fatty acid–sterol fractions were obtained by elution with hexane and ethyl acetate/benzene (7/3), respectively. Fatty acids were methylated with diazomethane. An aliquot of the fatty acid–sterol fraction was trimethylsilylated (TMS) with 25% *N,O*-bis(trimethylsilyl acetamide) acetonitrile solution, and sterol TMS derivatives were obtained. Hydrocarbons, fatty acid methyl esters, and sterol–TMS derivatives were analyzed by a JEOL JMS Automass 150 gas chromatograph–mass spectrometer equipped with a fused silica capillary column (DB5, 30 m × 0.25 mm i.d., film thickness 0.1 μm). The analytical uncertainty was within ±10%.

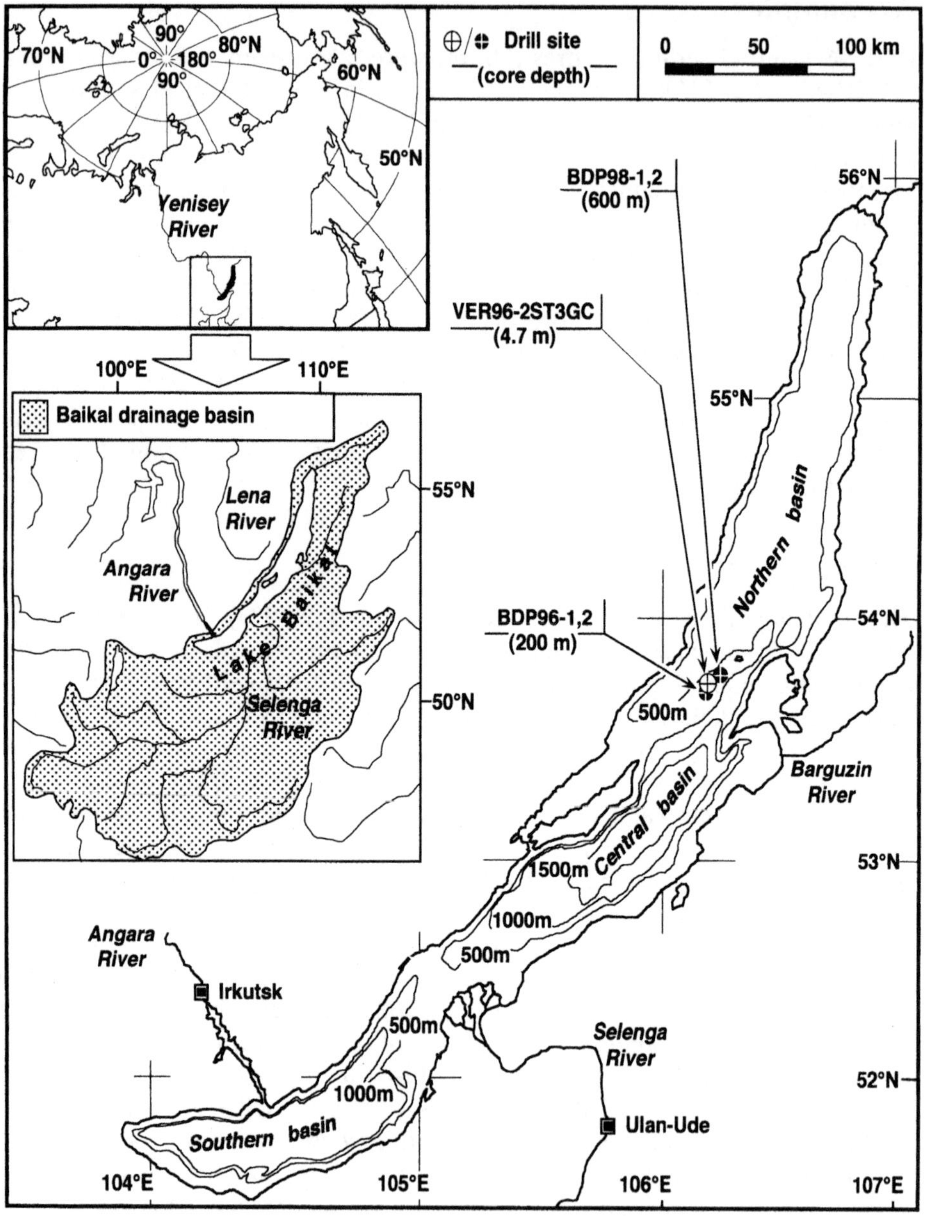

FIG. 1. Drilling sites at Academician Ridge in Lake Baikal, south-central Siberia, Russia

3 Results and Discussion

3.1 Sedimentation Rates and Ages

The BDP-96 and BDP-98 sediment cores were composed of clay-rich beds alternating with diatom frustule-rich beds throughout the samples. This rhythmic structure is mainly attributed to ecological changes caused by glacial and interglacial climate cycles (Baikal Drilling Project BDP-96 (Leg II) Members 1997; Baikal Drilling Project Group 2000). A paleomagnetic study of geomagnetic polarity reversals in the BDP-96-1 and -2 sediment cores revealed that the sedimentation rate was 3.8 cm/1000 years (kyr), with almost constant sedimentation (Sakai et al. 2000). The age of the bottom of the BDP-96-1 sediment core was estimated to be 5.2 Myr BP. The paleomagnetic study of the BDP-98-1 and -2 sediment cores showed that the sedimentation rates for depths from 0 to 277 m, from 290 to 352 m, and from 375 to 600 m were 4.1, 29.3, and 4.5 cm/kyr, respectively (Kashiwaya et al. 2001). The sedimentation rates of the BDP-98-1 and -2 sediment cores were similar to those of the BDP-96-1 and -2 sediment cores except for depths between 290 and 352 m (6.68–6.89 Myr BP). The reason for the sudden increase in sedimentation rate for these depths is not clear (Kashiwaya et al. 2001; H. Sakai et al., unpublished data 2002). The age of the bottom of the BDP-98-1 and -2 sediment cores was estimated to be 12 Myr BP.

3.2 Early Diagenesis of Organic Matter

Although approximately 90% of the organic matter formed by photosynthesis in the water column is degraded before sedimentation (Meyers and Ishiwatari 1993), TOC and TN contents and TOC/TN weight ratios appear to be insensitive to further diagenetic alteration after permanent burial in the bottom sediments. Thus, bulk organic parameters such as TOC, TN, and TOC/TN weight ratios are useful to estimate biological production, and to distinguish sources of organic matter in sediment cores (Meyers and Ishiwatari 1993; Krishnamurthy et al. 1986; Eartel and Hedges 1985; Ishiwatari and Uzaki 1987). Thus, they may be useful proxies of paleoclimatic changes.

It is well known that the present climate is similar to that in the marine $\delta^{18}O$ isotope stage 5e (121–123 kyr BP; e.g. Imbrie et al. 1992; NOAA Paleoclimatology Program 1997; Fig. 2). Biogenic silica records in Baikal sediment cores were well correlated with glacial and interglacial cycles, and were used as a proxy for paleoclimatic change (Colman et al. 1995). Biogenic silica contents in Holocene sediments (50%) are similar to those in sediments of stage 5e (52%) in Lake Baikal. This result shows that the climate in the 5e stage in Lake Baikal was similar to the present climate (Colman et al. 1995).

TOC contents in the surface sediments (0–2 cm) from Academician Ridge ranged from 2.0% to 3.1%, with an average of $2.6 \pm 0.36\%$ except for extreme values (Williams et al. 1993), which is 1.7 times higher than those (1.5%) in stage 5e of the BDP-96 and BDP-98 sediment cores (Fig. 2). In addition, the regres-

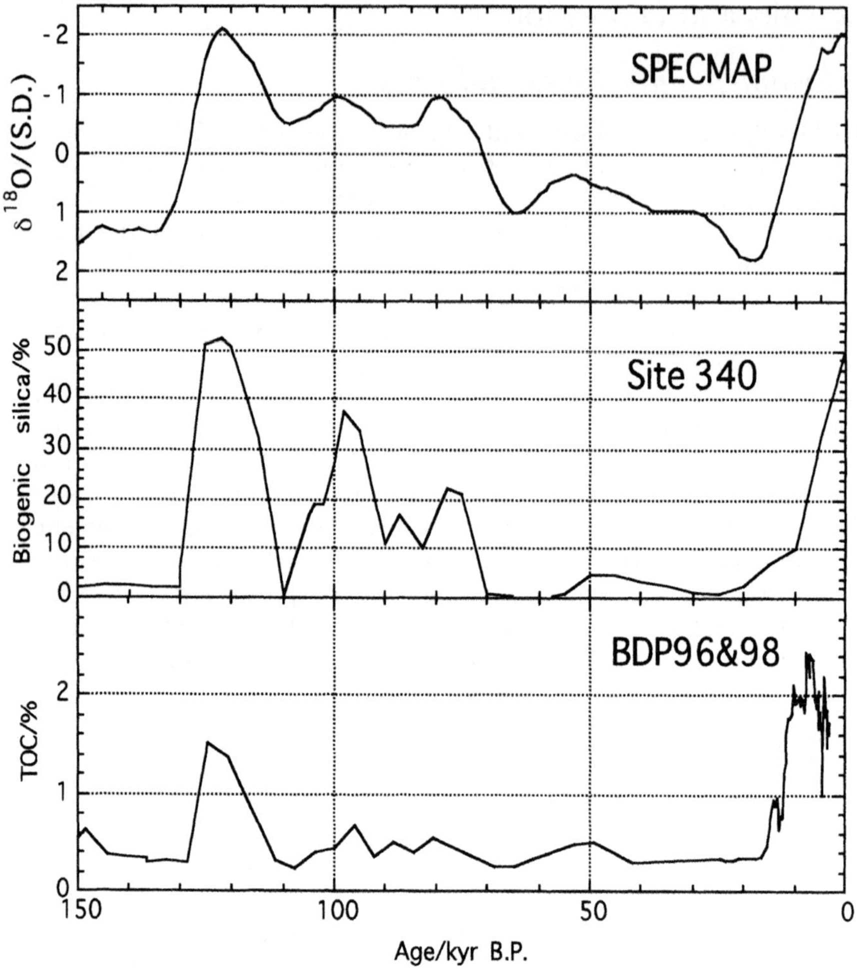

FIG. 2. Comparison of TOC and biogenic silica contents, with reference to SPECMAP. Biogenic silica data from Colman et al. (1995)

sion line between TOC and TN contents showed no excess nitrogen (inorganic nitrogen) in the surface sediments of Lake Baikal with a correlation coefficient (r^2) of 0.81 (revised from Williams et al. 1993; Fig. 3). However, the Y segment of the regression line for the relationship between the TOC and TN contents in the BDP-96 and BDP-98 sediment cores showed the presence of excess nitrogen of +0.052%, with a correlation coefficient of 0.41 (Fig. 4).

The degradation of organic matter yields carbon dioxide and ammonium salts. Carbon dioxide is easily removed from the sediment layer, but ammonium salts are preserved in clay layers and/or on silicate surfaces (e.g., Meyers and Ishiwatari 1993; Matsumoto et al. 2000). Excess nitrogen in the BDP-96 and BDP-98 sediment cores was therefore derived from the degradation of organic matter in the early diagenesis rather than directly supplied as inorganic salts.

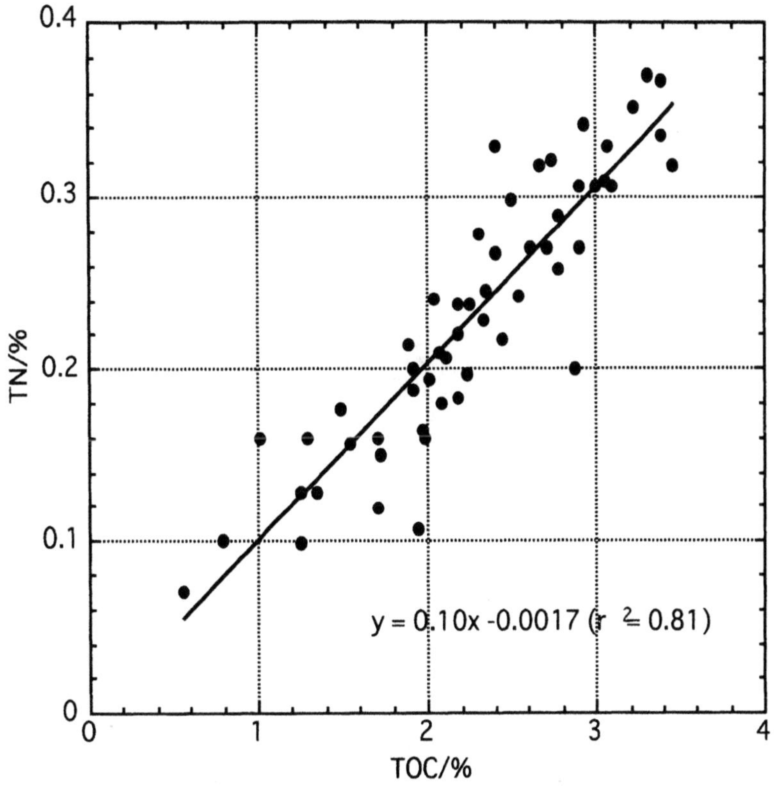

FIG. 3. Correlation between TOC and TN contents in surface sediments (0–2 cm) of Lake Baikal (data revised from Williams et al. 1993)

In addition, low TOC/TN weight ratios of these cores probably resulted from preferential losses of carbon and the preservation of nitrogen. TN contents can be well preserved in the sediment layers for long periods of time, rather than TOC contents.

According to a comparison of TOC and biogenic silica contents in the surface sediments and the stage 5e sediments, we calculated that approximately 40% of organic matter was degraded in the early diagenesis. TOC contents in Lake Baikal decreased in the early diagenesis of less than 10 kyr, as suggested by the BDP-93 sediment core from the Buguldeika of Lake Baikal (G.I. Matsumoto et al., unpublished data 1998). However, except for early diagenesis, the degradation of organic matter should be small, as reported previously (Meyers and Ishiwatari 1993; Krishnamurthy et al. 1986; Eartel and Hedges 1985; Ishiwatari and Uzaki 1987).

3.3 Changes in Biological Production

TOC and TN contents in sediment cores are markers of biological production in the drainage basin, and reflect the paleoclimate, i.e., high TOC and TN

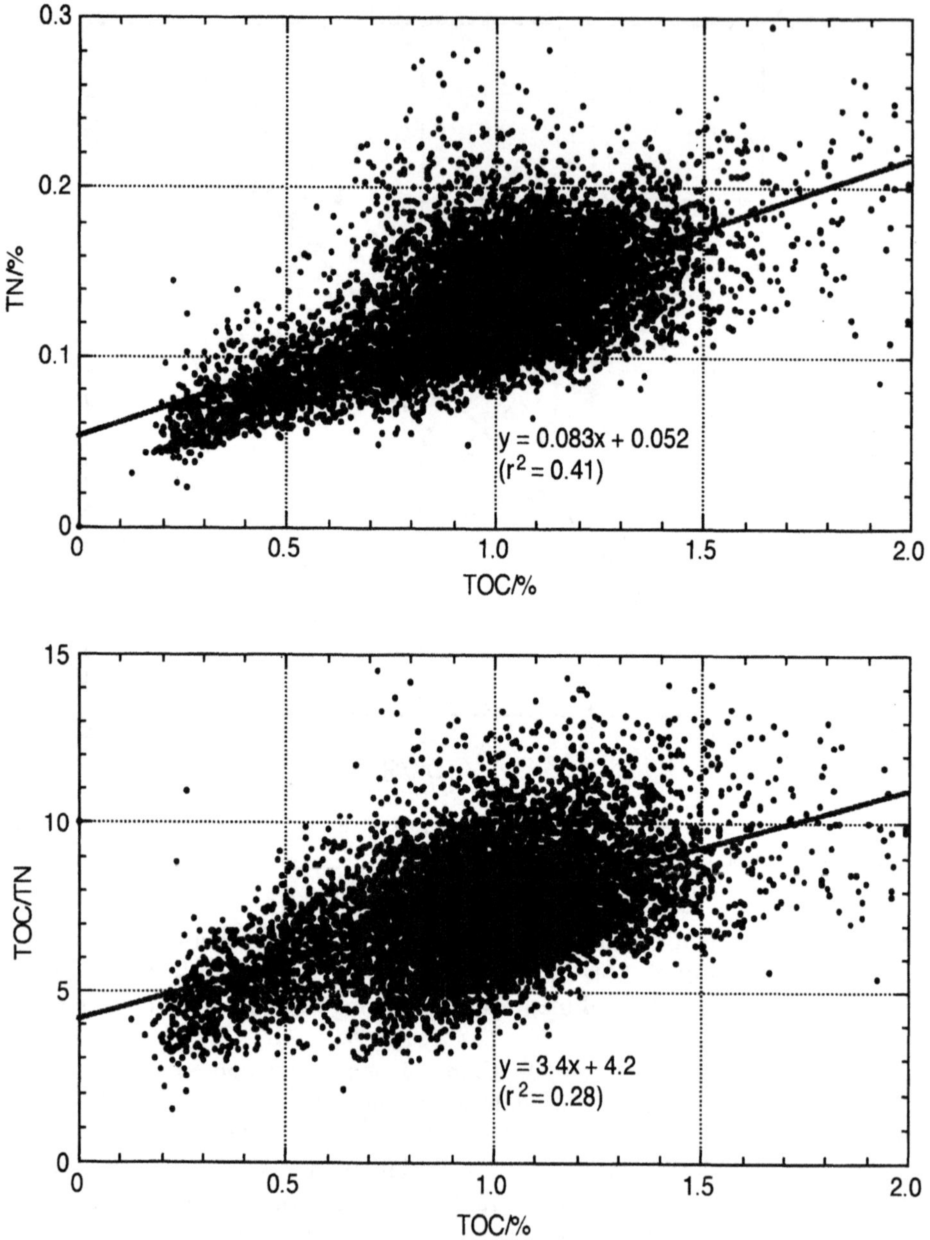

FIG. 4. Correlation between TOC and TN contents, and TOC contents and TOC/TN weight ratios for the BDP-96 and BDP-98 sediment cores ($n = 7710$)

contents indicate a warm climate, whereas low TOC and TN contents indicate the opposite (e.g., Krishnamurthy et al. 1986; Matsumoto et al. 2000). The TOC and TN contents in the BDP-96 and BDP-98 sediment cores, except for those younger than 10 kyr, ranged from 0.13% to 2.8%, with an average of 0.93%, and from 0.024% to 0.40%, with an average of 0.14%, respectively (Fig. 5). These values are similar to those in some oligotrophic lakes, such as Lake Michigan, USA (Meyers and Ishiwatari 1993), and Lake Biwa, Japan (Ishiwatari and Uzaki 1987), and thus the biological production in the Lake Baikal basin had been low throughout the sedimentation for 12 Myr. The TOC values fluctuated largely within a short period of time (less than 1.2 kyr; 5-cm sediment section), reflecting short-term environmental changes.

To clarify changes in long-term biological production and paleoenvironmental changes, a weighted mean value curve (2%) and a linear regression line are shown in Fig. 5 by bold solid lines, respectively. Relatively high regression lines correlation coefficients for TOC ($r^2 = 0.24$, $n = 7710$) and TN ($r^2 = 0.40$, $n = 7710$) demonstrated that TOC and TN contents decreased from 1.3% to 0.6% and from 0.18% to 0.07%, respectively, during 12 Myr (Fig. 5). These results indicated that biological production in the Lake Baikal basin has decreased by almost a half in the past 12 Myr. The correlation coefficient of the TN result was much higher than that of the TOC result, suggesting good preservation of TN values in the sediment layers.

Global cooling in the Cenozoic may have been caused by the uplift of the Tibetan plateau, and tectonically driven increases in chemical weathering has probably resulted in a decrease in atmospheric carbon dioxide concentration over the past 40 Myr (Raymo and Ruddiman 1992). The TOC and TN results from the BDP-96 and BDP-98 sediment cores clearly demonstrated that the cooling of the climate had also occurred in the Eurasian continental interior (see Fig. 5). Generally, the trend in the TOC contents in the BDP-96 and BDP-98 sediment cores for 5 Myr is similar to that in marine $\delta^{18}O$ records (e.g., Shackleton et al. 1990; Masuda 1991). These results revealed that climate changes in the Eurasian continental interior occurred simultaneously with changes in marine environments. This indicates that the TOC and TN profiles in the lake sediments are useful proxies for global paleoclimatic changes. These data showed fairly warm environmental conditions for 12–9.0, 7.7–7.0, 4.4–3.6, and 2.0–1.6 Myr BP, but cool environmental conditions for 8.6, 6.2–5.8, 2.8–2.2, and 1.2 Myr BP—the present (see Fig. 5). The warmest period during the past 12 Myr was 12–9.5 Myr BP. The major cooling period at about 8.6 Myr BP is thought to have been due to the influence of the initiation of the Indian monsoon caused by the uplift of the Himalayas and the Tibetan plateau (Prell and Kutzbach 1992). The major cooling at 6.2–5.8 Myr BP has not yet been reported, and can therefore be believed to reflect the local climate of the Baikal region. The major cooling at 2.8–2.2 Myr BP coincides with the formation of the ice sheet in the northern hemisphere. It is explained as the influence of the formation of the Panama uplift in late Neogene times, and the uplift of the Himalayan mountain chains (Masuda 1991; Driscoll and Haug 1998; Haug and Tiedemann 1998; Keigwin 1992).

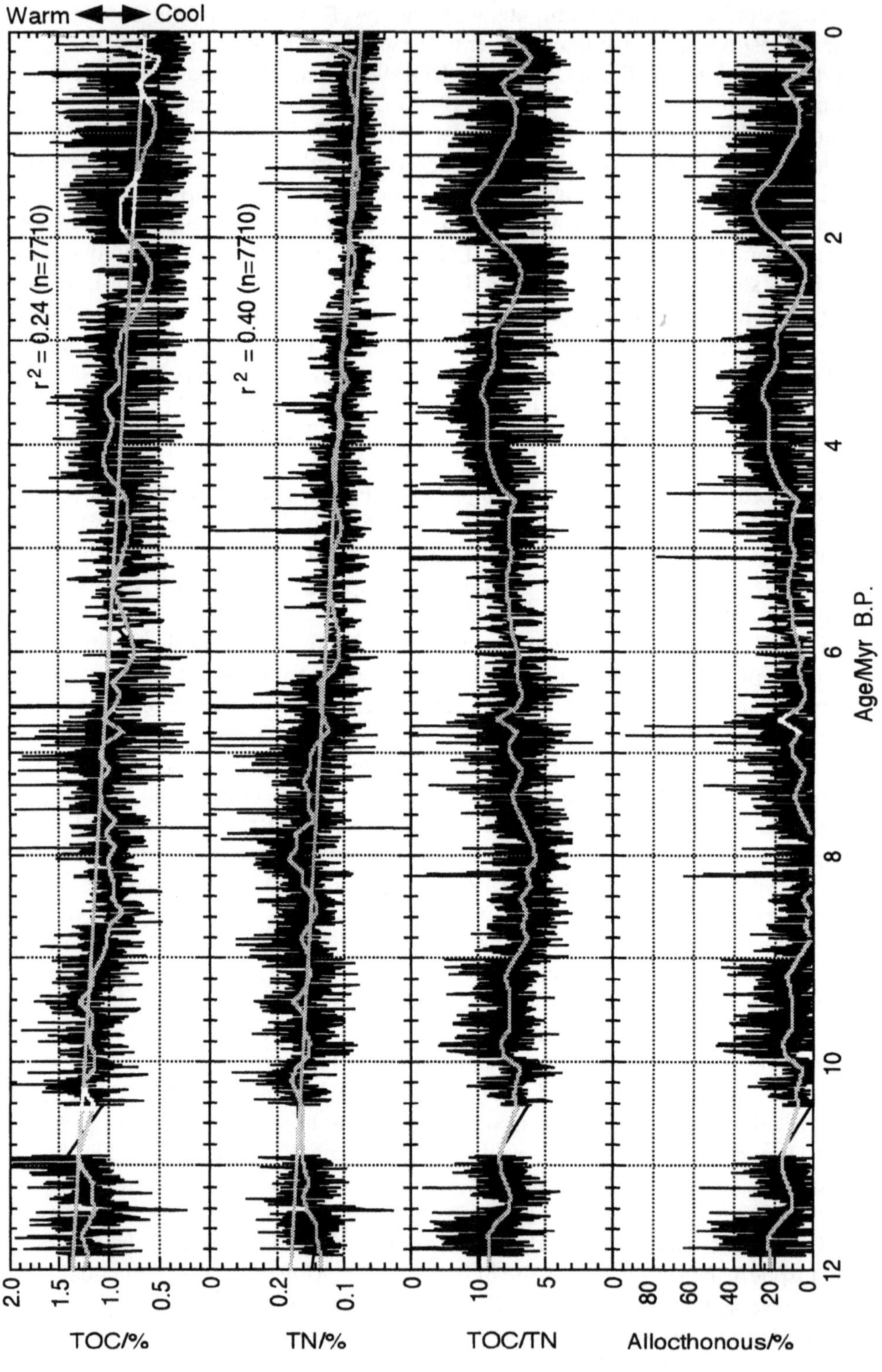
Warm
Cool
r 2 = 0.24 (n=7710)
r 2 = 0.40 (n=7710)
TOC/%
TN/%
TOC/TN
Allocthonous/%
Age/Myr B.P.

3.4 Allochthonous and Autochthonous Organic Matter

TOC/TN weight ratios of vascular land plant debris are 15 or more, whereas those of nonvascular aquatic plankton are as low as 4–10 (Meyers and Ishiwatari 1993; Krishnamurthy et al. 1986; Eartel and Hedges 1985). Thus, TOC/TN weight ratios reflect the relative contributions of vascular land plants (allochthonous organic matter) and plankton (autochthonous organic matter) in sediment cores. The TOC/TN weight ratios of the BDP-96 and BDP-98 sediment cores ranged from 3.0 to 13, with an average of 7.5 except for some extremely high and low values (see Fig. 5). The TOC/TN weight ratios were relatively low compared with those of other freshwater lakes, and were similar to those of marine sediments (Williams et al. 1993).

TOC contents were positively well correlated with TOC/TN weight ratios with a correlation coefficient (r^2) of 0.28 ($n = 7710$; see Fig. 4). These results clearly demonstrated that the higher biological production in the warm climate was mainly attributed to the contribution of allochthonous organic matter.

To estimate the relative contributions of allochthonous and autochthonous organic matter, we developed the following two simultaneous equations based on the TOC/TN weight ratios in the BDP-96 and BDP-98 sediment cores.

$$X + Y = 1.0$$
$$6.0X + 20Y = \mathrm{TOC/TN}$$

Here, we assumed that a TOC/TN ratio of 6.0 represented 100% contribution by autochthonous organic matter (X), and that a TOC/TN ratio of 20 represented 100% contribution by allochthonous organic matter (Y), based on the published results (Meyers and Ishiwatari 1993; Krishnamurthy et al. 1986; Eartel and Hedges 1985). The contribution of allochthonous organic matter ranged from 0% to 60% except for a few extremely high values (see Fig. 5). The pattern of changes in allochthonous organic matter is similar to that in the TOC results except for the period 8.4–7.8 Myr BP. Long-term observations (1950–1955, 1981–1984, and 1994–1997) on the input of nutrients and organic matter into Lake Baikal revealed that 89.9%–96.1% of organic matter as derived from river waters, that the contribution of atmospheric fallout was therefore small (Table 1; Solokovikova et al. 2000). Allochthonous organic matter in the sediment cores is therefore mainly supplied by river waters in warm, humid climatic conditions. Allochthonous organic matter was abundant in the periods 9.9–9.1, 4.4–3.2, and 2.0–1.5 Myr BP (see Fig. 5). In particular, the highest contribution of allochthonous organic matter was observed at the boundary of the Tertiary and Quaternary (1.6 Myr BP). These periods are believed to have been warm, with wet environmental conditions. In spite of the relatively high TOC contents in the period 8.4–7.8 Myr BP, the higher contribution of autochthonous organic matter

Fig. 5. Age distribution of TOC and TN contents, TOC/TN weight ratios, and allochthonous organic matter in the BDP-96 and BDP-98 sediment cores from Lake Baikal

TABLE 1. Input of organic matter into Lake Baikal

Years studied	Source	Organic matter	
		kt/y	%
1950–1955	River waters	584	96.1
	Atmospheric fall-out	24	3.9
1981–1984	River waters	734	92.9
	Atmospheric fall-out	56	7.1
1993–1997	River waters	595	89.9
	Atmospheric fall-out	67	10.1

Revised from Sorokovikova et al. (2000)

strongly suggests warm and dry climatic conditions because the contribution of land vascular plant biomass decreased considerably. In this period, higher temperatures accelerated biological production in the lake, and shortened the ice-covered periods during the year, but the biomass of land vascular plants decreased by aridification. Pollen analyses of the BDP-98-2 sediment core supports the idea that total pollen amounts decreased in the same period (Y. Hase et al., personal communication).

3.5 Molecular Fossil Information

3.5.1 Biomarkers in Plant Leaf and Soil Samples

The major hydrocarbons, fatty acids, and sterols in plant leaf and soil samples from the Lake Baikal basin are summarized in Table 2. Only long-chain (>C_{19}) *n*-alkanes having odd carbon numbers, ranging from *n*-C_{23} to *n*-C_{31}, were major hydrocarbons in all plant leaves analyzed. The most predominant *n*-alkane in *Abies sibirica* and *Larix dahurica* was *n*-C_{23} alkane, while that in *Pinus* spp., *Salix* spp., *Alnuster* sp., and *Betula* spp. was *n*-C_{27} alkane, except for a few cases. *Rhododendron dafurica* and *Agrostis clavata* contained *n*-C_{31} alkane as the predominant alkane. The major hydrocarbons in soil samples were *n*-alkanes ranging from *n*-C_{25} to *n*-C_{31} with odd carbon numbers, and were similar to those in plant leaf samples.

Major long-chain *n*-alkanoic acids were detected in carbon chain lengths ranging from C_{20} to C_{28} with even carbon numbers (see Table 2). Normal-C_{20} and *n*-C_{22} alkanoic acids were mainly detected in conifers and herbaceous plants, whereas *n*-C_{26} and *n*-C_{28} alkanoic acids were detected in broad-leaved tree, except for a few cases. The major fatty acids in soil samples were *n*-alkanoic acids ranging from *n*-C_{20} to *n*-C_{24} with even carbon numbers. 24-Ethylcholest-5-en-3β-ol (24-ethylcholesterol, C_{29}), 24-methylcholest-5-en-3β-ol (24-methylcholesterol, C_{28}), cholest-5-en-3β-ol (cholesterol, C_{27}), 24-ethylcholesta-5,22-dien-3β-ol (stigmasterol, C_{29}), and/or 24-methylcholesta-5,22-dien-3β-ol (brassicasterol, C_{28}) were detected in plant leaf samples (see Table 2). 24-Ethylcholesterol was the most predominant sterol in all the plant leaf and soil samples. No stanols were found in any plant leaf and soil samples.

TABLE 2. Major *n*-alkanes, *n*-alkanoic acids, and sterols found in plant leaf and soil samples from the Lake Baikal basin

Plants	*n*-Alkanes (%)[a]	*n*-Alkanoic acids (%)[b]	Sterols (%)[c]
Conifers			
Abies sibirica	23 (65), 27 (16), 25 (14)	22 (18), 24 (17)	Etc (62), Mec (27), Sti (10)
Larix dahurica 1	25 (30), 23 (27), 27 (15)	22 (36), 24 (31), 20 (5.0)	Etc (86), Mec (14)
Larix dahurica 2	23 (24), 25 (21)	22 (5.9), 24 (4.4)	Etc (85), Cho (8.3), Mec (6.5)
Picea obovata	—	20 (15), 24 (11), 28 (2.3)	Etc (86), Mec (14)
Pinus Pumila	—	20 (4.3), 22 (4.2)	Etc (88), Cho (12)
Pinus sibirica	29 (25), 27 (23), 25 (14)	22 (2.6), 20 (2.2)	Etc (69), Mec (31)
Pinus sylvestris	27 (33), 25 (21), 29 (15)	22 (11), 20 (5.2), 24 (2.4)	Etc (92), Mec (9)
Pinus sylvestris L.	27 (26), 29 (24), 25 (13)	22 (5.1), 20 (3.1), 24 (1.4)	Etc (100)
Pinus sp.	—	22 (23), 24 (11), 20 (6.3)	Etc (100)
Broad-leaved tree			
Alnuster fruticosa	27 (67), 29 (10)	28 (6.8), 22 (3.1), 20 (3.0)	Etc (94), Cho (5.7)
Betula ermani Cham	27 (33), 23 (28), 25 (26)	28 (4.8), 22 (3.6), 24 (2.5)	Etc (100)
Betula platyphylla Sukatcheu	27 (35), 25 (30), 23 (9.9)	28 (4.1), 20 (2.3), 22 (2.1)	Etc (100)
Populus maximowiczii Henny	27 (41), 25 (23), 29 (15)	26 (8.4), 24 (6.3), 28 (6.3)	Etc (100)
Populus suaveolens	29 (31), 27 (26), 25 (20), 23 (17)	20 (5.8), 22 (4.5)	Etc (98), Cho (2.0)
Rhododendron dafurica	31 (35), 27 (20), 33 (17), 29 (12)	20 (8.6), 28 (4.1), 30 (3.4)	Etc (71), Cho (29)
Salix sachalinensis Fr. Sohm	27 (35), 25 (33), 23 (11)	20 (7.4), 26 (7.3), 24 (5.1)	Etc (100)
Salix vulpina Anders	27 (43), 25 (30), 29 (11)	26 (8.1), 28 (6.2), 24 (3.9)	Etc (100)
Salix sp.	27 (50), 29 (42)	28 (6.7), 24 (3.9), 20 (3.8)	Etc (98), Cho (2.2)
Herbaceous plants			
Agrostis clavata Trinins	31 (62), 27 (10), 29 (9.4)	20 (2.6), 22 (2.4)	Etc (59), Mec (23), Sti (14)
Artemisia sp.	29 (40), 31 (30)	20 (30), 22 (6.0), 24 (3.6)	Etc (45), Sti (35), Bra (9.5)
Soil			
Soil 1	29 (20), 31 (16), 27 (16)	22 (9.8), 20 (8.9), 24 (6.9)	Etc (82), Mec (18)
Soil 2	27 (20), 31 (17), 25 (16)	22 (10), 20 (7.9), 24 (5.9)	Etc (52), Mec (26), Cho (22)

[a] mn (op), carbon chain length of *n*-alkanes (percentages in total *n*-alkanes (C_{13}–C_{35})); —, no data

[b] mn (op), carbon chain length of long-chain (>C_{19}) *n*-alkanoic acids (percentages in total *n*-alkanoic acids (C_{12}–C_{30}))

[c] abc (op), sterols (percentages in total sterols); Etc, 24-ethylcholesterol (C_{29}); Mec, 24-methylcholesterol (C_{28}); Sti, stigmasterol (24-ethylcholesta-5,22-dien-3β-ol, C_{29}); Cho, cholesterol (C_{27}); Bra, brassicasterol (24-methylcholesta-5,22-dien-3β-ol, C_{28})

3.5.2 Biomarkers in BDP-96 and BDP-98 Sediment Cores

Normal alkanes ranging from n-C_{15} to n-C_{35} were found in the BDP-96 and BDP-98 sediment cores with a predominance of odd carbon numbers, together with pristane, phytane, and squalane. The major hydrocarbons were n-C_{25}, n-C_{27}, n-C_{29}, and n-C_{31} alkanes, and/or squalane. Long-chain n-alkanes (C_{20}–C_{35}) originating from vascular plants were 94%–100% of total n-alkanes (C_{15}–C_{35}), and were well preserved in the sediment cores (Matsumoto et al. 2000). Normal alkanoic acids, ranging from n-C_{10} to n-C_{34}, were detected with a predominance of even carbon numbers. However, control experiment showed that most of the short-chain n-alkanoic and alkenoic acids (C_{12}–C_{19}) came from contamination during the experimental procedures. Thus only long-chain n-alkanoic acids (C_{20}–C_{34}) were discussed, as in the earlier article by Matsumoto et al. (2000). Sterols (C_{27}–C_{29}) were detected with an abundance of C_{27} and C_{29} stenols and stanols, while the contribution of C_{28} sterols was very small in all samples. Stanols were not found in plant leaf and soil samples (see Table 2), but were detected in all sediment samples as major sterols. Therefore stanols were derived from other sources, as discussed below.

Normal-C_{27} alkane was most predominant hydrocarbon in all the sediment layers (Fig. 6). Normal-C_{27} alkane is abundant in *Pinus* spp., *Alnuster* sp., *Betula* spp., *Populus* spp., and *Salix* spp. (Table 2). *Alnuster* spp. and *Populus* spp., however, were not predominant plants in the Lake Baikal basin (Y. Hase et al. personal communication). The abundance of n-C_{27} alkane therefore, reflects the predominance of *Pinus* spp., *Betula* spp., and/or *Salix* spp. in the Lake Baikal basin, although the contribution of *Betula* spp. may be greater than that of the other plant species (see Fig. 6, Table 2). The abundance of n-C_{27} alkane at 7.2, 6.3, and 2.0 Myr was due to the contribution of *Betula* spp. (see Fig. 6). In contrast, the increase in n-C_{29} alkane from 1 Myr ago to the present can be explained by the increase in conifers in the Lake Baikal basin.

Normal-C_{20} alkanoic acid was less than 2% from 12 to 6.8 Myr ago, but increased abruptly to about 5% or more from 6.8 Myr to the present (see Fig. 6). Normal-C_{20} alkanoic acid is abundant in *Pinus* spp. *Populus* spp., and *Rhododendron* spp. (see Table 2). This may be due to the increase in conifers in the Lake Baikal basin because *Populus* spp. and *Rhododendron* spp. were not predominant plants in the Lake Baikal basin (Y. Hase et al. personal communication).

Normal-C_{31} alkane is abundant in *Rhododendron dafurica,* as well as some herbaceous plants such as *Agrosis clavata* (see Table 2). However, *Rhododen-*

FIG. 6. Age distribution of selected biomarker compounds in the BDP-96 and BDP-98 sediment cores from Lake Baikal. Normal-C_{27}, n-C_{29} alkane, or squalane as percentage in total alkanes (n-C_{15}–n-C_{35}). Normal-C_{20} alkanoic acid as percentage in long-chain n-alkanoic acid (n-C_{20}–n-C_{34}). C_{27} sterol: cholesterol and cholestanol. C_{29} sterol: 24-ethylcholesterol and 24-ethylcholestanol

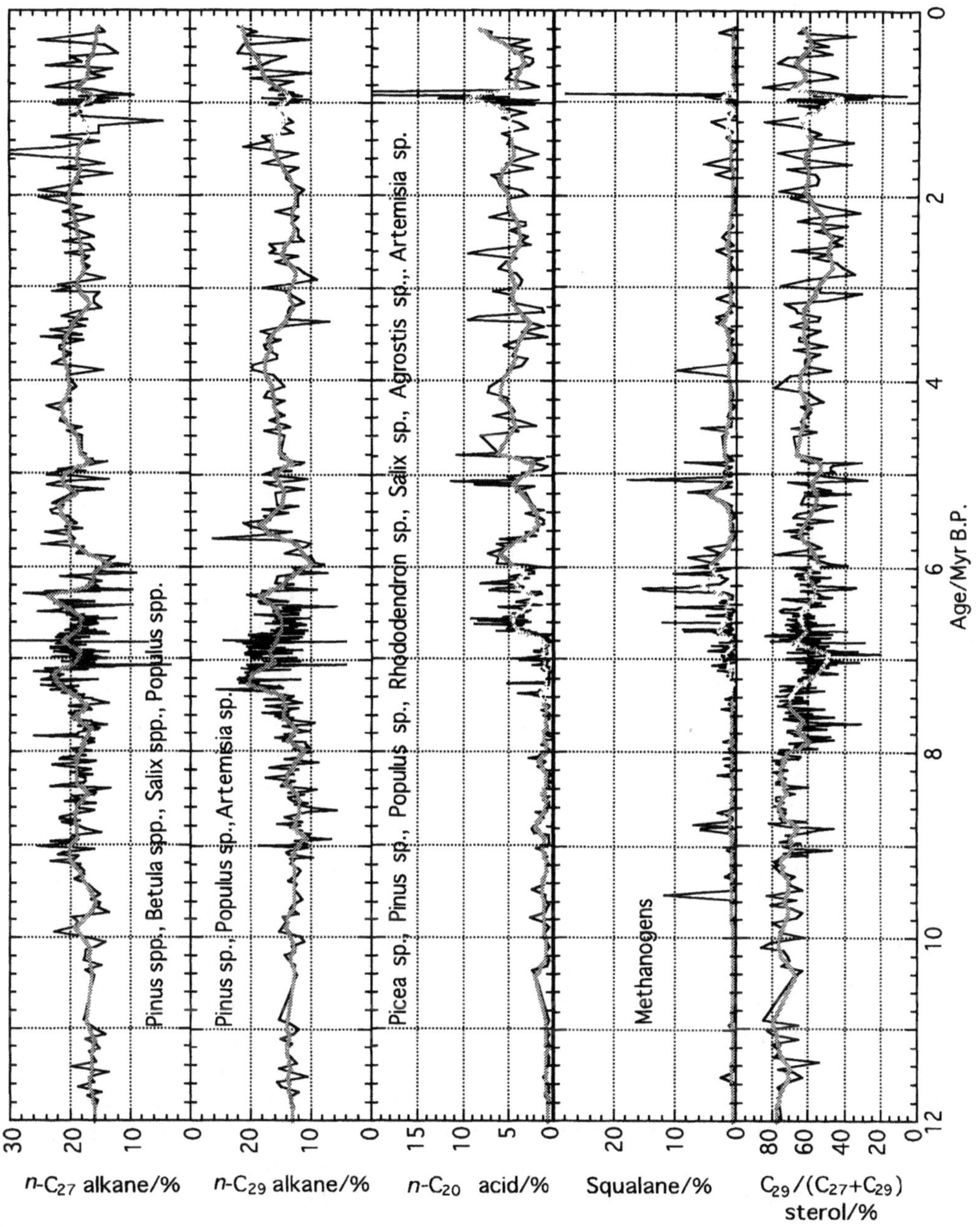
Pinus spp., Betula spp., Salix spp., Populus spp.
Pinus sp., Populus sp., Artemisia sp.
Picea sp., Pinus sp., Populus sp., Rhododendron sp., Salix sp., Agrostis sp., Artemisia sp.
Methanogens
Age/Myr B.P.
n-C27 alkane/%
n-C29 alkane/%
n-C20 acid/%
Squalane/%
C29/(C27+C29) sterol/%

dron dafurica was not abundant in the surroundings of Lake Baikal. Thus, n-C_{31} alkane could be a biomarker for herbaceous plants, and this suggests that herbaceous plants increased from 1 Myr ago to the present in the drainage basin of the lake due to aridification of the climate. These organic components are mainly derived from river waters in particulate and dissolved forms, including fallen leaves and soil particles, with some influence from atmospheric fall-out (see Table 1).

Squalane was found in many sediment layers, and was abundant 9.5, 6.3, 5.1, and 0.9 Myr ago (see Fig. 6). Squalane is produced by methanogens, and is believed to be a biomarker for methanogens (Langworthy 1985), although it is present in some vascular plants in small percentages (G.I. Matsumoto et al., unpublished data, 2000). Methanogens are widely distributed in the surface sediments of Lake Baikal (B. Namsaraev et al., personal communication). Isotope studies of methane detected in the BDP-96-1 and -2 sediment cores showed that methane originated from methanogens (N. Yoshida et al., personal communication). Thus, squalane in the sediment core probably reflects the abundance of methanogens.

The major sterols found in the BDP-96 and BDP-98 sediment cores were cholesterol, cholestanol, 24-ethylcholesterol, and 24-ethylcholestanol. As stated above, stanols were not detected in plant leaf and soil samples. Stanols are often found in microalgae such as dinoflagellates, diatoms, and raphidophytes (e.g., Volkman et al. 1998). They are also formed by bacterial reduction of stenols (e.g., Nishimura 1982). Stanols, including cholestanol and 24-ethylcholestanol, are present in lake and marine sediments (Pearce et al. 1998). Stanols detected in the BDP-96 and BDP-98 sediment cores therefore originated from microalgae in the lake, in addition to the microbial reduction of stenols.

Generally, 24-ethylcholesterol (C_{29}) is the most predominant sterol in vascular plants (see Table 2), while cholesterol (C_{27}) is a major sterol of plankton in the lake (e.g., Matsumoto et al. 1982; Volkman et al. 1998). It is known, however, that many kinds of microalgae synthesize C_{29} sterols, including 24-ethylcholesterol (Matsumoto et al. 1982; Volkman et al. 1998). Here we assumed that the $C_{29}/(C_{27} + C_{29})$ sterol ratio (%) is a marker of allochthonous organic matter. These ratios, ranging mainly from 40% to 80%, were much greater than those of allochthonous organic matter calculated from TOC/TN weight ratios, although their patterns resemble each other (see Figs. 5 and 6). This result strongly suggests that algal C_{29} sterols contributed considerably to the BDP-96 and BDP-98 sediment. Hence, the $C_{29}/(C_{27} + C_{29})$ sterol ratios (%) reflect the contribution of vascular plants in addition to the contribution of microalgae in the lake.

Stanol/sterol ratios depend on the microalgal sources, the relative stability of stanols and stenols, and the reduction of stenols into stanols. The rate of microbial degradation of stenols is greater than that of stanols, and thus the stanol/sterol ratios increase with the degradation of organic matter (Nishimura 1977; Pearce et al. 1998). Stanols are also formed by the reduction of stenols (Volkman et al. 1998). The stanol/sterol ratios may be a marker of the redox conditions in the sedimentary environments of the lake. In particular,

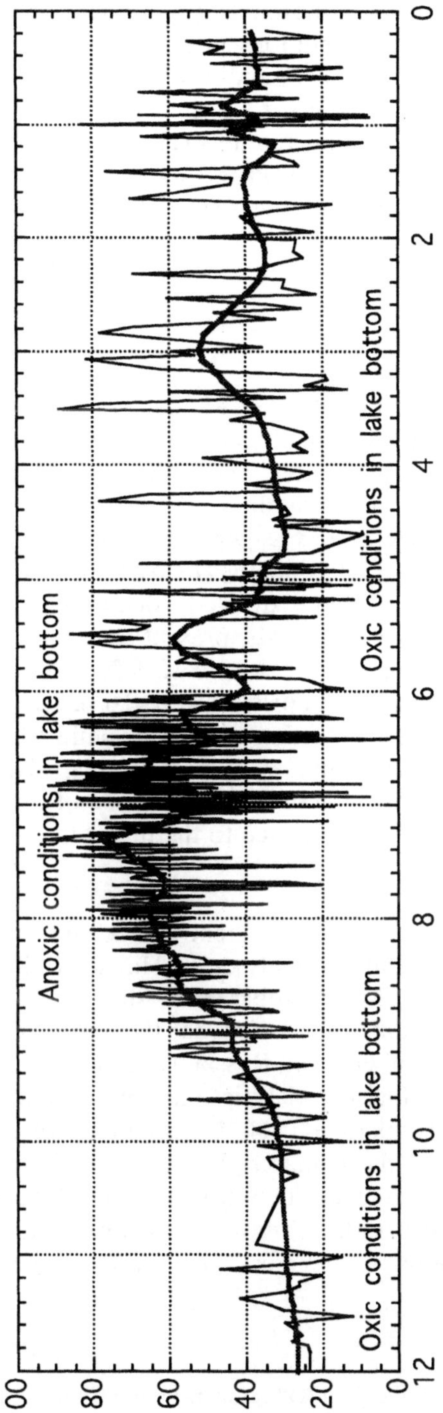

FIG. 7. C_{27} stanol/sterol ratios for the BDP-96 and BDP-98 sediment cores from Lake Baikal

cholesterol mainly originates from plankton in the lake, and thus the (cholestanol/(cholesterol + cholestanol) (%) ratios) (STA/STE ratios (%)) could directly reflect redox conditions in the sedimentary environments. The STA/STE ratios (%) in the BDP-96 and BDP-98 sediment cores were very low 12 Myr ago (25%), increased toward 7.5 Myr ago (80%), and decreased again 5–4 Myr ago (30%), with large fluctuations (Fig. 7). The redox conditions in the sedimentary environments depend on the water depth, biological production, and water circulation. Currently, the reason for these changes is not clear, and requires further study.

4 Conclusions

The organic components in the BDP-96 and BDP-98 sediment cores from Academician Ridge in Lake Baikal were studied to elucidate biological production, sources of organic matter, and paleoenvironmental change in the Eurasian continental interior and in the world during the last 12 Myr. Biological production in the drainage basin was generally low throughout the sedimentation layers. The TOC and TN contents are useful proxies for global climate change. The TOC and TN contents revealed a decrease in biological production to less than half from 12 Myr ago to the present caused by global cooling of the climate. The major warming episodes were 12–9.5, 7.7–7.0, 4.4–3.6, and 2.0–1.6 Myr ago, while the major cooling episodes were 8.6, 6.2–5.8, 2.8–2.2, and 1.2 Myr ago. The higher biological production in warm, humid climatic conditions was mainly attributed to the contribution of land vascular plants. Biomarkers suggested that there was an increase in the contribution of conifers in the drainage basin of the lake from about 1 Myr ago to the present.

Acknowledgments. The authors are greatly indebted to M. Kanadani and H. Kusaka for the TOC and TN analyses, K. Hirose, M. Iizuka, Y. Ishibashi, S. Kaminaga, A. Komoriya, M. Matsubara, M. Matsuo, K. Mine, H. Shima, T. Shinoda, N. Suzuki, M. Tajima, and T. Oya for their biomarker analyses, and the BDP-96 and BDP-98 members from Russia, Japan, the USA, and Germany who worked on the drilling, logistics, descriptions, and sampling. We thank Dr. Y. Soma, University of Shizuoka, and Prof. M. Nishimura, Tokai University, for their useful suggestions about the manuscript. This research was supported by the Science and Technology Agency (now Ministry of Education and Science) of Japan, the Siberian Branch of the Russian Academy of Science, US NSF, and the Japan Baikal International Center for Ecological Research.

References

Baikal Drilling Project BDP-96 (Leg II) Members (1997) Continuous paleoclimate record recovered for last 5 million years. EOS 78(51):597–604

Baikal Drilling Project Group (2000) Paleoclimate record in the late Cenozoic sediments of Lake Baikal (by 600 m deep-drilling data) (in Russian). Russ Geol Geophys 41:3–32

Colman SM, Peck JA, Karabanov EB, Carter SJ, Bradbury JP, King JW, Williams DF (1995) Continental climate response to orbital forcing from biogenic silica records in Lake Baikal. Nature 378:769–771

Driscoll NW, Haug GH (1998) A short circuit in thermohaline circulation: a cause for northern hemisphere glaciation? Science 282:436–438

Eartel JR, Hedges JI (1985) Sources of sedimentary humic substances: vascular plant debris. Geochim Cosmochim Acta 49:2097–2107

Haug GH, Tiedemann R (1998) Effect of the formation of the Isthmus of Panama on Atlantic Ocean thermohaline circulation. Nature 393:673–676

Hutchinson DR, Golmshtok AJ, Zonenshain LP, Moor TC, Scholz A, Klitgord KD (1992) Depositional and tectonic framework of the rift basins of Lake Baikal from multichannel seismic data. Geology 20:589–592

Imbrie J, Berger A, Boyle EA, Clements SC, Duffy A, Howard WR, Kukula G, Kutzbach J, Martinson DG, McIntyre A, Mix AC, Molfino B, Morley JJ, Peterson LC, Pisias NG, Prell WL, Raymo ME, Shackleton NJ, Toggweiler JR (1992) On the structure and origin of major glaciation cycles. 1. Linear responses to Milankovitch forcing. Paleoceanography 7:701–738

Ishiwatari R, Uzaki M (1987) Diagenetic changes of lignin compounds in a more than 0.6 million-year-old lacustrine sediment (Lake Biwa, Japan). Geochim Cosmochim Acta 51:321–328

Karabanov EB, Kuzmin MI, Prokopenko AA, Williams DF, Khursevich GK, Bezrukova EV, Kerber EV, Gvozdkov AN, Gelety VF, Weil D, Schwab M (2000) Glaciations of central Asia in the late Cenozoic according to the sedimentary record from Lake Baikal. In: Minoura K (ed) Lake Baikal. Elsevier, Amsterdam, p 71–84

Kashiwaya K, Ochiai S, Sakai H, Kawai T (2001) Orbit-related long-term climate cycles revealed in a 12-Myr continental record from Lake Baikal. Nature 410:71–74

Kawamuro K, Shichi K, Hase Y, Iwauchi A, Minoura K, Oda T, Takahara H, Sakai H, Morita Y, Miyoshi N, Kuzmin MI (2000) Forest–desert alternation history revealed by the pollen record in Lake Baikal over the past 5 million years. In: Minoura K (ed) Lake Baikal. Elsevier, Amsterdam, p 101–107

Keigwin LD (1992) Isotopic paleoceanography of the Caribbean and east Pacific: role of Panama uplift in late Neogene times. Science 217:350–353

Krishnamurthy RV, Bhattacharya SK, Kusumgar S (1986) Palaeoclimatic changes deduced from $^{13}C/^{12}C$ and C/N ratios of Karewa lake sediments, India. Nature 323:150–152

Langworthy TA (1985) Lipids of archaebacteria. In: Woese CR, Wolfe RS (eds) The bacteria, vol VIII. Archaebacteria. Academic Press, Orlando, p 459–497

Masuda F (1991) Evaluation of the present on the history of paleoclimatic change (in Japanese). J Geol 100:976–987

Mats VD (1993) The structure and development of the Baikal rift depression. Earth-Sci Rev 34:81–118

Matsumoto GI, Takamatsu N (1998) Analyses of environmental change by biomarkers (in Japanese). In: Matsumoto GI, Minoura K, Kashiwaya K (eds) Science of global environmental change. Kokon Shoin, Tokyo, p 149–161

Matsumoto GI, Watanuki K (1992) Geochemical features of organic components in an extremely acid crater lake (Yugama) of the Kusatsu–Shirane Volcano in Japan. Geochem J 26:117–136

Matsumoto G, Torii T, Hanya T (1979) Distribution of organic constituents in lake waters and sediments of the McMurdo Sound region in the Antarctic. Mem Natl Inst Polar Res Spec Iss 13:103–120

Matsumoto G, Torii T, Hanya T (1982) High abundance of algal 24-ethylcholesterol in Antarctic lake sediment. Nature 299:52–54

Matsumoto GI, Kosaku S, Takamatsu N, Akagi T, Kawai T, Ambe Y (2000) Estimation of paleoenvironmental changes in the Eurasian continental interior during the past 5 million years inferred from organic components in the BDP-96/1 sediment core from Lake Baikal. In: Minoura K (ed) Lake Baikal. Elsevier, Amsterdam, p 119–126

Meyers PA, Ishiwatari R (1993) Lacustrine organic geochemistry: an overview of indicators of organic matter sources and diagenesis in lake sediments. Org Geochem 20:867–900

Nikolaev VG, Vanyakin LA, Kalinin VV, Milanovsky VY (1985) The sedimentary section beneath Lake Baikal. Int Geol Rev 27:449–459

Nishimura M (1977) Origin of stanols in young lacustrine sediments. Nature 270:711–712

Nishimura M (1982) 5β-isomers of stanols and stanones as potential markers of sedimentary organic quality and depositional paleoenvironments. Geochim Cosmochim Acta 46:423–432

NOAA Paleoclimatology Program (1997) Paleoceanographic data file 1. File 17 (http://www.ngdc.noaa.gov/paleo/paleodat.html)

Pearce GES, Harradine PJ, Talbot HM, Maxwell JR (1998) Sedimentary sterols and steryl chlorin esters: distribution differences and significance. Org Geochem 28:3–10

Prell WL, Kutzbach JE (1992) Sensitivity of the Indian monsoon to forcing parameters and implication for its evolution. Nature 360:647–652

Raymo ME, Ruddiman WF (1992) Tectonic forcing of late Cenozoic climate. Nature 359:117–122

Sakai HS, Nomura M, Horii M, Kashiwaya K, Tanaka A, Kawai T, Kravchinsky V, Peck J, King J (2000) Paleomagnetic and rock magnetic studies on Lake Baikal sediments: BDP-96 borehole at Academician Ridge. In: Minoura K (ed) Lake Baikal. Elsevier, Amsterdam, p 35–52

Shackleton NJ, Berger A, Peltier WR (1990) An alternative astronomical calibration of the lower Pleistocene timescale based on ODP site 677. Trans R Soc Edinburgh Earth Sci 81:251–261

Sorokovikova LM, Khodzher T, Sinyukovich VN, Golobokoba LP, Bashenkhaeva ND, Netsvetaeva OG (2000) Atmospheric and riverine input of nutrients and organic matter into Lake Baikal. In: Minoura K (ed) Lake Baikal. Elsevier, Amsterdam, p 236–246

Takamatsu N, Matsumoto GI, Kato N, Kawai T (2000) Paleoenvironmental changes in the Eurasian continent interior inferred from chemical elements in sediment cores (BDP-96/1,BDP-96/2) from Lake Baikal. In: Minoura K (ed) Lake Baikal. Elsevier, Amsterdam, p 127–135

Volkman JK, Barrett SM, Blackburn SI, Mansour MP, Sikes EL, Gelin F (1998) Microbial biomarkers: a review of recent research developments. Org Geochem 29:1163–1179

Williams DF, Qui L, Karabanov E, Gvozdkov A (1993) Geochemical indicators of productivity and sources of organic matter in surficial sediments of Lake Baikal. Russ Geol Geophys 33:111–125

Paleoenvironmental Changes during the Last 12 Million Years in the Eurasian Continental Interior Estimated by Chemical Elements in Sediment Cores (BDP-96 and BDP-98) from Lake Baikal

NOBUKI TAKAMATSU[1], GENKI I. MATSUMOTO[2], NAOYUKI KATO[3], and TAKAYOSHI KAWAI[4]

Summary. The content of 29 chemical elements (Na, K, Mg, Ca, Al, Fe, Li, Be, Sc, Ti, V, Cr, Mn, Co, Ni, Cu, Zn, Ga, Ge, As, Rb, Sr, Mo, Cs, Ba, W, Pb, Th, and U) in 589 subsamples of BD-P96 cores and 703 subsamples of BDP-98 cores from the Academician Ridge of Lake Baikal were determined using inductively coupled plasma-mass spectromety (ICP–MS) in order to estimate paleoenvironmental changes during the last 12 million years in the Eurasian continental interior. The content of terrigenous elements (Na, K, Mg, Ca, Al, Ti, etc.) increased gradually from about 6 Ma BP to the present, with repeated major fluctuations after 3.6 Ma BP, indicating a significant climate change at 3.6 Ma BP in the Eurasian continental interior. The abrupt change may be mainly associated with the uplift of the Tibetan plateau and Himalayas. This result corresponds with the rapid increase in the eolian dust accumulation rate in the central north Pacific sediment core (Rea et al. 1998). The data from BDP98 cores indicated that some big changes in the sources of chemical elements occurred at about 7 Ma BP, when the sedimentation rate in the core suddenly increased. We also estimated from the chemical index of alteration (CIA) that the variation in the climate from 8 to 12 Ma BP was small, and the rapid climate change started at 3.6 Ma BP.

Key words. Paleoenvironment, Lake Baikal, Sediment core, Chemical element

[1] Department of Chemistry, Faculty of Science, Toho University, Funabashi, Chiba 274-8510, Japan
[2] Department of Environmental Information Science, School of Social Information Studies, Otsuma Women's University, Tama, Tokyo 206-8540, Japan
[3] Department of Chemistry, Toho University School of Medicine, Ota-ku, Tokyo 143-8540, Japan
[4] National Institute for Environmental Studies, 16-2 Onogawa, Tsukuba 305-8506, Japan (present address: Department of Earth and Environmental Sciences, Graduate School of Environmental Studies, Nagoya University, Furo-cho, Chikusa-ku, Nagoya 464-8602, Japan)

1 Introduction

Studies of the bottom sediments of lakes are very important for understanding the paleoclimate changes, as well as those of seas. Lake Baikal is located in south Siberia, and is the deepest (1643 m) and the oldest (30 million years (Myr)) freshwater lake on Earth. The maximum thickness of its sediment is 8 km (Scholz et al. 1993). Therefore, Lake Baikal gives us a great opportunity to examine the paleoenvironmental changes in the Eurasian continental interior during the last 30 Myr (e.g., Nikolaev et al. 1985; Hutchinson et al. 1992; Mats 1993; Baikal Drilling Project Leg II Members 1997). Four long cores, BDP-96/1 and 2 and BDP-98/1 and 2, were drilled near Academician Ridge in 1996 and 1998 by members of the international Baikal Drilling Project (BDP) to interpret paleoclimatic history from various proxy data (e.g., Karabanov et al. 2000; Kawamuro et al. 2000; Matsumoto et al. 2000). It has been found that significant signals relating to changing paleoclimates are contained in the chemical element contents and their ratios in Lake Baikal sediments (Takamatsu et al. 2000; Goldberg et al. 2000; Phedorin et al. 2000; Edgington et al. 1996). Some elements, i.e., K, Rb, Cs, Ba, Nd, etc., negatively correlate with the biogenic silica which are the proxies of the productivity in the lake, and hence may directly reflect paleoenvironmental changes. Conversely, Br and U correlate positively with biogenic silica.

Here, we report the results of chemical element analyses of the BDP-96 and 98 sediment cores to clarify the paleoenvironmental changes during the last 12 Myr in the Eurasian continental interior.

2 Materials and Methods

2.1 Samples

BDP-96/1 and 2 sediment cores were taken in 1996 at Academician Ridge, a submerged topographic high between the north and central basins of Lake Baikal (53°41′48″N, 108°21′06″E, water depth 321 m, Fig. 1). The drilling methods are given in detail elsewhere (Williams et al. 1997). BDP-98/1 and 2 (600 m) sediment cores were taken in 1998 at Academician Ridge (53°44′48″N, 108°24′34″E, water depth 333 m, Fig. 1) near the BDP-96/1 and 2 sediment core site, with a core recovery of 95% (Baikal Drilling Project Group, 2000).

2.2 Analytical Methods

We analyzed the chemical elements content of 142 samples of BDP-96/1 at approximately 1.4 m intervals, and 447 samples of the BDP-96/2 core at intervals of approximately 20 cm. We also analyzed 703 samples of the BDP-98/1 and 2 cores. The digestion method for the core samples, and the analytical methods for the chemical elements, are given in detail elsewhere (Takamatsu et al. 2000). A 100-mg powdered subsample of each sediment sample was digested with a

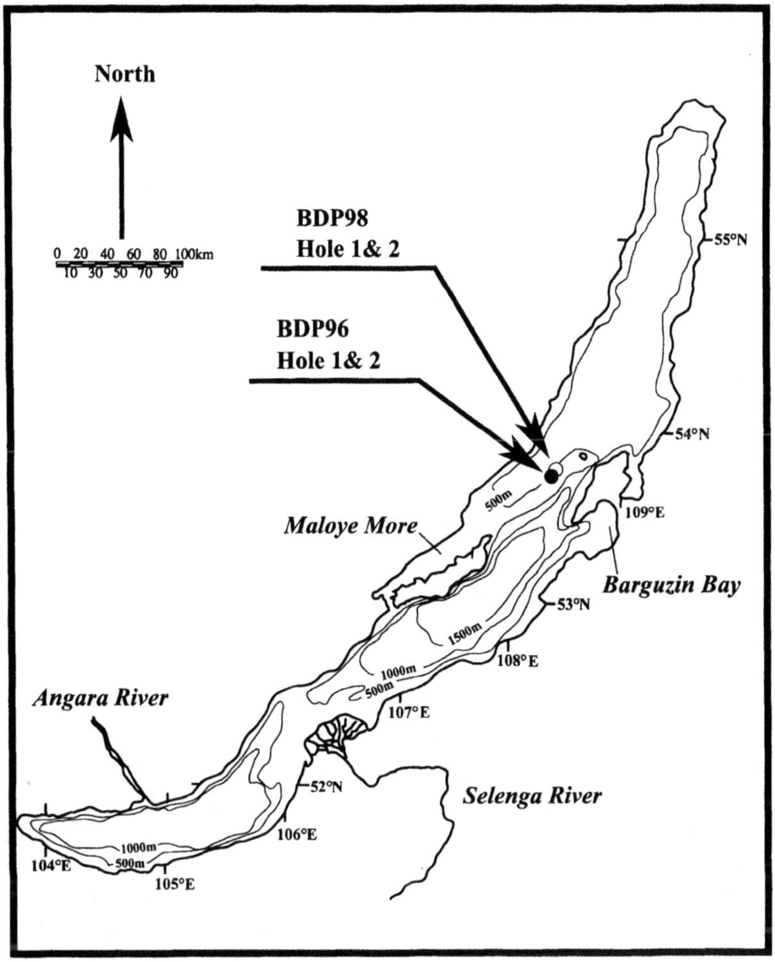

FIG. 1. Drilling sites at the Academician Ridge at Lake Baikal

mixture of nitric acid (3 ml), hydrogen peroxide (2 ml), and hydrofluoric acid (4 ml) in a Teflon vessel. The vessel was then exposed for 90 min to microwaves generated by a magnetron in a microwave-assisted decomposition apparatus (Model 7165, OI. Analytical, College Station, TX, USA). After drying the digested sample solution, the residue was dissolved in 1 ml nitric acid and transferred to a 100-ml volumetric flask with distilled water. All acids and reagents used in this experiment were of very pure grade (TAMAPURE-AA-100, Tamakagaku Kogyo, Kawasaki, Kanagawa, Japan). After appropriate dilution, the content of the elements (Li, Be, Al, Sc, Ti, V, Cr, Mn, Co, Ni, Cu, Zn, Ga, Ge, As, Rb, Sr, Mo, Cs, Ba, W, Pb, Th and U) in the sediment samples were determined with an inductively coupled plasma-mass spectrometer (Perkin Elmer

ELAN 5000, Perkin Elmer, Wellesley, MA, USA) using rhodium as the internal standard. The total Fe_2O_3 content was determined by a colorimetric method. The CaO, MgO, Na_2O, and K_2O content was determined by atomic absorption spectrometry. The precision and accuracy of all the element data were assessed by a comparison with the results of an analysis of JLk-1 (lake sediment) issued by the Geological Survey of Japan (Tsukuba, Ibaragi, Japan). The uncertainties of precision and accuracy for all elements were less than 10%.

3 Results and Discussion

3.1 Sediment Core

The depths along the BDP-96 and BDP-98 cores were converted into ages using age scales based on paleomagnetic epochs. The sedimentation rate of BDP-96/1 and 2 was 3.8 cm/kyr (Sakai et al. 2000). The sedimentation rates at depths of 0–277 m, 290–352 m, and 375–600 m for BDP-98/1 and 2 were 4.1, 29.3, and 4.5 cm/kyr, respectively (Kashiwaya et al. 2001). A sudden increase in the sedimentation rate was found at 290–352 m depth.

3.2 Variations in Content of Chemical Elements

To get the overall profiles of the elements investigated, we combined the data of the BDP-96/2 core (0–2.5 Ma), the BDP-98/1 core (2.5–5.0 Ma), and the BDP-98/2 core (5.0–12 Ma). The variations in the content of some elements are shown in Figs. 2–4. The general trend shows a gradual increase from about 6 Ma to the present, with repeated major fluctuations, and relatively constant values between 6 and 12 Ma, except between 7 and 8 Ma. However, each element has its own behavior, reflecting its chemical properties. An abnormally high Ni, Co, As, and Pb content in all cores was occasionally observed, as shown in BDP-96/1 (Takamatsu et al. 2000). In particular, the As content sometimes showed extremely high values compared with the base line value (Fig. 5). The high content may be reflecting low redox conditions caused by a chemical reaction such as sulfate reduction after sedimentation (Takamatsu 1998). We can conclude that the behaviors of these elements are associated with the redox condition changes after the sedimentation events, and cannot be used as climate proxies.

3.3 Chemical Elements as Paleoclimate Proxies

It is well accepted that the siliceous microfossil record is a paleoclimate proxy (Baikal Drilling Project Leg II Members 1997). It has also been shown that biogenic silica and diatoms were abundant during the interglacial periods, but rare during the glacial periods. We therefore, examined in detail which elements could be correlated with biogenic silica, using the data from the BDP-96/2 core. The

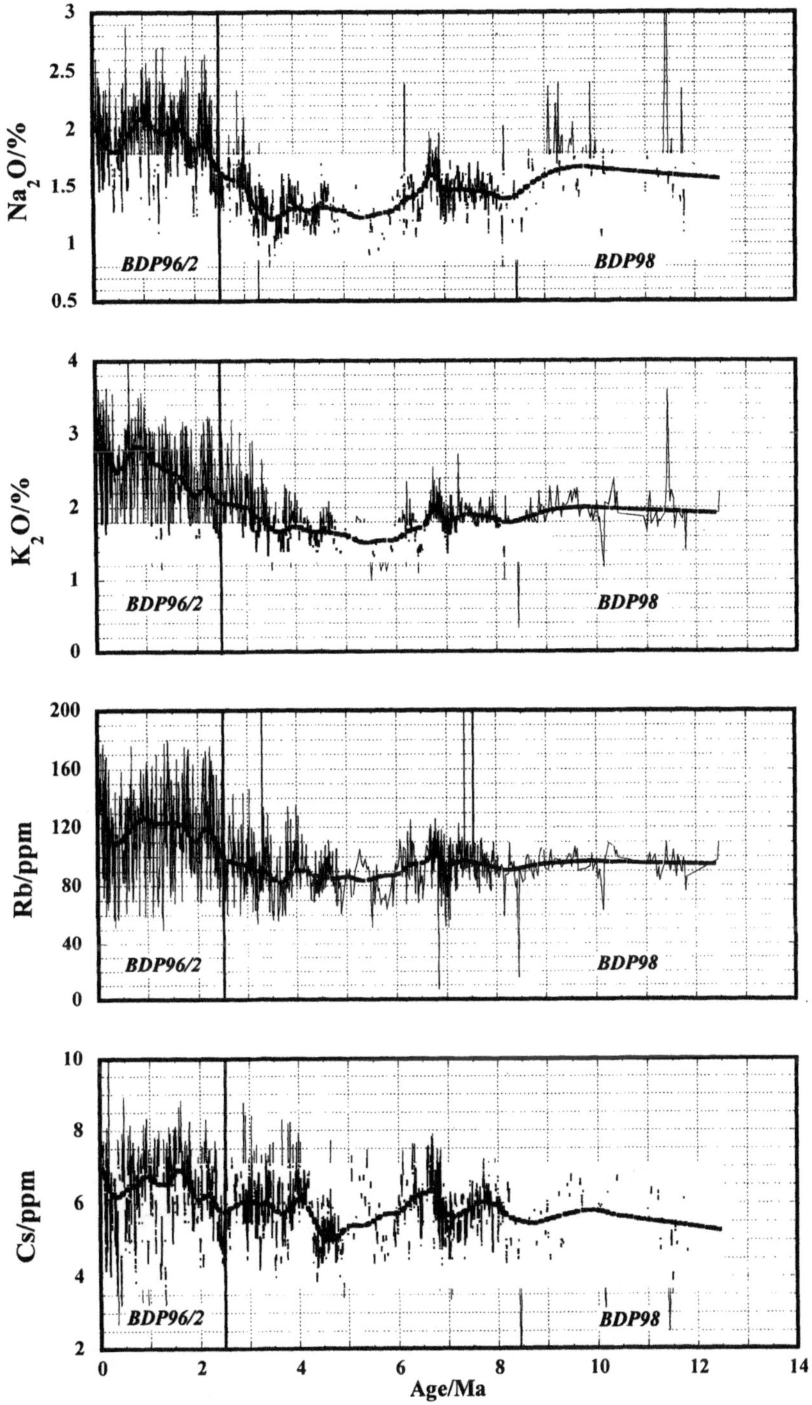

Fig. 2. Changes in the content of alkali metal elements in BDP-96/2 and BDP-98/1 and 2 cores. *Broken lines* are the weighted average curves with an 8% smoothing coefficient

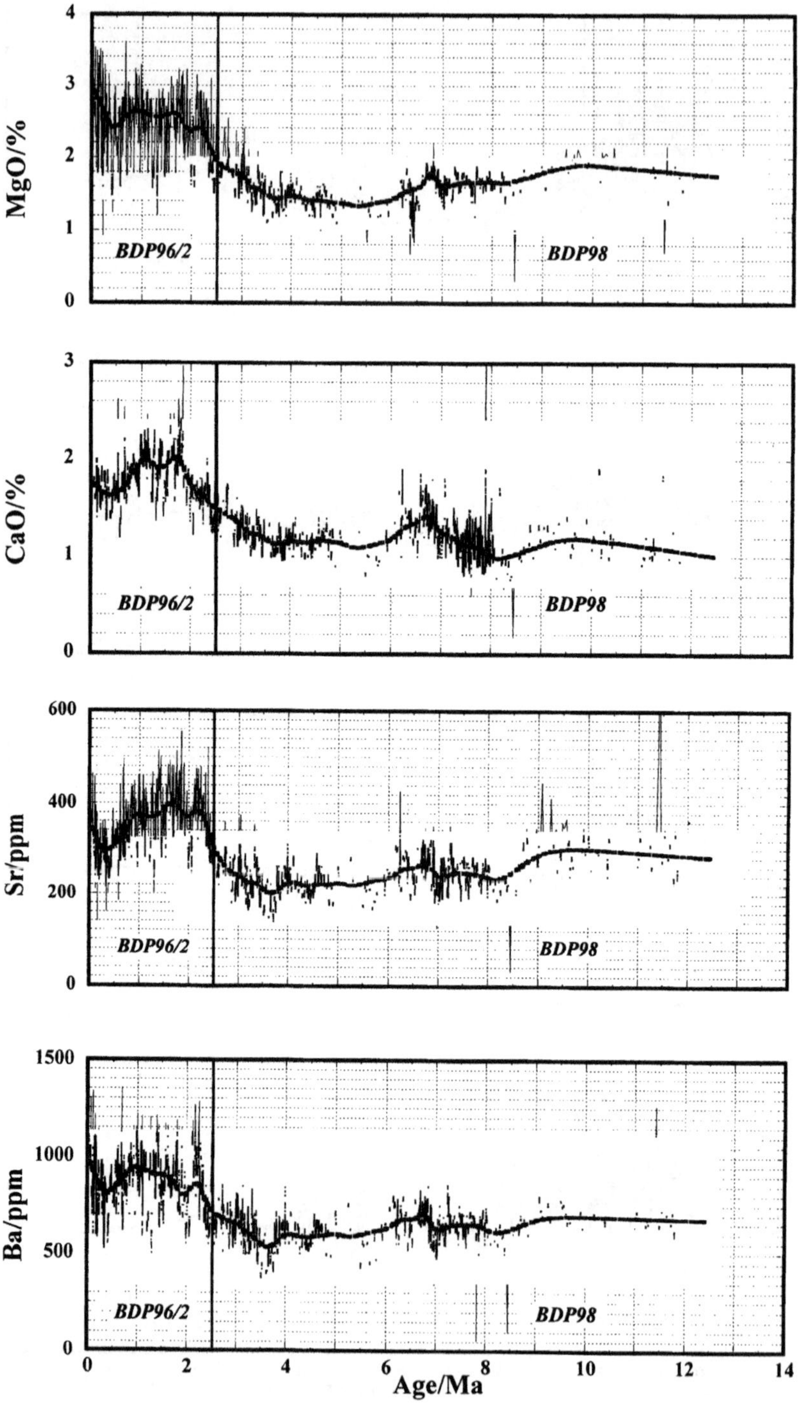

FIG. 3. Changes in the content of alkaline earth metal elements in BDP-96/2 and BDP-98/1 and 2 cores. *Broken lines* are the weighted average curves with an 8% smoothing coefficient

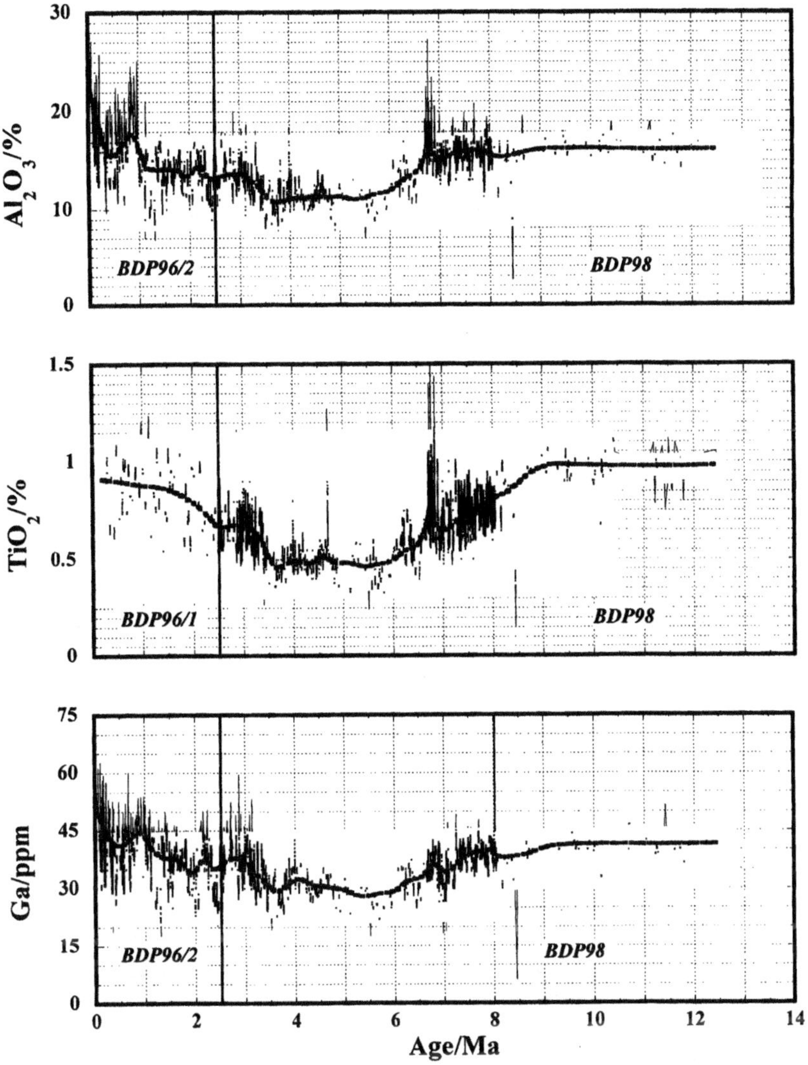

FIG. 4. Changes in the content of Al_2O_3, TiO_2, and Ga in BDP-96/2 and BDP-98/1 and 2 cores. *Broken lines* are the weighted average curves with an 8% smoothing coefficient

K_2O content in the BDP-96/2 core negatively correlated with the diatom frustule content of the core, as shown in Fig. 6, suggesting that K_2O may also be a proxy for major cooling or warming episodes in southern Siberia. The K_2O content strongly correlates with those of MgO ($R = 0.91$), Na_2O ($R = 0.80$), Al_2O_3 ($R = 0.78$), Rb ($R = 0.88$), and Ga ($R = 0.91$). These results indicate that these elements may also be paleoclimate proxies, at least in sediments younger than 2.5 Ma, although we have no available data according to the diatom frustule content of the BDP-98/1 and 2 cores.

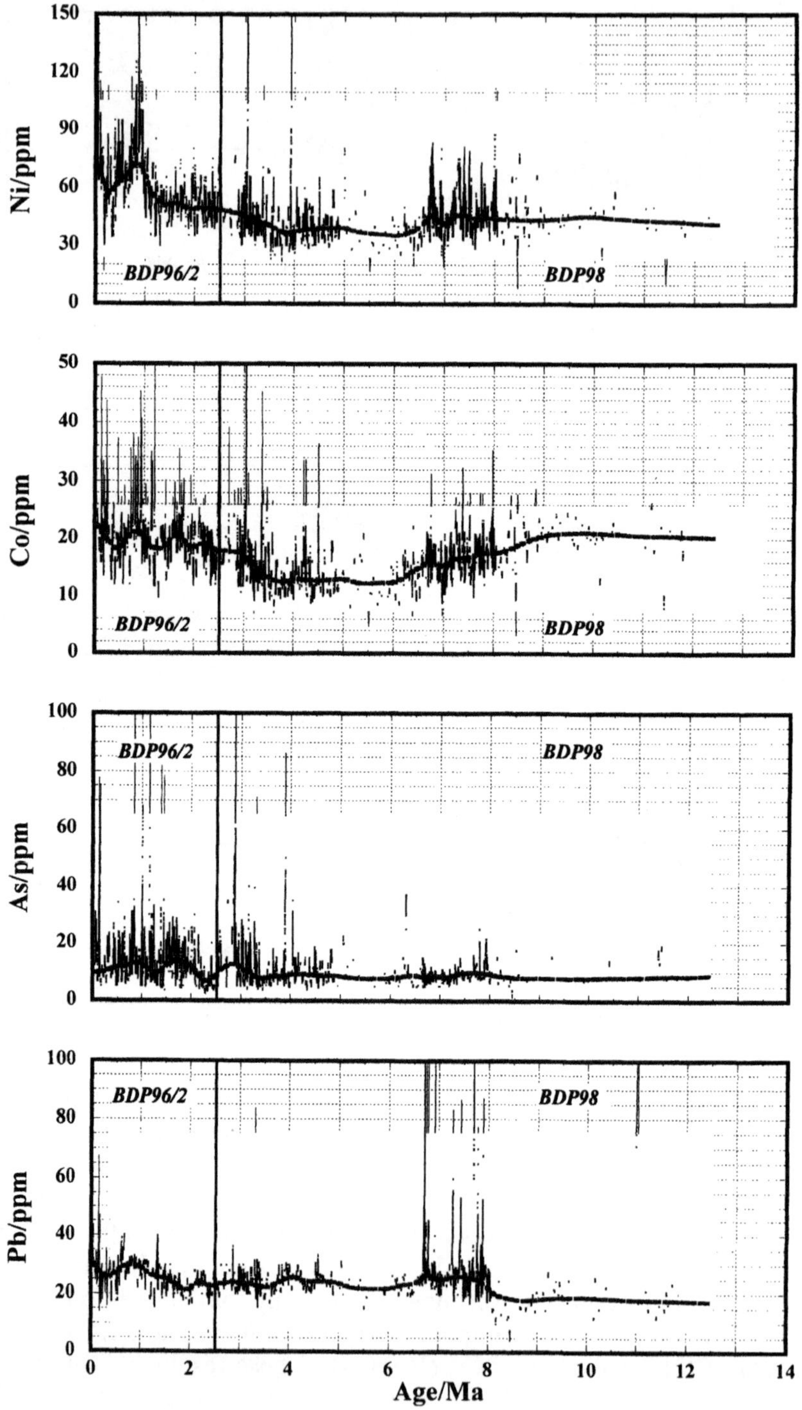

FIG. 5. Changes in the content of Ni, Co, As, and Pb in BDP-96/2 and BDP-98/1 and 2 cores. *Broken lines* are the weighted average curves with an 8% smoothing coefficient

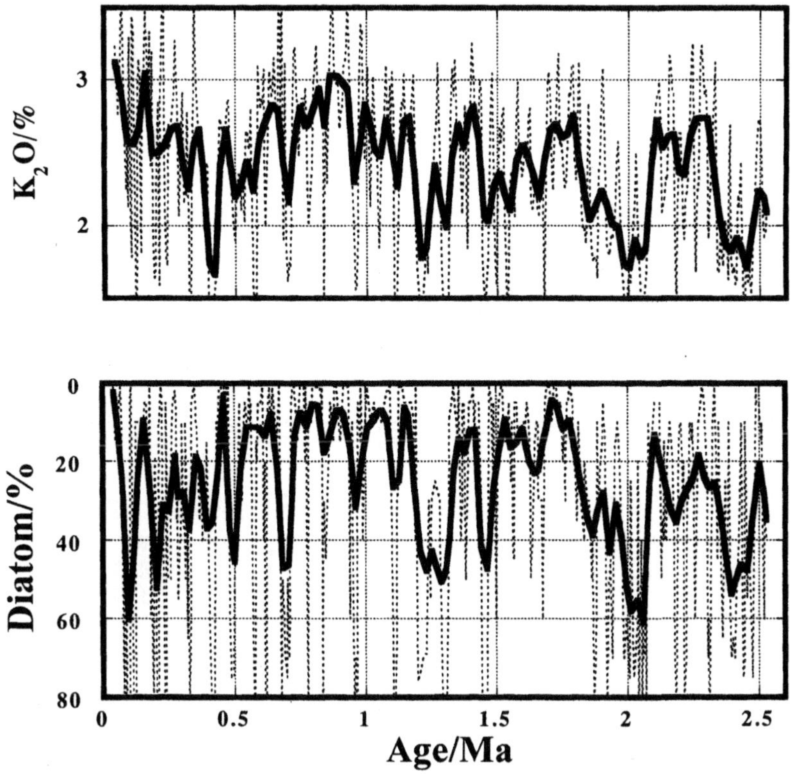

FIG. 6. Reverse relationship between K_2O content and diatom abundance in the BDP-96/2 core. *Solid lines* are the weighted average curves with a 3% smoothing coefficient

3.4 Correlation with Geological Events

Although there were some differences in the content of alkali metals (Na, K, Rb, and Cs) and alkaline earth metals (Ca, Mg, Sr, and Ba), they were roughly similar, as shown in Figs. 2 and 3. Takamatsu et al. (2000) applied principal component analysis to the ratios of 28 chemical elements to Al in the BDP-96/1 core, and estimated that a significant climate change occurred at 3.5 Ma, although the analytical intervals were rough. In this work, it should be noted that all variations of alkali metals, alkaline earth metals, Al_2O_3, TiO_2, and Ga in the BDP-98/1 core clearly showed that the significant climate change occurred at 3.6 Ma BP in the Eurasian continental interior. This result corresponds with the rapid increase in the eolian dust accumulation rate in the central north Pacific sediment core (ODP site 885/886; Rea et al. 1998). Figure 7 shows a comparison between the variation in Al_2O_3 content of BDP cores and the variation in the mass accumulation rate of eolian dust at Site 885/886 in the north Pacific. Rea et al. (1998) showed that dust deposition in the central north Pacific increased fairly rapidly by one order of magnitude at 3.6 Ma, and they associated this fact with the uplift

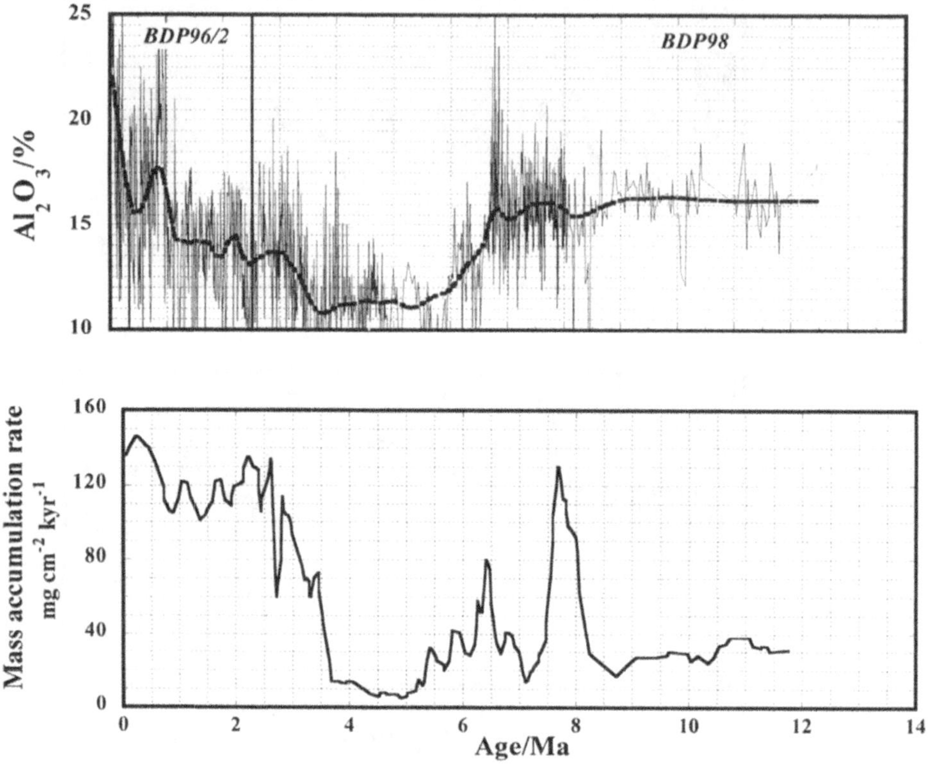

FIG. 7. Comparison of the Al_2O_3 content in BDP cores (the *broken line* is the weighted average curve with an 8% smoothing coefficient) with the mass accumulation rate of eolian dust at site 885/886 in the north Pacific (from Rea et al. 1998; the data were cited from URL: http://www.ngdc.noaa.gov/paleo)

of the Tibetan plateau, which shut off the Indian Ocean moisture source to central and western China. Rapid rifting began in Lake Baikal at about the time of the early–late Pliocene boundary at 3.5 Ma (Mats 1993). These results suggest that the sediments in Lake Baikal were strongly influenced by the tectonic activity in northern China and southern Siberia, which led to the drying of Asia in the late Cenozoic period.

The data of the BDP-98/1 and 2 cores show that the chemical element contents were low between 4 and 6 Ma, and considerably higher between 6 and 7 Ma BP, when the sedimentation rate in the core suddenly increased. The reason for the sudden increase is not clear at present. The chemical element contents were higher, and variations in amplitude were lower, in sediment older than 8 Ma.

3.5 Chemical Index of Alteration

The chemical element contents are largely influenced by the content of biogenic silica. Therefore, we have to compensate for the dilution effect of the biogenic silica when considering the changes in the composition of chemical elements which reflect the difference in source and/or lithology. We cannot discuss the lithology of BDP sediments directly because we have no available data for mineral components, such as the primary and the secondary minerals, in the subsamples we analyzed. Here, we calculate the chemical index of alteration (CIA) using molecular proportions (Andrew et al. 1996).

$$CIA = (Al_2O_3 \times 100)/(Al_2O_3 + CaO + Na_2O + K_2O)$$

Thus, CIA values approaching 100 are typical of materials formed under heavily leached conditions where soluble calcium, sodium, and potassium have been removed. It is known that kaolinite and chlorite attain CIA values of nearly 100, whereas illites and smectite have CIA values around 75–85. In comparison, unleached feldspars have CIA values around 50. Therefore, the CIA values in BDP sediment cores may roughly predict the degree of alteration of source rocks, because they show a relative abundance of minerals. Yuretich et al. (1999) found illite, interstratified illite–smectite, chlorite, and kaolinite as the major minerals in the BDP-93 cores. Horiuchi et al. (2000) found that late Quarternay sediment cores exhibit higher smectite contents from warm periods with a humid climate. The proportions of these minerals, especially kaolinite, which are indicative of a moist environment, may be related to the weathering regimes in the source region (Rea et al. 1998). A high CIA value corresponds to high humidity. The inverse variation in the CIA values of BDP sediment cores is shown in Fig. 8. This figure indicates that the significant climate change in the Lake Baikal region started at 3.6 Ma BP, the climate between 3.6 and 6 Ma was warm and wet, and the climate between 6 and 7 Ma was considerably cooler and drier.

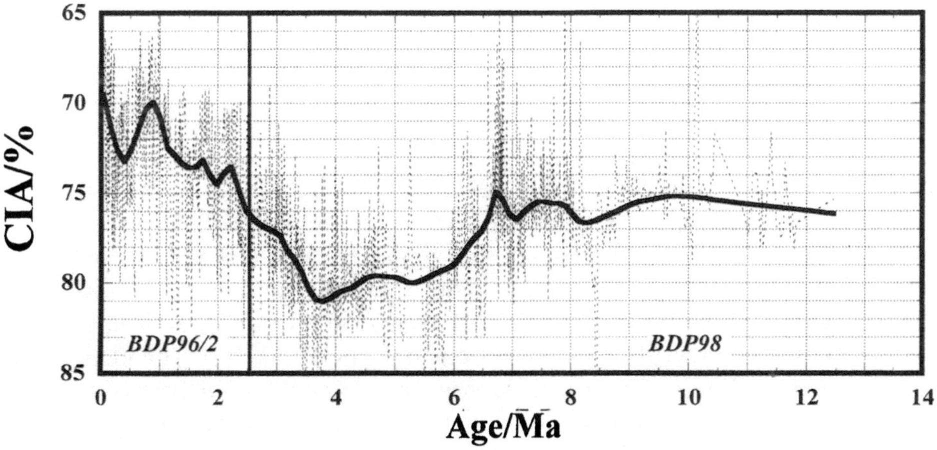

Fig. 8. Change in the chemical index of alteration (CIA) = $(Al_2O_3 \times 100)/(Al_2O_3 + CaO + Na_2O + K_2O)$ in BDP-96/2 and BDP-98/1 and 2 cores. The *solid line* is the weighted average curve with an 8% smoothing coefficient

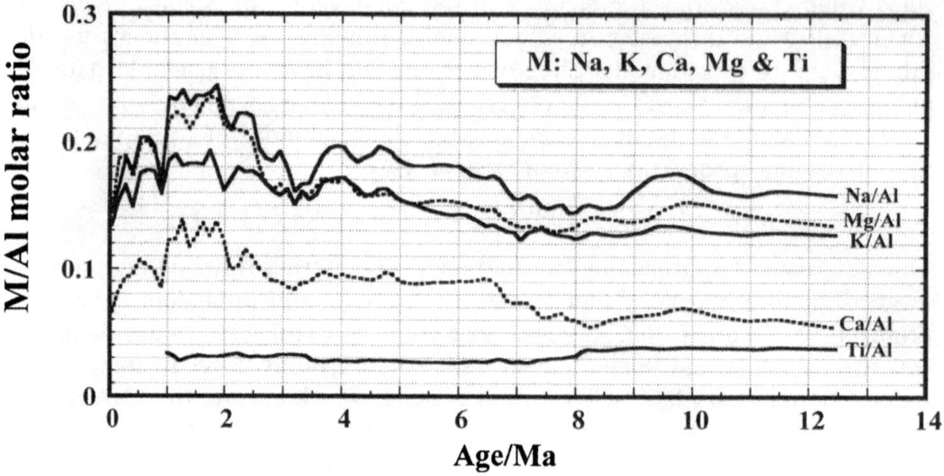

FIG. 9. Changes in element/aluminum molar ratios in BDP-96/2 and BDP-98/1 and 2 cores. All lines are weighted average curves with a 3% smoothing coefficient

The molar ratios of some major elements to Al in the BDP-96/2 and BDP-98/1 and 2 cores are shown in Fig. 9 to illustrate the degree of alteration of the originating rocks. These ratios, especially the Na/Al and Ca/Al molar ratios, suggest that the aridification started at 3.6 Ma BP, and that some big changes in the sources of chemical elements may have occurred at about 7 Ma BP, when the sedimentation rate in the core suddenly increased. This is more clearly shown by applying principal component analysis to our data. Figure 10 shows the variation in the first principal component score obtained by an analysis of data sets for 28 element/aluminum weight ratios. This may indicate that the sediments have gradually changed from distal deltaic facies to fine hemipelagic sediments (Moore et al. 1997).

4 Conclusion

The terrigeneous element contents (alkali metals and alkaline earth metals, Ti, Al, etc.) increased gradually from 6 Ma BP to the present, with repeated major fluctuations after 3.6 Ma BP, indicating a significant climate change at 3.6 Ma BP (the onset of aridification in Asia) in the Eurasian continental interior. This abrupt change may be associated with the change in the atmospheric circulation caused by the uplift of the Tibetan plateau and the Himalayas. The element/Al molar ratios in BDP-98 cores, and the principal component analysis of the data set of the element/Al weight ratios, indicated that some big changes in the sources of chemical elements occurred at about 7 Ma BP, when the sedimentation rate in the core suddenly increased. We also estimated from the data that the variation in the climate from 8 to 12 Ma BP was small. The chemical index

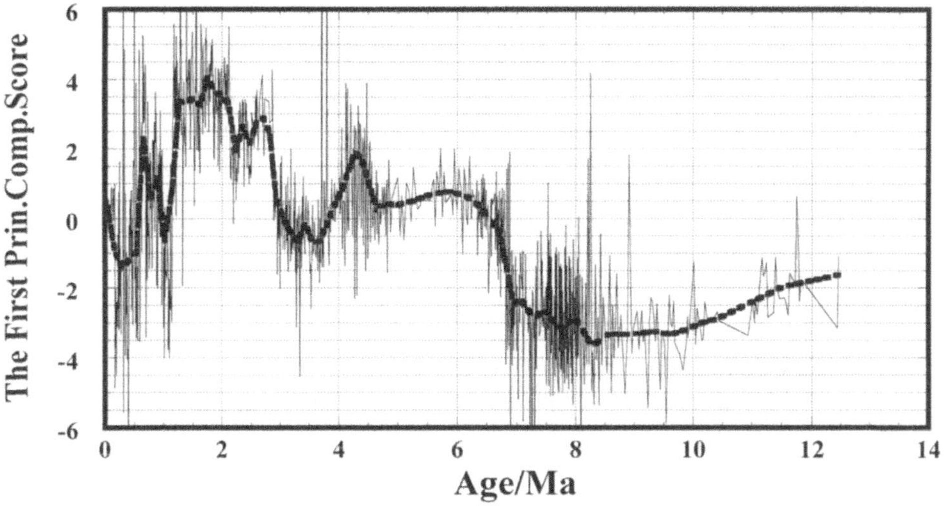

FIG. 10. Changes in the first principal component score obtained by principal component analysis of data sets for 28 element/aluminum weight ratios in the BDP-96/2 and BDP-98/1 and 2 cores. The *broken line* is the weighted average curve with a 3% smoothing coefficient

of alteration in the BDP-96 and 98 cores suggested that the humidity during 3.6–5.5 Ma BP was relatively high in the last 12 M years, and became rapidly lower after 3.6 Ma BP.

Acknowledgments. The authors thank T. Kawamura, J. Yamamoto, T. Kato, R. Kiuchi, M. Kodama, K. Kondo, A. Oshitanai, and M. Ono for their chemical analyses. We thank the BDP-96 and 98 members from Russia, Japan, the USA, and Germany who worked in drilling, logistics, descriptions, sampling, etc. This research was supported by the Science and Technology Agency (now the Ministry of Education and Science) of Japan, the Siberian Branch of the Russian Academy of Science, the US National Science Foundation, and the Japan Baikal International Center for Ecological Research.

References

Andrews JE, Brimblecombe P, Jickells TD, Liss PS (1996) An introduction to environmental chemistry. Blackwell Science, London, p 79–84

Baikal Drilling Project Group (2000) Paleoclimate record in the late Cenozoic sediments of Lake Baikal (by 600-m deep-drilling data) (in Russian). Russ Geol Geophys 41:3–32

Baikal Drilling Project Leg II Members (1997) Continuous paleoclimate record recovered for the last 5 million years. EOS 78:597–604

Edgington DN, Robbins JA, Colman SM, Orlandinid KA, Gustina MP (1996) Uranium-series disequilibrium, sedimentation, diatom frustules, and paleoclimate change in Lake Baikal. Earth Planet Sci Lett 142:29–42

Goldberg EL, Phedorin MA, Grachev MA, Bobrov VA, Dolbuya IP, Khlystov OM, Levina OV, Ziborova GA (2000) Geochemical signals of orbital orcing in the records of paleoclimates found in the sediments of Lake Baikal. Nucl Instrum Methods Phys Res A 448:384–393

Horiuchi K, Minoura K, Hoshino K, Oda T, Nakamura T, Kawai T (2000) Paleoenvironmental history of Lake Baikal during the last 23000 years. Paleogeogr Paleoclimatol Paleoecol 157:95–108

Hutchinson DR, Golmshtok AJ, Zonenshain LP, Moor TC, Scholz CA, Klitgord KD (1992) Depositional and tectonic framework of the rift basins of Lake Baikal from multichannel seismic data. Geology 20:589–592

Karabanov EB, Kuzmin MI, Prokopenko AA, Williams DF, Khursevich GK, Bezrukova EV, Kerber EV, Gvozdkov AN, Gelety VF, Weil D, Schwab M (2000) Glaciations of central Asia in the late Cenozoic according to the sedimentary record from Lake Baikal. In: Minoura K (ed) Lake Baikal. Elsevier, Amsterdam, p 71–84

Kashiwaya K, Ochiai S, Sakai H, Kawai T (2001) Orbit-related long-term climate cycles revealed in a 12-Myr continental record from Lake Baikal. Nature 410:71–74

Kawamuro K, Shichi K, Hase Y, Iwauchi A, Minoura K, Oda T, Takahara H, Sakai H, Morita Y, Miyoshi N, Kuzmin M (2000) Forest–desert alternation history revealed by the pollen record in Lake Baikal over the past 5 million years. In: Minoura K (ed) Lake Baikal. Elsevier, Amsterdam, p 101–107

Mats VD (1993) The structure and development of the Baikal rift depression. Earth Sci Rev 34:81–118

Matsumoto GI, Kosaku S, Takamatsu N, Akagi T, Kawai T, Ambe Y (2000) Estimation of paleoenvironmental changes in the Eurasian continental interior during the past 5 million years inferred from organic components in the BDP96/1 sediment core from Lake Baikal. In: Minoura K (ed) Lake Baikal. Elsevier, Amsterdam, p 119–126

Moore TC, Klitgord KD, Golmstok AJ, Weber E (1997) Sedimentation and subsidence patterns in the central and north basins of Lake Baikal from seismic stratigraphy. GSA Bull 109:746–766

Nikolaev VG, Vanyakin LA, Kalinin VV, Milanovsky Y (1985) The sedimentary section beneath Lake Baikal. Int Geol Rev 27:449–459

Phedorin MA, Goldberg EL, Grachev MA, Levina OL, Khlystov OM, Dolbnya IP (2000) The composition of biogenic silica, Br and Nd distributions in the sediments of Lake Baikal as proxies of changing paleo-climates of the last 480 kyr. Nucl Instrum Methods Phy Res A 448:400–406

Rea DK, Snoeckx H, Joseph LH (1998) Late Cenozoic eolian deposition in the north Pacific: Asian drying, Tibetan uplift, and cooling of the northern hemisphere. Paleoceanography 13:215–224

Sakai H, Nomura S, Horii M, Kashiwaya K, Tanaka A, Kawai T, Kravchinsky V, Peck J, King J (2000) Paleomagnetic and rock magnetic studies on Lake Baikal sediments: BDP96 borehole at Academician Ridge. In: Minoura K (ed) Lake Baikal. Elsevier, Amsterdam, p 35–52

Scholz AC, Klitgord KD, Hutchinson DH, Ten Brink US, Zonenshain LP, Golmshtok AY, Moore TC (1993) Results of 1992 seismic reflection experiment in Lake Baikal. EOS Trans Am Geophys Univ 74:469–470

Takamatsu T (1998) Science of global environmental change: Baikal Drilling Project (in Japanese). In: Matsumoto GI, Kashiwaya K, Minoura K (eds) Kokon Shoin, Tokyo, p 137–144

Takamatsu N, Matsumoto GI, Kato N, Kawai T (2000) Paleoenvironmental changes in the Eurasian continent interior inferred from chemical elements in sediment cores (BDP96/1, BDP96/2) from Lake Baikal. In: Minoura K (ed) Lake Baikal. Elsevier, Amsterdam, p 127–135

Williams DF, Peck J, Karabanov EB, Prokopenko AA, Kravchinsky V, King J, Kuzmin MI (1997) Lake Baikal record of continental climate response to orbital insolation during the past 5 million years. Science 278:1114–1117

Yuretich R, Melles M, Sarata B, Grobe H (1999) Clay minerals in the sediments of Lake Baikal: a useful climate proxy. J Sediment Res 69:582–596

Pliocene–Quaternary Vegetation and Climate History of the Lake Baikal Area, Eastern Siberia

E.V. Bezrukova[1], N.V. Kulagina[2], P.P. Letunova[1], E.B. Karabanov[3,4], D.F. Williams[3], M.I. Kuzmin[4], S.M. Krapivina[1], K.E. Vershinin[1], and O.N. Shestakova[2]

Summary. A palynological study was carried out on the BDP-96 hole 1 core. We distinguished six major large stages (A1-6) characterized by vegetation changes in the Lake Baikal catchment area. The existence of the most abundant amount of pollen, *Tsuga*, is a specific feature of the pollen spectra of stage A6 from 3.1 to 2.6 Ma. The pollen complex of dark coniferous arboreal plants such as *Tsuga*, *Abies*, and *Picea*, which are sensitive to air humidity, serves as an indicator of the amelioration of climatic conditions. There is a process of replacement of *Tsuga* pollen by *Abies* in the pollen spectra of stage A5. The age of this stage was determined to be from 2.6 to 2.35 Ma. These are the first reliable palaeobotanical data obtained from continuous cross sections, which reflect the response of regional vegetation and flora to the first well-known significant cooling in the northern hemisphere—the Praetiglian. During stage A4 larch appears and is widely distributed. This is probably connected with the start of a temporary cold continental, climate. At the beginning of stage A3, about 1.85 Ma, a new amelioration of climatic conditions takes place. The beginning of stage A2, about 1.5 Ma, was characterized by an abrupt change in the composition of regional vegetation and flora. *Tsuga* disappears from the flora composition, and *Larix* and *Betula* are again widely distributed. *Pinus pumila* appears, and continues to exist. During stage A1, vegetation which was similar to that of the present started to persist in the region, and the climate became sharply continental, and moderately cold

Key words. Palynology, Pliocene, Baikal region, Flora and vegetation reconstruction, Paleolandscapes, Paleoclimate

---

[1] Limnological Institute, Siberian Branch of the Russian Academy of Sciences, Ulanbatorskaya str. 3, Irkutsk 664033, Russia
[2] Institute of Earth Crust, Siberian Branch of the Russian Academy of Sciences, Lermontova str., 126, 664033, Irkutsk, Russia
[3] Baikal Drilling Project, Department of Geological Sciences, University of South Carolina, Columbia, SC 29208, USA
[4] Vinogradov Institute of Geochemistry, Siberian Branch of the Russian Academy of Sciences, Favorsky str. 1A, Irkutsk 664033, Russia

1 Introduction

Complex investigations of Lake Baikal sediments obtained from the top of the underwater Academician Ridge have indicated that these sediments contain a continuous record of paleoclimates of the upper Pleistocene (Bradbury et al. 1994; Colman et al. 1995; Grachev et al. 1997). The characteristic feature of the record is the high content of diatoms in layers formed in warm climatic conditions, and insignificant diatom concentrations in layers belonging to the periods of cooling (Bezrukova et al. 1991; Grachev et al. 1998). The biogenic silica (SiObio) concentration, water content, and density of sediments, as well as some geochemical and biogeochemical indicators, correlate well with the diatom content. The study of two cores, BDP-96-1 (200 m length) and BDP-96-2 (100 m), obtained from Academician Ridge, showed that the alternation of layers of clays containing no diatoms and diatom silt is found down to a depth of at least 200 m (Grachev et al. 1998; Kuzmin et al. 1997, 1999; Williams et al. 1997; Baikal Members 1997). The cores obtained via drilling are well dated. The Brunhes–Matuyama boundary (0.78 Ma) is found at a depth of 34 m, and the Matuyama–Gauss boundary (2.58 Ma) is observed 105 m from the sediment surface. The average sedimentation rate at the drill site during the whole studied period was constant at 4–5 cm/ky. Spectral analyses of the distribution of biogenic silica and the water content (Grachev et al. 1998) indicated that for the last 500 ky periods 100, 41, 23, and 19 ky are well observed in the Baikal archive. This leads to the conclusion that the climate of Eastern Siberia within the interval indicated changed in line with global changes found in the SPECMAP profile (Colman et al. 1995; Kuzmin et al. 1999). Since the vegetation is very sensitive to climate, changes in the palynological records can thus be used to elucidate the origin of continental climates far removed from the influence of oceans and ice sheets.

The first results of the palynological analysis of the BDP-96-1 core (200 m), ascertained by time resolution to be 45–50 ky (Bezrukova et al. 1999), revealed three long periods of cooling in the region. These periods were dated at 2.5–2.8, 1.5–1.8, and 0.6–0.8 Ma. The results of the palynological investigations of the upper 34 m of the BDP-96-1 core, characterizing the changes in the climate and vegetation during the time from 200 ky to the Brunhes–Matuyama (0.78 Ma) boundary, and with a time resolution of 4–5 ky, were published later (Bezrukova and Letunova 2001). This chapter considers the fluctuations in the vegetation composition recorded in pollen spectra from the upper 122-m section of the BDP-96-1 core, spanning about 3.1 Myr. The sampling interval was 2.5 ky on average. Such a detailed pollen analysis was being performed for the first time for both Lake Baikal sediments and Asian continental deposits. The data obtained can be used as the basis for a through correlation of sediments from Lake Baikal proper, and lacustrine and continental sediments from other depressions in the Baikal Rift Zone. These data are regarded as being suitable to obtain a reference palynological sequence for the Baikal Region.

2 Materials and Methods

The sediments from the core were sampled using a small plastic box of known volume (≈ 300 mm^3), and were then prepared for pollen analysis using the procedure of Faegri and Iversen (1989). The method was supplemented by processing the sediment sample in 5% sodium pyrophosphate to remove any clays (Bates et al. 1978). The pollen residue obtained was mixed with a glycerine–gelatine mixture and put on a slide. A routine palynological analysis was conducted using a Karl Zeiss microscope with a magnitude of 650x. At least 300 pollen grains of arboreal taxa were counted for samples with a rich pollen content. If a sample had very little pollen, then 150 grains and spores of each plant group were counted. Pollen types were identified using the reference collection of the Institute of the Earth Crust SB RAS (Irkutsk, Russia), and identification keys for classification and verification (Grana Palynologica 1961; Moore et al. 1991; Markgraf and Di'Antoni 1978; Kupriyanova 1965). The pollen diagram for the late Pliocene–middle Pleistocene (Fig. 1) was not divided into the traditional regional pollen zones, as the number of samples used for plotting was 800, which would make the diagram barely readable. The diagram shows the main stages of the changes in the vegetation and flora composition in the Lake Baikal catchment area. These stages result primarily from changes to the global climate. The divisions of stages A1–6 are based on the change in the ratio of pollen between arboreal and nonarboreal plants, and trends in the changing percentages of taxa (for instance, *Tsuga*, *Abies*, *Picea*, *Pinus Haploxylon*, *Pinus Diploxylon*, and *Betula*) which dominate the late Pliocene–Quaternary of Eastern Siberia. The spectra of all samples with *Tsuga* pollen contain pollen of such moderate thermophillous taxa as *Carpinus*, *Corylus*, *Ulmus*, *Quercus*, and *Tilia*. The spectra of samples of stage A6 include the pollen of *Fagus* and Juglandaceae. Because the pollen of the last two taxa is found as single grains, these are not shown on the pollen diagram. The sum percentage content of all the thermophillous taxa mentioned does not exceed 3%. The data on spore–pollen spectra (SPS) of surface samples, areas of recent forest-forming species and their pollen productivity, and vertical zoning in vegetation distribution were considered to reconstruct an image of the vegetation cover. The features forming recent SPS in both the upper layer of Baikal sediments and the continental sediments of the shoreline have been reported (Bezrukova 1994, 1998; Ogura et al. 1998). The age determination of major stages in the evolution of the vegetation is based on an age model obtained from studies of the paleomagnetic features of core sediments. The paleomagnetic age model is exact and detailed, and is independent of the correlation from climatic indicators (Williams et al. 2001).

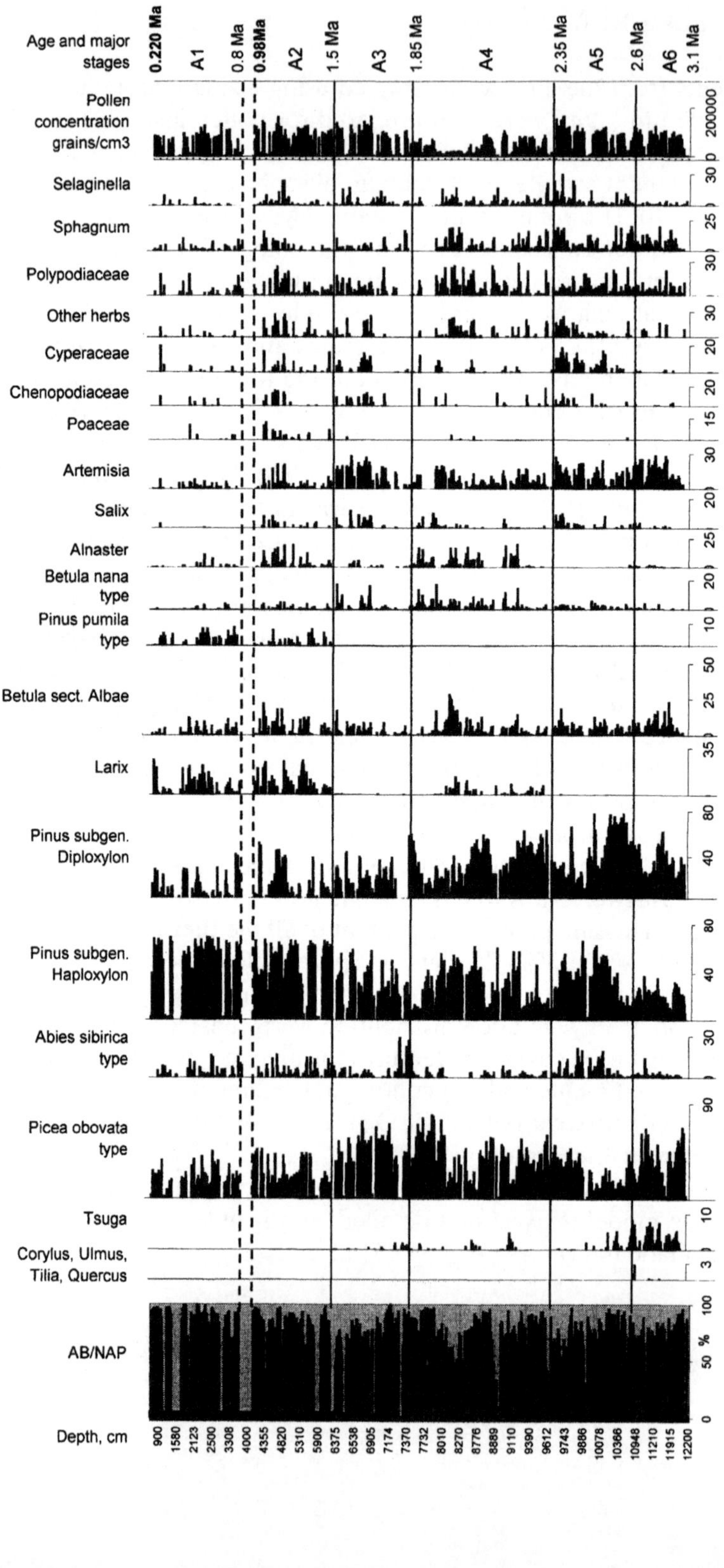

FIG. 1. Simplified pollen diagram of the bottom sediments of Lake Baikal (BDP-96-1 core). The determination of the relative frequency of pollen was calculated based on the total tree pollen. *Black*, arboreal pollen; *Gray*, nonarboreal pollen

3 Recent Vegetation and Climate

The changes in the flora composition, landscape, and climate can be better represented by a comparison of recent vegetation patterns and climate with those of the past. Therefore, we will briefly describe the recent vegetation and climate in the region. The mountain-taiga landscapes, which occupy about two-thirds of the area, are predominant at present in the Baikal watershed. Dark coniferous forests containing Siberian cedar (*Pinus sibirica* Du Tour), fir (*Abies sibirica* Ledeb.), and spruce (*Picea obovata* Ledeb.) prevail on the western, wetter macroslopes of the mountain ranges. The eastern macroslopes, with a drier continental climate, are characterized by light coniferous forest containing common pine (*Pinus sylvestris* L.) and larch (*Larix sibirica* Ledeb.). On the higher slopes, the forests are gradually replaced by a sparse growth of trees in the subalpine mountain belt. The zone of high mountain vegetation is found above this. Most of the bare mountain zone is occupied by mountain tundra. The steppes of the Baikal basin do not form an independent belt. They are fragmentary, and are located in the Priolíkhonie, on Olíkhon Island, and on the eastern slopes of ridges with the least precipitation (Molozhnikov 1986).

The specific features of the recent climate in the region result from its high mountain relief, its submeridional location in the lacustrine basin, the intracontinental location of the Asian anticyclone effect, and the influence of the constant temperature of the Baikal water column on the offshore zone. The average annual temperature is negative everywhere (from –0.5°C in the southern basin to –3.5°C in the north). The average January temperature varies in the same direction from –17.3 to –22.7°C, and the average July temperature varies from +17 to +12°C. Precipitation is scarce and irregular. The typical maximum precipitation is 630 mm, in the southern basin, and the minimum, 340 mm, is common in the middle area (Baikal Atlas 1993).

4 Results: Description of the Stages

Figure 1 shows six stages in the changes of the vegetation and flora composition in the region investigated.

4.1 Stage A6: 3.1–2.6 Ma

Pollen of *Picea* (to 60% maximum), *Pinus* subgenus *Diploxylon* (to 45%), *Pinus* subgenus *Haploxylon* (to 35%), and *Artemisia* (to 30%) are predominant in the pollen spectra. This stage is also characterized by the maximum *Tsuga* content (to 8%), as well as pollen of temperate taxa such as *Carpinus*, *Corylus*, *Ulmus*, *Quercus*, and *Tilia* (up to 2.8% in total).

4.2 Stage A5: 2.6–2.35 Ma

The pollen spectra in this stage show the predominance of *Pinus* subgenus *Diploxylon* (to 78%) and *Pinus* subgenus *Haploxylon* (to 67%). *Picea* pollen abundance decreases significantly (35% on average). However, *Abies* pollen increases to 25%. Cyperaceae and Chenopodiaceae (to 18% and 15%, respectively) are predominant among herb pollen. The *Tsuga* pollen content decreases to 4%–5%, and is found only at the beginning of this stage.

4.3 Stage A4: 2.35–1.85 Ma

Picea (to 80%), and *Pinus* subgenus *Haploxylon* and *Pinus* subgenus *Diploxylon* (to 65% and 75%, respectively) dominate the pollen spectra in this stage. The *Alnaster* pollen content markedly increases (to 19%), while the *Artemisia* content decreases to 15%–18%. The characteristic feature of the spectra in this stage is a significant amount of *Larix* pollen. The spectra of the two previous stages contain only minute amounts of this taxa, but the spectra of stage 4 contain up to 10% of *Larix* pollen. The *Abies* pollen concentration decreases markedly (from 25% to 8%), and *Tsuga* pollen occurs sporadically, but does not exceed 4%.

4.4 Stage A3: 1.86–1.5 Ma

The pollen spectra at the beginning of this stage contain abundant *Abies*, *Picea*, and *Pinus* subgenus *Haploxylon*. Then the pollen of *Picea* (70% on average), *Pinus* subgenus *Haploxylon*, *Artemisia*, Chenopodiaceae, and mesophyte herbs, and Polypodiaceae spores becomes predominant. *Larix* pollen is found as solitary grains. *Tsuga* pollen does not exceed 1.5%–2% and is found in four spectra only at the beginning of the stage.

4.5 Stage A2: 1.5–0.98 Ma

Tsuga pollen is not found in the spectra of this stage. The *Picea* content decreases to an average 40%. The pollen of *Abies* (to 15%) and *Pinus* subgenus *Haploxylon* (to 68%) is abundant. The *Larix* (to 25% on average) and *Alnaster* (to 15%) contents increase sharply. A constant component in the spectra of this stage is the pollen of *Pinus pumila* (to 7%). Poaceae, Chenopodiaceae, mezophyte herbs, spores of Polypodiaceae, and *Selaginella* prevail among herb pollen and spores.

The palynological data have not been obtained for the 0.98–0.88 Ma interval because the samples were not available. No number is given to this interval in Fig. 1.

4.6 Stage A1: 0.88–0.22 Ma

Tsuga and temperate taxa pollen are absent in the spectra of this stage. The pollen spectra are dominated by *Abies* (12% on average), *Pinus* subgenus

Haploxylon (65% on average), and *Larix* (20% on average). The pollen of *Pinus pumila* (to 8%) is predominant in the group of shrubs, while the herb pollen content decreases sharply. Poaceae and Chenopodiaceae are predominant among these pollen.

5 Discussion

A comparison of features of the local flora evolution with these from adjacent areas is of great importance for understanding the evolution of the flora in the Pliocene, and for a robust interpretation of the data. Upper Pliocene slope sediments from the Transbaikalia (Rezanov 1988) show spectra with a predominance of *Pinus* subgenus *Diploxylon*, *P.* subgenus *Haploxylon*, and *Betula* pollen. For *Abies*, *Picea*, *Alnus*, *Ulmus*, and *Corylus*, only a single pollen grain of each was found. The late Pliocene flora, reconstructed from the palynological data from the ancient Manzurka river deposits collected from the Buguldeika River valley (Trofimov et al. 1995), is characterized by a complicated combination of coniferous, coniferous and broad-leaved, and steppe-like forests, indicating a vertical zoning of the vegetation cover. Typical components of coniferous forests included some species of pine (subgenus *Diploxylon* and *Haploxylon*), *Abies*, *Picea*, and *Tsuga*. The broad-leaved group contained *Corylus*, *Carpinus*, *Ulmus*, *Juglans*, *Quercus*, *Fagus*, and *Tilia*. The total amount did not exceed 3%. The age of those floras, from the absolute dating of basalts, is younger than 3.2 Ma (Devyatkin et al. 1989). The late Pliocene flora from the alluvial deposits of the Pribaikalia characterizes the vegetation of mixed coniferous and broad-leaved forests. These forests were composed of *Picea*, *Abies*, *Pinus* subgenus *Haploxylon*, and *P.* subgenus *Diploxylon*. Pollen from taxa typical of the temperate flora, such as *Ulmus*, *Tilia*, *Corylus*, and *Carpinus*, is found as single grains.

Volkova (1977) concluded that the species now growing in Western Siberia were predominant in the composition of the flora of the late Pliocene in Western Siberia. The vegetation was mainly composed of conifers such as *Picea*, *Abies*, *Pinus* subgenus *Haploxylon*, and *P.* subgenus *Diploxylon*, and of small-leaved shrub trees such as *Betula* and *Alnus*, accompanied by the broad-leaved trees *Ulmus*, *Tilia*, *Quercus* subgenus *Lepidobalanus*, and *Carpinus*. Using data from paleocarpological studies of Siberian Pliocene flora samples, Dorofeev (1956) came to the conclusion that this flora was intermediate in character between the Miocene and recent times. The fossil flora includes *Picea*, *Pinus*, *Larix*, *Betula*, *Alnus*, and *Tsuga*.

The pollen of evergreen conifers such as *Tsuga*, *Abies*, and *Picea* indicates humid climatic conditions. The minimum possible winter temperature can be estimated by the environmental requirements of *Tsuga*. In the Asian part of the area, it grows in humid conditions caused by the proximity of the Himalayas and Tibet. In Japan, *Tsuga* grows mainly in the subalpine zone and the upper part of the warm temperate forest zone. In Northern America, *Tsuga* grows in the mountain taiga of the western shore, and in broad-leaved and conifer forests of the mid-West in regions with cool and wet climates. It also grows in forests with *Abies*, *Picea*, and *Pinus*. The same complex of predominant conifers existed in the Pribaikalia mountain range. The conditions necessary for the existence of vegetation containing *Tsuga* as well as *Corylus*, *Tilia*, and *Quercus* include high precipitation (not less than 600–800 mm). The summer temperature was probably similar to the present day. At the same time, the abundance of *Pinus* subgenus *Diploxylon* and *Betula* pollen, and that of herbs such as *Artemisia* and Chenopodiaceae, indicates a wide distribution of steppe and forest–steppe landscapes. A similar modern pollen composition is found in the steppes and forest–steppe of Priolkhonie. This can be verified by the ratio of arboreal and nonarboreal plants in the upper layer of the bottom sediments from the central Baikal basin (Bezrukova et al. 1999), showing that the most recent spectra reflect the modern vegetation pattern, apart from some differences in the offshore zones of the three basins. The largest quantities of herb pollen are found in sites where the lake bottom is located close to steppe and forest–steppe landscapes on the shore. The combined existence of wet, dark coniferous, dry forest–steppe, and steppe vegetation complexes suggests contrasting environmental conditions, and thus vegetation occurrence, in this region of alternating mountains and valleys.

The palynocomplexes obtained for the late Pliocene from the lacustrine sediments of BDP-96-1, which contain pollen of various taxa growing on different types of environment, do not necessarily indicate significant Pliocene warming and climate aridization, although traces of this are found in the region (Vorobieva et al. 1995). The paleosoils of the late Pliocene do contain such information. The peculiarities of their structure confirm that the palaeosoils were formed in a subtropical climate with a wet cool winter and dry hot summer. Such paleosoils (slitozems) occur widely on Olíkhon Island, Priolíkhonie, South Priangarie. The structure of their profiles indicates a significant drying of the climate when the soils were formed. Nevertheless, the pollen spectra in BDP-96-1 core samples do not contain the pollen of representatives of semidesert, steppe, and forest–steppe landscapes such as *Amygdalus* and *Arteniaca.* A wide distribution of forest–steppe and steppe landscapes was reconstructed using the high values of the herb pollen in the spectra of BDP-96-1, in particular *Artemisia* (about 30% on average), and a predominance of *Betula* and *Pinus* subgenus *Diploxylon* pollen. A constant, small amount of temperate flora pollen, such as *Tilia*, *Corylus*, *Carpinus*, and *Quercus* subgenus *Lepidobalanus*, indicates a warm climate in the first half of the late Pliocene. Today these trees do not grow in this region because they require warmer climatic conditions. Thus, the role of temperate and subtropical flora elements in spectra from a large lake such as Baikal can be significantly underestimated. The pollen spectra in BDP-96-1 sediments show both the complicated climatic conditions and the diversity of the vegetation habitat in the uplifted relief of the region as a whole, although local features have not been taken into consideration. However, these do not contradict the paleontological and paleopedological data.

5.1 Features of Vegetation and Climate Evolution

At the beginning of the late Pliocene, i.e., the start of the origins of the sediments in stage A6, the vegetation of the mountains framing the Lake Baikal depression included a complex combination of mountain–taiga forest and submontane–taiga forest, steppe, and steppe landscapes. The height differences in the vegetation cover indicated the extremes in height in the terrain which existed at that time (Mats 1993). Dark coniferous forests, consisting of *Abies*, *Picea*, *Tsuga*, and *Pinus* subgenus *Haploxylon*, grew on the wet slopes of the upper mountain levels (Belova 1985). *Alnus*, *Ulmus*, and *Carpinus* formed the valley vegetation assemblages. *Quercus*, *Acer*, and *Tilia* compared part of the structure of the forest assemblages of the lower slopes (Rezanov 1988; Trofimov et al. 1995). The steppe and forest–steppe landscapes occurred in the submontane valleys, on Olíkhon Island, and in Priolkhonie (Bezrukova et al. 1999). The climate of the area had clear, contrasting seasonal differentiation. The average annual temperature could vary from +3 to +10°C. The average July temperature could vary from +22 to +15°C, and the average January temperature could vary from 0 to +10°C (Vorobieva et al. 1995). The upper slopes around the lake were marked by high precipitation (not less than 1000 mm/year). Priolkhonie, which was in the rain shadow of the mountain framing the western shore, was characterized by lower precipitation, mainly during the winter season. The Asian anticyclone started to influence the climate in the region at the same time (Borisov 1965). About 2.6 Ma, a significant cooling and less precipitation led to the predominance of a thinned-out forest consisting of *Picea*, *Abies*, *Pinus*, cold steppes, and mossy bogs on the shore. Forest landscapes predominate after 2.6 Ma. However, there is markedly less *Tsuga* and fewer broad-leaved trees (stage A5). One more significant cooling period between 1.86 and 1.5 Ma resulted in a further decrease in the forest landscapes (Belova 1985; Bezrukova et al. 1999). After that, forest–tundra landscapes developed with a significant number of *Larix*, *Picea*, *Alnaster*, and *Salix*. These landscapes were also the result of the increase in height in the mountains surrounding lake Baikal (Mats 1993). That cooling is close to the boundary of the Pliocene and Quaternary. The Quaternary landscapes included taxa now growing in Eastern Siberia. The coniferous forests contain *Picea*, *Abies*, *Pinus* subgenus *Haploxylon*, and *P.* subgenus *Diploxylon*. The small-leaved forests contain *Betula* and *Alnaster*. Broad-leaved taxa (*Ulmus, Tilia*, *Quercus*, *Carpinus*, and *Corylus*) were scarce. In the Pribaikalia mountains as well as in Mongolia and the Transbaikalia mountain ranges, *Tsuga* had almost disappeared from the flora composition. This and other genera typical of the humid flora of the late Pliocene did not return (Devyatkin et al. 1989). The third significant cooling, recorded in lacustrine sediments, happened about 0.8 Ma (Bezrukova et al. 1999). The start of stage A2 (about 1.5 Ma) is marked by an abrupt change in the structure of the vegetation and flora composition. *Tsuga* disappeared, while *Larix*, *Pinus* subgenus *Haploxylon*, and *Pinus pumila* became abundant. Vegetation which was similar to that of the present day started to appear from that time boundary. The average annual temperatures were negative, and the average annual precipitation changed markedly depending on height and slope exposure.

6 Conclusions

The palynological data have allowed a study of the main features and major stages of the changes in the vegetation and flora of the Eastern Siberia in this continuous record. The main reasons for the changes in the flora composition and landscapes were gradual climate cooling in the Miocene, and sharp climate variations in the Quaternary. A significant climate change caused by the change in global ice volume (Prueher and Rea 1998) happened in this region at the end of the early Pliocene, about 2.8–2.6 Ma. After that, the present-day taxa became predominant in the flora composition. The main forest-forming species were *Picea*, *Pinus* subgenus *Haploxylon*, *Pinus* subgenus *Diploxylon*, *Abies*, *Tsuga*, and *Betula*. During the second cooling at the boundary of the Pliocene and Pleistocene, at about 1. 5 Ma, forest–steppe landscapes became abundant, which indicates the existence of several glaciations centers in the mountains. The third cooling gave rise to sharp variations in the climate and a wide distribution of the boreal–arctic elements of the flora (*Larix* and *Pinus pumila*) during the late Pleistocene. The cold epochs are observed about every 1 Myr (2.6–2.8, 1.5–1.8, and about 0.8 Ma). A comparative analysis of the Pliocene pollen spectra from several land-based cores (Belova 1985) and from the lacustrine sediments showed that the former contained abundant pollen from moderately warm-loving small-leaved and broad-leaved trees, shrubs, and herbs in varied quantities, while the lacustrine sediments mainly contained the pollen of coniferous trees. This conclusion indicates the importance of sediments from large lakes to our acquisition of paleogeographic information at a wide regional level.

Acknowledgments. This study was partly supported by the Russian Foundation for Basic Research (RFBR grant 01-05-97206), the US National Science Foundation (grant EAR09614770), and the Science and Technology Agency, Japan. The authors would like to thank the crew of the research vessel *Ulan-Ude*. We are grateful to Dr. V.A. Misharina for the help in pollen identification. We would also sincerely like to thank Prof. V.D. Mats and Dr. A.V. Belov for valuable discussions and comments on an earlier draft of this chapter.

References

Baikal Atlas (1993) Russian Federal Service of Cartography, Moscow, p 160
Bates CD, Coxon P, Gibbard PL (1978) A new method for the preparation of clay-rich sediment samples for palynological investigations. New Phytol 81:459–463

Baikal Drilling Project Members (1997) Continuous paleoclimate record of the last 5 Ma from Lake Baikal, Siberia. Baikal Drilling Project BDP-96 (Leg II). EOS Am Geophys Union Trans 78:597–604

Belova VA (1985) Vegetation and climate of south-eastern Siberia in the late Cenozoic (in Russian). Nauka Press, Novosibirsk, p 158

Bezrukova EV (1994) Composition peculiarities of modern spore-and-pollen spectra of Pribaikalie and bottom sediments of Lake Baikal. In: Baikal as a natural laboratory for global change. Abstracts. Irkutsk Russia 2:5–6

Bezrukova EV (1998) Peculiarities of sub-recent spore–pollen spectra composition of pre-Baikal (in Russian). Geogr Nat Resource 1:142–147

Bezrukova EV, Letunova PP (2001) A high-resolution record of east Siberian paleoclimates in the early and middle Pleistocene by palynological studies of Baikal sediments from the deep borehole BDP-96-1 (in Russian). Russ Geol Geophys 42:98–107

Bezrukova EV, Bogdanov YA, Williams DF, Granina LZ, Grachev MA, Ignatova NV, Karabanov EB, Kuptzov VM, Kurylev AB, Letunova PP, Likhoshway EV, Chernyaeva GP, Shimaraeva MK, Yakushin AO (1991) Deep changes of the north Baikal ecosystem in the Holocene (in Russian). Dokl AN SSSR 321:1032–1037

Bezrukova EV, Kulagina NV, Letunova PP, Shestakova ON (1999) Climate and vegetation changes in the Baikal region for the last 5 Ma. Russ Geol Geophys 40:722–731

Borisov AA (1965) Paleoclimates on USSR territory (in Russian). Leningrad State University Press, Leningrad, p 111

Bradbury JP, Bezrukova EV, Chernyaeva GP, Colman SM, Khursevich GK, King J, Likhoshway YV (1994) A synthesis of post-glacial diatom records from Lake Baikal. J Paleolimnol 10:213–252

Colman SM, Peck JA, Jones GA, Karabanov EB, Carter SJ, Bradbury JP, King JW, Williams DF (1995) Continental climate response to orbital forcing from biogenic silica records in Lake Baikal. Nature 378:769–771

Devyatkin EV, Malaeva EM, Murzaeva VE (1989) The late Cenozoic of Mongolia—stratigraphy and paleogeography (in Russian). Nauka Press, Moscow, p 241

Dorofeev PI (1956) Pliocene flora of lower Kama. In: Gerasimov WG (ed) For the 75th anniversary of academic V.N. Sukachev (in Russian). AN SSSR Press, Moscow, Leningrad, 171–182

Faegri K, Iversen J (1989) Textbook of pollen analysis, 4th ed (revised by Faegri K, Kaland PE, Krzywinski K). Wiley, Chichester

Grachev MA, Likhoshway EV, Vorobyova SS, and others (1997) Signals of the paleoclimates of the upper Pleistocene in the sediments of Lake Baikal. Russ Geol Geophys 38:957–980

Grachev MA, Vorobiova SS, Likhoshway YV, Goldberg EL, Ziborova GA, Levina OV, Khlystov OM (1998) A high-resolution diatom record of the palaeoclimates of east Siberia for the last 2.5 My from Lake Baikal. Q Sci Rev 17:1101–1106

Grana Palynologica (1961) Erdtman G. (ed), vol 3, Stockholm

Kupriyanova LA (1965) The palynology of the Amentiferae (in Russian). Nauka Press, Moscow, Leningrad, p 163

Kuzmin MI, Grachev MA, Williams DF, Kawai T, Horie T, Oberhaensli H (1997) The continuous record of paleoclimates of last 4.5 Ma from Lake Baikal, 1997. Russ Geol Geophys 38:1021–1023

Kuzmin MI, Karabanov EB, Prokopenko AA, Williams DF, Gvozdkov AN, Gelety VF (1999) Climatic events in Siberia during the upper Brunhes according to the Lake Baikal sedimentary record. Berliner Geowiss Abh E 30:315–323

Markgraf V, Di'Antoni HL (1978) Pollen flora of Argentina. University of Arizona Press, Tuscon, p 208

Mats VD (1993) The structure and development of the Baikal rift depression. Earth Sci Rev 34:81–118

Molozhnikov VN (1986) Plant communities of pre-Baikal (in Russian). Nauka Press, Novosibirsk, p 270

Moore PD, Webb JA, Collinson ME (1991) Pollen analysis, 2nd edn. Blackwell, Oxford, p 215

Ogura A, Takahara H, Krivonogov SK, Bezrukova EV, Morita Y, Shinomiya T, Kawamuro K (1998) Pollen–tree abundance relationship from the Khamar-Daban Mountains, in the southeastern area of Lake Baikal. In: BICER: BDP and DIWPA. Joint International Symposium on Lake Baikal. Abstract 83

Prueher LM, Rea DK (1998) Rapid onset of glacial conditions in the sub-arctic North Pacific region at 2.67 Ma: clues to causality. Geology 25:1027–1030

Rezanov II (1988) Cenozoic sediments and morphostructure of eastern pre-Baikal (in Russian). Nauka Press, Novosibirsk, p 126

Trofimov AG, Malaeva EM, Kulikov OA, Kulagina NV, Ufimtzev GF (1995) Manzurka alluvium (in Russian). Institute of Earth Crust SB RAS Press, Irkutsk, p 50

Volkova VS (1977) Stratigraphy and vegetation history development of western Siberia in the Late Cenozoic (in Russian). Nauka Press, Moscow, p 237

Vorobieva GA, Mats VD, Shimaraeva MK (1995) Paleoclimates of the late Cenozoic of the Baikal region (in Russian). Russ Geol Geophys 36:82–96

Williams DF, Peck J, Karabanov EB (1997) Lake Baikal record of continental climate response to orbital insolation during the past 5 million years. Science 278:1114–1117

Williams DF, Karabanov EB, Prokopenko AA, Kuzumin MI, Khursevich GK, Gvozdkov AN, Bezrukova EV, Solotchina EP (2001) Late Miocene–Pleistocene records from Baikal sediments: markers for paleoclimatic and biostratigraphic reconstructions (in Russian). Russ Geol Geophys 42:35–48

Vegetation Changes in the Baikal Region during the Late Miocene Based on Pollen Analysis of the BDP-98-2 Core

TAKESHI MAKI[1], YOSHITAKA HASE[2], KIMIYASU KAWAMURO[3], KOJI SHICHI[4], KOJI MINOURA[5], TAKEFUMI ODA[6], and NORIO MIYOSHI[7]

Summary. A palynological study was conducted on samples taken from the bottom sediments of Lake Baikal. The pollen records from 600-m to 400-m depth of the BDP-98 core show that the vegetation consisted mainly of broad-leaved tree forests composed of Fagaceae, Ulmaceae, and Juglandaceae in the lower part of the core. In the middle and upper parts of the core, the vegetation consisted mainly of coniferous tree forests and deciduous broad-leaved tree forests. The coniferous tree forests were composed of *Pinus* and *Picea* accompanied by *Tsuga* and *Larix*, and the deciduous broad-leaved tree forests were mainly composed of *Betula* and *Alnus*. A major reduction in palynological diversity and abundance was found at 8.6 Ma. In addition, a drastic change in the Steppe/forest index (SFI; Traverse 1988) was detected, reflecting the intensification of dry conditions. After the change, *Betula* and *Alnus* became more important elements of the deciduous broad-leaved tree forests, suggesting that a cool continental climate prevailed.

Key words. BDP-98 core, Pollen analysis, Late Miocene, Vegetation change

[1] Graduate School of Social and Cultural Studies, Kyushu University, 4-2-1 Ropponmatsu, Chuo-ku, Fukuoka 810-0044, Japan
[2] Department of Earth Sciences, Kumamoto University, 2-39-1 Kurokami, Kumamoto 860-8555, Japan
[3] Kansai Branch Center, Forestry and Forest Products Research Institute, Momoyama-cho, Fushimi-ku, Kyoto 612-0855, Japan
[4] Tohoku Branch Center, Forestry and Forest Products Research Institute, 92-25 Nabeyashiki, Shimokuriyagawa, Morioka 020-0123, Japan
[5] Department of Geoenvironmental Science, Tohoku University, Aramaki, Aoba-ku, Sendai 980-8578, Japan
[6] Nagoya University Center for Chronological Research, Furo-cho, Chikusa-ku, Nagoya 464-8602, Japan
[7] Department of Biosphere–Geosphere System Science, Faculty of Informatics, Okayama University of Science, 1-1 Ridai-cho, Okayama 700-0005, Japan

1 Introduction

Kabanova and Filimonova (1978) published their pollen record from the Anadyr Basin next to the northern part of the Bering Sea in the Paleogene and early Neogene, although the late Miocene was not mentioned. Kulchitskii et al. (1993) described the vegetation and climate changes in the late Cenozoic from palynological and carpological studies on the northwestern coast of Lake Baikal. The floral changes recorded in long cores from Lake Baikal are important to a further understanding of the vegetation history not only of the Baikal area, but also of northeastern Eurasia in the late Cenozoic era. In this study, we reconstructed the long-term vegetation and climate changes within the Baikal region during the late Miocene (12–7.3 Ma), based on pollen analyses of samples between the 600-m and 400-m depth sections of the BDP-98-2 core, which was taken from Academician Ridge in Lake Baikal (Fig. 1), and was well preserved.

2 Materials and Methods

2.1 Grain-Size Analysis

The BDP-98-2 core samples are mainly composed of clay and silt with a small amount of sand. Four hundred and fifty-six samples were taken from the core at 40-cm intervals between 600-m and 400-m depth, and were analysed for grain sizes using the laser-beam diffraction method. Figure 2 shows the composition of the sand, silt, and clay rates based on grain-size analysis. The sand, silt, and clay contents vary in the range 0.3%–1.5%, 67%–84%, and 8%–32%, respectively. The sand content increases slightly between 530 m and 480 m, and the clay content shows a slight increase between 480 m and 400 m. With the exception of a sandy layer at 583 m, the grain-size composition between 600 m and 400 m is steady and does not show any long-term change in depositional regimes.

2.2 Age Model for the BDP-98-2 Core

The age model of the BDP-98-2 core is constructed by a study of paleomagnetic stratigraphy (Sakai et al. 2003). The age from 600-m to 400-m depth was found to be 12 Ma to 7.3 Ma. The average age interval of each sample at 40-cm intervals for grain-size and pollen analyses is ca. 8900 years.

2.3 Pollen Analysis

We conducted pollen analyses by the standard method of a KOH–HF–acetolysis mixture on each sample (Faegri and Iversen 1989). A known concentration of plastic microspheres (25 μm diameter) in suspension (Ogden 1986) was

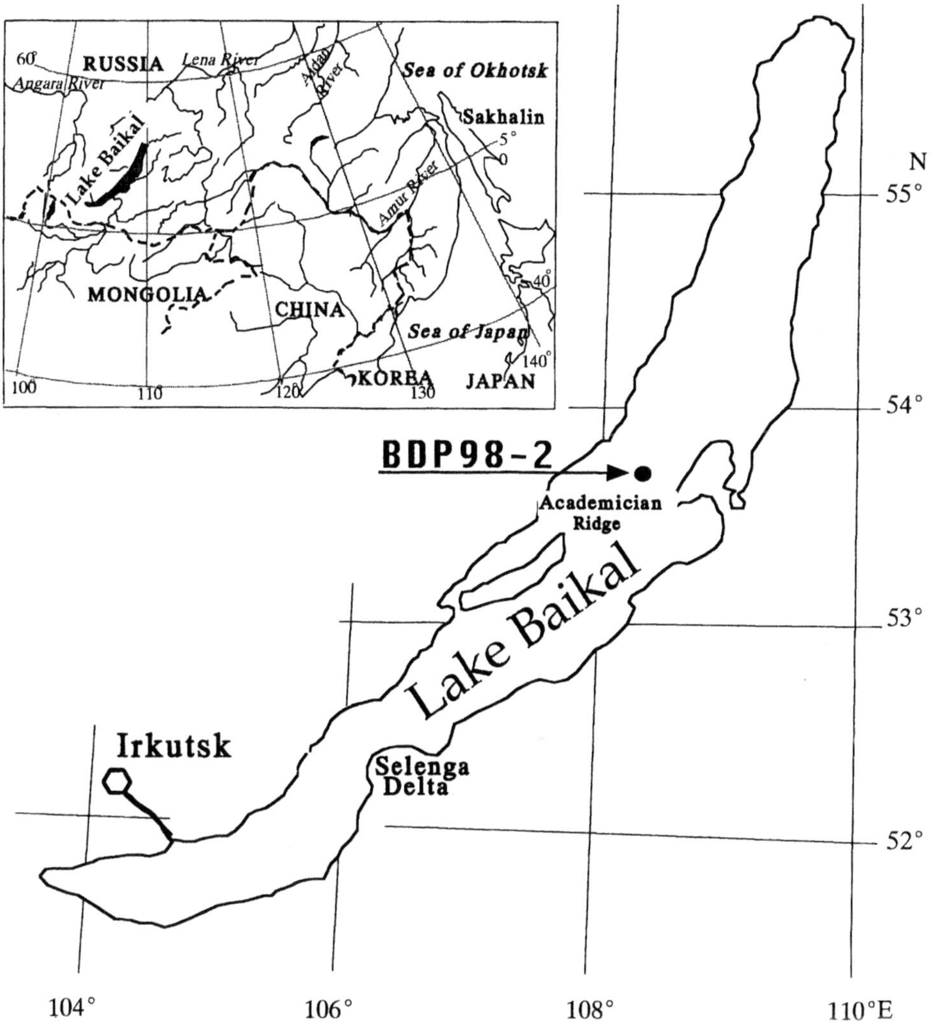

FIG. 1. Location map of the BDP-98-2 drilling sites

also added to each sample to determine the pollen concentration. Prepared samples were mounted in glycerin jelly on microscope slides.

The percentage values of each taxon in the pollen diagram were calculated based on the sum of the arboreal pollen.

2.3.1 Description of Pollen Zones

The major pollen taxa are given in the pollen diagram in Fig. 3. Based on differences in the characteristics of pollen assemblages, the diagram is divided into pollen zones labeled A to J. The ages of each pollen zone were estimated by a

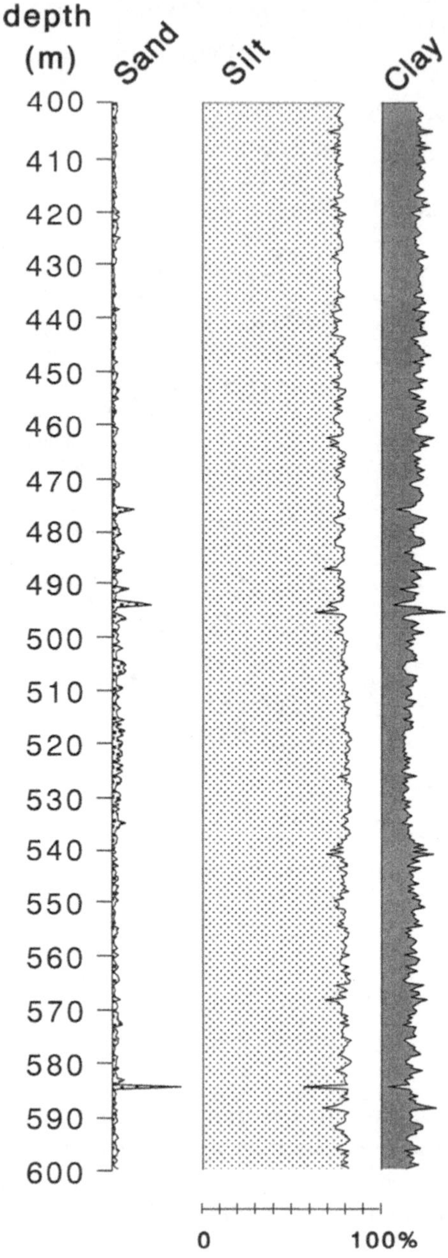

FIG. 2. Grain-size distribution vs depth between the 600-m to 400-m depth sections of the BDP-98-2 core

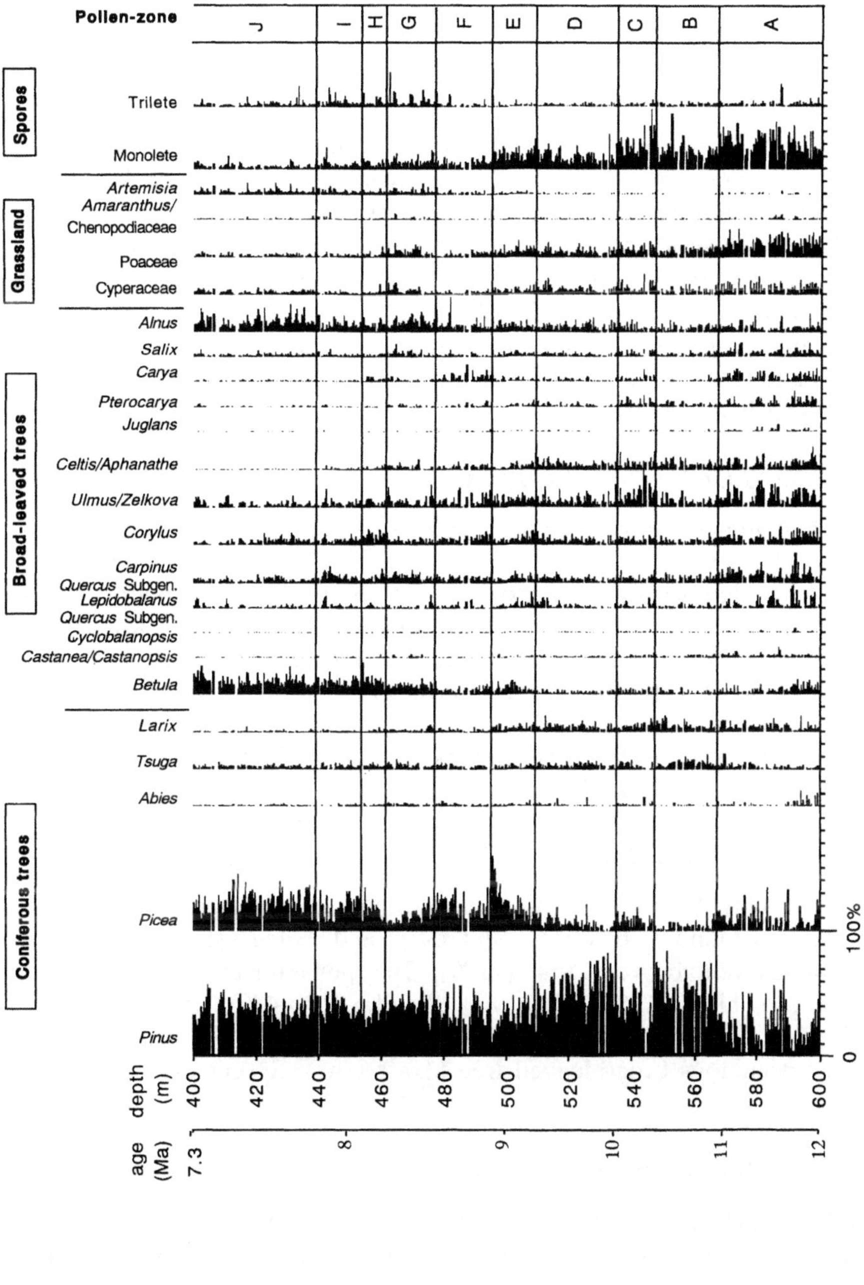

FIG. 3. Pollen and spores diagram of the BDP-98-2 core

paleomagnetic chronological method (Sakai et al. 2003) as follows: zone A, 12–11.2 Ma; zone B, 11.2–10.5 Ma; zone C, 10.5–10.2 Ma; zone D, 10.2–9.8 Ma; zone E, 9.8–9.3 Ma; zone F, 9.3–8.6 Ma; zone G, 8.6–8.3 Ma; zone H, 8.3–8.1 Ma; zone I, 8.1–7.9 Ma; zone J, 7.9–7.3 Ma. *Pinus* was dominant in almost all pollen assemblages, although no distinction was made between Subgen. *Diploxylon* and *Haploxylon* types.

2.3.1.1 Zone A (600–567 m; 12–11.2 Ma)

Pinus oscillates between 10% and 50%; *Picea* occurs between 5% and 20%, and sometimes reaches 30%. Relatively high percentages of *Abies* are found in the lower part of this zone (<3%), and are followed by *Tsuga* (1%–3%) and *Larix* (2%–5%), which show low but stable frequencies. Broad-leaved tree taxa such as *Carpinus* (5%–10%), *Corylus* (5%–10%), *Ulmus* and/or *Zelkova* (4%–12%), and Juglandaceae (*Juglans* 2%–5%, *Pterocarya* 0.5%–8%, and *Carya* 1%–7%) show relatively high values in this zone. Herbaceous taxon percentages (Poaceae 5%–20% and Cyperaceae 3%–10%) are relatively high. Monolete spores range from 10% to 30%, and trilete spores are rare (<3%).

2.3.1.2 Zone B (567–547 m; 11.2–10.5 Ma)

Relatively high percentages of *Pinus* (50%–70%) and low percentages of *Picea* (1%–5%) characterize this zone. *Abies* is rare, and *Tsuga* occurs continuously, but with low values (5%–8%). *Larix* (3%–5%) occurred in the upper part of the zone. The broad-leaved tree taxa occurred in low values overall, with a reduction in *Carya*. Ulmaceae (*Ulmus* and/or *Zelkova* 2%–6%, *Celtis* and/or *Aphanathe* 1%–7%) and Betulaceae (*Carpinus* 2%–6%, *Corylus* 1%–4%, *Alnus* 2%–5%) still occur constantly. The percentages of herbaceous pollen (Cyperaceae 5%–9%, Poaceae 5%–9%) and spores (11%–20%) level out throughout this zone, but are less than those in zone A.

2.3.1.3 Zone C (547–534 m; 10.5–10.2 Ma)

Pinus is usually found at 20%–40%, although it occasionally increases to 60%. *Picea* is found at 6%–12%, and *Larix* and *Tsuga* values are low, while *Abies* values remain very low (<1%). The percentages of *Ulmus* and/or *Zelkova* are high (10%–30%), but *Betula* (1%–4%), *Alnus* (2%–5%), and *Quercus* Subgen. *Lepidobalanus* (0.3%–5%) remain low, while the values for other deciduous broad-leaved tree taxa recover. Spores occur throughout this zone.

2.3.1.4 Zone D (534–509 m; 10.2–9.8 Ma)

Pinus shows high values (50%–75%), while *Picea* percentages are low (2%–8%). Generally, the percentages of other coniferous tree taxa are also low. In deciduous broad-leaved tree taxa, *Ulmus* and/or *Zelkova* and *Alnus* show occasional high values. Herbaceous taxa and spores still show low percentages.

2.3.1.5 Zone E (509–495 m; 9.8–9.3 Ma)

The percentages of *Pinus* show a decrease (50%–25%), while those of *Picea* gradually increase through this zone (up to 37%). In deciduous broad-leaved tree taxa, *Ulmus* and/or *Zelkova* (5%–10%), *Carpinus* (3%–6%), *Corylus* (4%–9%), *Quercus* Subgen. *Lepidobalanus* (1%–9%), and *Alnus* (3%–6%) are constant. In the upper part of this zone, *Betula* occurs at 3%–6%, while most other percentages decrease. Even herbaceous taxa and spores are low.

2.3.1.6 Zone F (495–476 m; 9.3–8.6 Ma)

This zone is characterized by high percentages of *Picea* (20%–50%) and *Betula* (10%–25%), and by relatively high percentages of *Carya* (2%–7%). Herbaceous taxa and spores still show low values.

2.3.1.7 Zone G (476–463 m; 8.6–8.3 Ma)

Pinus shows a substantial increase (35%–50%), while *Picea* percentages are low (6%–16%) and *Tsuga* increases slightly. The percentages of *Betula* (6%–8%) and *Carpinus* (3%–7%) are low but consistent. *Ulmus* and/or *Zelkova* and *Corylus* fluctuate unsteadily. *Alnus* occurs continuously, with some fluctuations (8%–15%). Occasional increases in both Poaceae and *Artemisia* were noted (6%–9%). Monolete spore values are steady (5%–7%), while those of trilete spores are relatively high.

2.3.1.8 Zone H (463–453 m; 8.3–8.1 Ma)

Pinus is steady (30%–40%) and *Picea* has recovered (13%–20%). The other coniferous tree taxa are poorly represented except for a slight increase in *Tsuga*. *Betula* occurs steadily at slightly higher values (7%–12%), while *Carpinus* (1.5%–5%), *Corylus* (4%–9%), and *Alnus* (4%–7%) are also important elements in this zone.

2.3.1.9 Zone I (453–440 m; 8.1–7.9 Ma)

Pinus (30%–50%) and *Picea* (20%–30%) maintain high values, while other coniferous tree taxa have low values. *Betula* and *Alnus* occur commonly, with some fluctuations, while *Carpinus* shows a slight increase in the upper part of this zone (2.5%–5%). *Corylus* and *Ulmus* and/or *Zelkova* are minor elements. Herbaceous taxa are present at low values without any marked fluctuations, while spores have only sporadically high values.

2.3.1.10 Zone J (440–400 m; 7.9–7.3 Ma)

The percentages of *Pinus* (30%–40%) and *Picea* (20%–30%) are still high throughout this zone. *Betula* (7%–15%) and *Alnus* (5%–10%) also maintain high values, while the percentages of other deciduous broad-leaved tree taxa are low. Herbaceous taxa remain low except for slightly higher values for Cyperaceae (1%–5%) and *Artemisia* (2%–5%).

2.3.2 Change of Occurrences of Pollen and Spores

Figure 4a shows percentage diagrams of coniferous trees, broad-leaved trees, herbaceous pollen, and spores. Between 600 and 573 m, the percentage of coniferous tree pollen shows repeated increases and decreases, but at 573 m depth it became high, and after that it maintains high values, suggesting that coniferous vegetation became prevalent in the Baikal region from 11.2 Ma. The percentage of broad-leaved tree pollen is even higher than that of coniferous tree pollen between 600 and 573 m, but after that it shows moderate values. The percentage of herbaceous pollen is steady between 600 m and 400 m, and does not show any significant changes. Spores show a gradual decrease toward the upper part of the core.

2.3.3 Pollen Concentration

The concentration diagram of arboreal pollen grains is shown in Fig. 4b. The concentration of arboreal pollen grains is steady between 600 m and 517 m, and shows sporadic high values between 517 m and 478 m. At 478 m depth, the concentration of arboreal pollen shows a significant decrease, and levels out below 439 m. It then shows a slight increase, followed by consistent values.

Changes in the pollen concentration are influenced by the sedimentation rate as well as pollen deposition. Consequently, if the rate is high, pollen concentration might decrease because of the dilution effect of additional sediment particles, and if the rate is low, more pollen could be contained in fewer deposited particles. Although this point should be considered, the paleomagnetic data did not show a change in sedimentation rate in this section of the core (Sakai et al. 2003). Also, the fluctuation rate of the grain size (see Fig. 2) scarcely changed, except at 583 m. These data coincide with a steady sedimentary condition on Academician Ridge in Lake Baikal. Therefore, it can be concluded that the fluctuation in the pollen concentration values mainly represents variations in the pollen load of the surrounding vegetation. Consequently, it is suggested that the abundance of the forests surrounding Lake Baikal changed as follows, with the ages having been derived from magnetostratigraphic studies (Sakai et al. 2003). Forests were well established until 9.8 Ma, and then expanded between 9.8 Ma and 9.3 Ma. In the next period, 9.3 Ma to 9.0 Ma, the forests returned to the same level as before 9.8 Ma. The forests recovered again between 9.0 Ma and 8.5 Ma, after which a marked decrease took place. This lower level persisted until 7.8 Ma, after which there was a substantial forest recovery.

2.3.4 Steppe/Forest Index

The steppe/forest index (SFI) ratio (Traverse 1988) was adapted to visualize the relationship between forest and steppe taxa as an indicator of precipitation levels (Fig. 4c). The SFI is given by the ratio of the sum of *Artemisia*, *Amaranthus* and/or Chenopodiaceae to that of arboreal genera. There is a drastic shift at

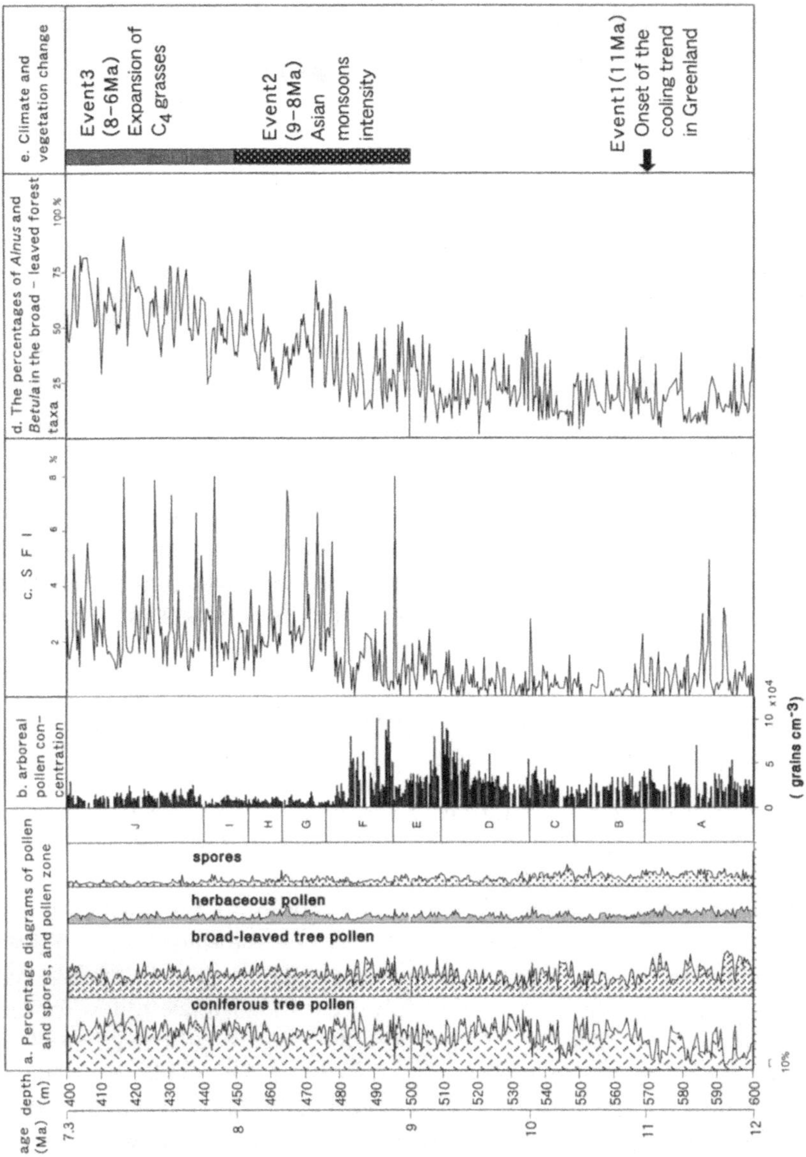

FIG. 4 **a** Percentage diagrams of broad–leaved trees, coniferous trees, grass pollen, and spores. **b** Total count of arboreal pollen grains. **c** Percentages of steppe elements (steppe/forest index (SFI), Traverse 1988). SFI = (*Artemisia* + *Amaranthus* and/or Chenopodiaceae) / (the above + tree genera) × 100. **d** Percentages of *Alnus* and *Betula* in broad – leaved tree taxa. **e** Some climatic events reported previously. Event 1 (11 Ma) is known as the time of the initiation of the cooling trend in Greenland (Helland and Holmes 1997). Event 2 (9–8 Ma) is known as the time of increasing monsoon intensity (An et al. 2001). Event 3 (7.3 Ma) is known as the time of expansion of C_4 grasses (Cerling et al. 1997)

approximately 8.6 Ma, which is the boundary between zones F and G. This indicates that the onset of the increase in the steppe element occurred in zone G, reflecting the dry conditions which became prevalent over the Baikal region from that time.

2.3.5 *Betula* and *Alnus*: as an Indication of Climatic Changes

Vegetation changes shown in the BDP-98 core and described above showed significant fluctuations in the percentages of broad-leaved tree taxa, especially *Betula* and *Alnus*. Figure 4d shows the percentages of *Betula* plus *Alnus* in the broad-leaved tree taxa. In Fig. 4d, a drastic shift at 478 m is clearly seen. Before that level, other broad-leaved elements fluctuate without any systematic change. Above 478 m, *Betula* and *Alnus* show a significant increase. Aboreal pollen concentrations decreased abruptly at 478 m depth (Fig. 4b), after which the percentage of *Betula* and *Alnus* increased in the broad-leaved tree taxa (Fig. 4d). It is considered that the vegetation changes represented by these fluctuations of the broad-leaved tree taxa were mainly a response to climatic changes, such as climatic cooling and drying. That is to say, in the Baikal region, the cooling trend and associated dry conditions accelerated the decrease of broad-leaved trees. However, *Betula* and *Alnus* adapted to the climatic changes and increased their populations.

2.3.6 Late Miocene Vegetation and Climatic Changes in the Baikal Area

The history of the local vegetation and climate of the Lake Baikal area in the late Miocene from 12 to 7.3 Ma has been clarified by this study. The forests were composed of broad-leaved trees and conifers from 12 to 11.2 Ma. Kulchitskii et al. (1993) had already briefly revealed the vegetation of the northeastern part of Lake Baikal from the late Miocene to the Pleistocene, and showed that the vegetation consisted of conifers and many broad-leaved trees in the late Miocene (Table 1). The prevailing taxa varied from that in existence today: for example, the conifers *Podocarpus* and *Dacrydites*, and the broad-leaved trees *Myrica*, *Tilia*, *Magnolia*, and *Liriodendron* also grew there. The evidence of this vegetation indicates that the conditions were warm, unlike the boreal taiga that exists today. This palynological study of zone A seems to coincide with the results of Kulchitskii et al. (1993), as the broad-leaved trees found in this study were the same kind found by Kulchitskii et al. (1993). In this period, some occasional expansion of coniferous tree pollen occurred instead of the dominance of the broad-leaved tree pollen. This may reflect the global cooling suggested by marine cores collected off the shore of southeast Greenland (Larsen et al. 1994; Helland and Holmus 1997).

Forests dominated by *Pinus* became well established, and *Picea* and other coniferous tree taxa were present as minor elements in the forests between 11.2

TABLE 1. Late Miocene to early Pliocene floral list from the drilling core of the northwest coast of Lake Baikal (after Kulchitskii et al. 1993)

Conifers	Broad-leaved trees	Grassland taxa	Fern
Picea sect. *Eupicea*	*Betula* sp.	Cyperaceae	*Sphagnum* sp.
Picea sect. *Omorica*	*Alnus* sp.	Polygonaceae	Polypodiaceae
Picea sect. sp.	*Salix* sp.		*Lycopodium* sp.
Tsuga	*Corylus* sp.		*Osmumda* sp.
Pinus subgen. *Haploxyrlon*	Ericales		
Pinus subgen. *Diploxylon*			
Abies sp.	(Turgai Flora)		
Pinus pumila	*Juglans polyporate*		
Podocarpus sp.	*Myrica* sp.		
Dacrydites	*Pterocarya* sp.		
	Carya sp.		
	Ulmus sp.		
	Tilia sp.		
	(rare genera)		
	Fagus sp.		
	Celtis sp.		
	Moraceae		
	Magnolia sp.		
	Liriodendron sp.		
	Platanaceae		
	Ilex sp.		

and 10.5 Ma. This indicates that the climate became dry, but not very cool. Between 10.5 and 9.8 Ma, the forests dominated by *Pinus* became well established, and deciduous broad-leaved forests are assumed to have occupied a small area. The SFI for this period is maintained at a low level, suggesting that steppe vegetation had not yet expanded. Therefore, the climate was moderate during this period, and provided increased humidity over this region.

Between 9.8 and 9.3 Ma, *Picea* became an important element replacing *Pinus*, while *Betula* increased gradually and other deciduous broad-leaved tree taxa were commonly present, suggesting that the climate became cooler than that of the previous age. Between 9.3 and 8.6 Ma, *Carya* was dominant in zone A, and was strikingly abundant throughout this period. In addition, an abundance of pollen grains indicates that the climate became slightly warmer, leading to rising precipitation.

Between 8.6 and 8.3 Ma, deciduous broad-leaved forests had little variety, and in a kind of steppe taxa, *Artemisia* replaced Poaceae and Cyperaceae, suggesting that a continental climate prevailed. This vegetation change may correspond with the global expansion of C_4 grasses (8–6 Ma; Cerling et al. 1997), and may also relate to the intensification of the Asian monsoons (9–8 Ma; An et al. 2001). During this period, the relatively high percentages of grassland taxa indicate the vegetation inhabiting the drainage basin. Between 8.3 and 8.1 Ma, the climate remained cool. Judging from the decline in the pollen concentration and the low percentages of grassland taxa and spores, the cool climate was accompanied by dry conditions.

The climatic trend that prevailed became strong between 8.1 and 7.9 Ma. During this period, *Pinus* and *Picea* occurred in almost the same volumes. There were many secondary forests of *Betula*, and many stands of *Alnus*. Between 7.9 Ma and 7.3 Ma, the forests dominated by *Pinus* and *Picea* mainly covered the highlands, and shrubs, mainly consisting of *Alnus*, occupied the drainage basins and areas around ponds, but grassland vegetation was restricted.

Acknowledgments. We thank Dr. Takayoshi Kawai, Professor of Nagoya University, Dr. Mikhail I. Kuzmin Director of the Vinogradov Institute of Geochemistry Siberian Division, Russian Academy of Science, Prof. Douglas Williams, South Carolina University, and other members of the Baikal Drilling Project for their strong support. We are also indebted to Dr. Alfred Traverse, Emeritus Professor of Pennsylvania University, who kindly reviewed the manuscript and gave us many suggestions, and to Dr. Hikaru Takahara, Professor of Kyoto Prefectural University, for his useful advice. This work was supported by the international cooperative project of STA Japan, the US NFS, and the Russian Academy of Science.

References

An Z, Kutzbach JE, Prell WL, Porter SC (2001) Evolution of Asian monsoons and phased uplift of the Himalaya–Tibetan plateau since late Miocene times. Nature 411:62–66

Cerling TE, Harris JM, MacFadden BJ, Leakey MG, Quade J, Eisenmann V, Ehleringer JR (1997) Global vegetation change through the Miocene/Pliocene boundary. Nature 389:153–158

Faegri K, Iversen J (1989) Textbook of pollen analysis. Wiley, New York, 328 p

Helland PE, Holmes MA (1997) Surface textual analysis of quartz sand from ODP Site 918 off the southeast coast of Greenland suggests glaciation of southern Greenland at 11 Ma. Palaeogeogr Palaeoclim Palaeoecol 135:109–121

Kabanova GKh, Fillimonova LA (1978) Palynological characteristic of the Paleogene-Neogene deposits of the Lower Andyr Basin. Russ Geol Geophys 19:78–81

Kulchitskii AA, Krivonogov SK, Misharina VA, Chernyaeva GP (1993) Key section of Northern-Baikalian Upper-Cenozoic deposits. Russ Geol Geophys 34:1–7

Larsen HC, Saunders AD, Clift PD, Beget J, Wei W, Spezzaferri S, ODP Leg152 Scientific Party (1994) Seven million years of glaciation in Greenland. Science 264:952–955

Ogden J (1986) An alternative to exotic spore or pollen addition in quantitative micro fossil studies. Can J Earth Sci 23:102–106

Sakai H, Nomura S, Horii M, Kashiwaya K, Kawai T, Kravchinsky V, Peck J, King J (2003) Paleomagnetism and paleo-environmental magnetism studied on BDP-98 sedimentary cores from Lake Baikal. In: Kashiwaya K (ed) Long continental records from Lake Baikal. Springer, Berlin

Traverse A (1988) Paleopalynology. Unwin Hyman, Boston, 600 p

Residual Photosynthetic Pigments in the Sediment of Lake Baikal as Indicators of Phytoplankton History

Mitsuyuki Soma[1], Yuko Soma[1], Yukinori Tani[1], Nobuyasu Itoh[1], Kota Kurihara[1], Fumiko Nara[1], Atsushi Tanaka[2], and Takayoshi Kawai[2*]

Summary. Phytoplankton photosynthetic pigments preserved in lake sediments provide important information about the past environment of the lake as it affected the distribution of phytoplankton. This review chapter demonstrates that the pigments can represent the changes in the algal community in Lake Baikal during both millennial-scale and mega-annum-scale periods.

In a sediment core (3.6 m) from the southern basin representing the past 25 000 yr, the increase in biogenic silica (BGS) which started after 10 000 yr BP was preceded by an increase in total organic carbon (TOC), chlorophyll derivatives, and carotenoids which started as early as 15 000 yr BP. This indicates that algae other than diatoms first proliferated after the last glacial maximum. Diatoms significantly increased late in the Holocene, reaching a maximum after 5000 yr BP.

In a 0- to 100-m section of the 600-m long BDP-98 core, which spans over 2.8 million years, steryl chlorin esters (SCEs) were the major chlorophyll derivatives, and accounted for a maximum of 90% of the total chlorophyll derivatives. Depth profiles of TOC, BGS, and SCEs show a glacial/interglacial cycle, where they increase during the same periods. The sterol composition of the SCEs showed a remarkable variation among warm periods, which means that the taxonomic composition of the phytoplankton community from which the SCEs originated also varied. We suggest that green algae and dinoflagellates, as well as diatoms, were abundant in Lake Baikal until 1.2 million years ago.

Key words. Photosynthetic pigments, Lake Baikal, Sediment, Phytoplankton, Carotenoids, Steryl chlorin esters

---

[1] Institute for Environmental Sciences, University of Shizuoka, 52-1 Yada, Shizuoka 422-8526, Japan

[2] National Institute for Environmental Studies, 16-2 Onogawa, Tsukuba 305-8506, Japan (*present address: Department of Earth and Environmental Sciences, Graduate School of Environmental Studies, Nagoya University, Furo-cho, Chikusa-ku, Nagoya 464-8602, Japan)

1 Introduction

To reconstruct the paleoenvironment of Lake Baikal, many indicator parameters in the sediment have been under intensive study (Baikal Drilling Project Group 2000; Minoura 2000). Diatoms have been the dominant phytoplankton in Lake Baikal from the Plio-Pleistocene until today (Horie 1992; Kozhova and Izmest'eva 1998; Khursevich et al. 2000), and depth profiles of biogenic silica (BGS) originating from fossil diatoms in the sediment reflect continental climate change in the Northern Hemisphere (Colman et al. 1995; Baikal Drilling Project Group 2000; Khursevich et al. 2000). To date, however, the importance of other phytoplankton groups in the paleoenvironment of Lake Baikal has not been assessed. Recent developments in high-performance liquid chromatography have made it possible to analyze photosynthetic pigments readily with precision (Mantoura and Llewellyn 1983; Wright 1984; Goodwin and Britton 1988; Wright and Jeffery 1997), and these phytoplankton pigments, notably carotenoids, provide direct information about the phytoplankton population from which the pigments originated.

We have analyzed the photosynthetic pigments in the sediments of Lake Baikal, and been able to detect chlorophyll derivatives and carotenoids from recent (Soma et al. 1996; Tani et al. 2001) and surficial (Soma et al. 2001a) sediments of this deep, oxic, and oligotrophic lake. In this chapter, we first describe the distribution of pigments in a sediment core that should be representative of the pelagic southern basin of Lake Baikal, covering the period of the last glacial/postglacial transition (Tani et al. 2002). It will be shown that the sedimentary profiles of photosynthetic pigments, as well as that of total organic carbon (TOC), throughout this period are significantly different from those of BGS, which essentially reflect the diatom population, and that other phytoplankton have also been abundant.

We then describe the depth profiles of steryl chlorin esters (SCEs) as a major group of the residual photosynthetic pigments, BGS, and TOC in the 0- to 100-m section of the BDP-98 core (Soma et al. 2001b).

SCEs are a class of chlorophyll derivatives in which the sterol moiety is esterified, mainly to pyropheophorbide *a*. As the SCE analogue of chlorophyll *b* has been identified (Talbot et al. 1999a), we will use SCE-*a* and SCE-*b* as appropriate. Grazing of phytoplankton by zooplankton produces SCEs, which are egested in fecal pellets (King and Repeta 1991; King and Wakeham 1996; Harradine et al. 1996; Talbot et al. 1999a, b, 2000). SCEs are observed widely in marine and lacustrine sediments (King and Repeta 1991; Prowse and Maxwell 1991; Eckardt et al. 1991, 1992; Pearce et al. 1998; Naylor and Keely 1998), are relatively stable with respect to diagenetic transformation, and can persist over millions of years (Prowse and Maxwell 1991). Sterols are well known as biomarkers for phytoplankton (Volkman 1986; Patterson 1991). However, the distribution of sterols in SCEs more directly reflects the original algal source than do the free sedimentary sterols, as esterification of the chlorin nucleus seems to protect the distribution from selective biotransformation (Harradine et al. 1996; Pearce et al. 1998).

The result will indicate the alteration of phytoplankton community structures in Lake Baikal for the last 2.8 million years, and show that SCEs retained in the sediment can usefully serve as paleolimnological indicators of the structures of phytoplankton communities in lake ecosystems.

2 Materials and Methods

2.1 Sediment Cores

A gravity core, Ver.99 G-6 (360 cm), was collected in the southern basin of Lake Baikal (52°05′20″N, 105°50′24″E, water depth 200 m) on July 7, 1999. The sample core was sliced into 3- to 5-cm sections, and the sliced sediment samples were stored at −80°C until analysis.

A 0- to 100-m section of the 600-m long BDP-98 sediment core (Baikal Drilling Project Group 2000), obtained as a part of the Baikal Drilling Project (BDP), was used for analysis. The drilling point was located at 53°44′45″N, 108°24′33″E in Academician Ridge at 335 m below water level. The sedimentation rate for the 0- to 100-m section of the BDP-98 core was estimated to be 4.1 cm kyr^{-1} on average (Sakai et al. 2002, private communication; Kashiwaya et al. 2002). However, the detailed paleomagnetic analysis showed that the sedimentation rate between 50 and 100 m was slower (Sakai et al. 2002, private communication; Kashiwaya et al. 2002), and the layer at 100-m depth corresponded to 2.8 Ma. Part (0.5–6.0 g) of the 2-cm-thick sections from every 10-cm length of the core was used for analysis, and the samples were kept frozen at −20°C until analysis.

2.2 Analysis of Photosynthetic Pigments

Photosynthetic pigments in the sediment samples were analyzed by high-performance liquid chromatography (HPLC) under reversed-phase conditions with photodiode array absorbance detection, as described previously (Soma et al. 1996). A typical HPLC chromatogram of an extract of the G-6 core (42- to 45-cm depth) sediment detected by absorbance at 400 nm is shown in Fig. 1. The assignment of pigments was based on the HPLC retention time and the absorption spectrum of standard samples, or from literature data (King and Repeta 1991; Harris et al. 1995; Harradine et al. 1996) when the standards were not available. Four pyropheophorbide *a* steryl esters were synthesized according to the method of King and Repeta (1994) and used for identification. The sterols used were cholesterol, stigmasterol, campesterol, and *β*-sitosterol, which were commercially available. SCEs derived from pyropheophytin *a* were quantified assuming that the specific absorbance was the same as that of pyropheo-

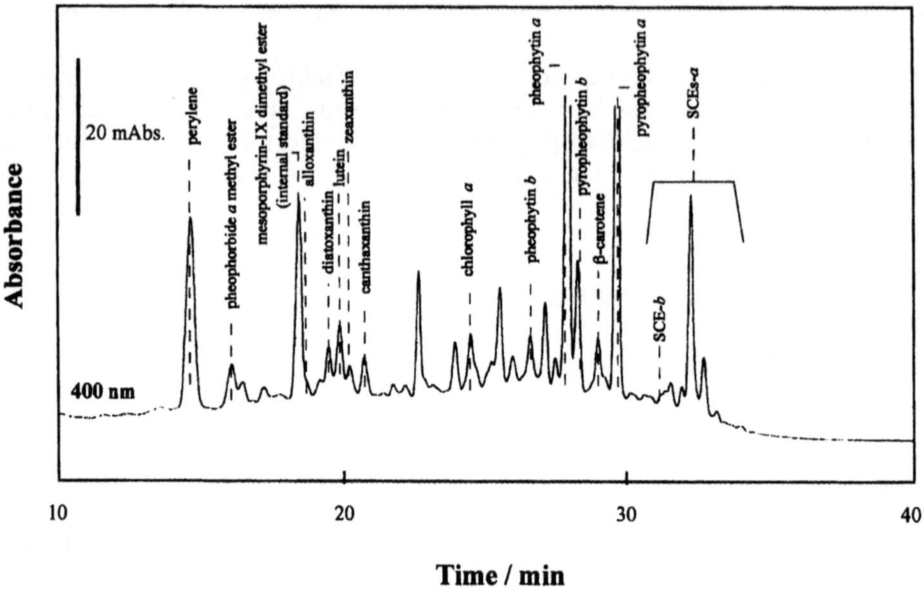

FIG. 1. Typical HPLC chromatogram of extracts from the G-6 core at a depth of 42–45 cm detected by absorbance at 400 nm. *SCE-b*, steryl chlorin ester of pyropheophorbide *b*; *SCEs-a*, steryl chlorin esters of pyropheophorbide *a*

phytin *a*. The SCE derived from pyropheophytin *b* was assumed to have the same specific absorbance as pheophytin *b*. The methyl ether of chlorophyllone *a* and the epimer of chlorophyllone *a* were assumed to have the same absorbance as pheophytin *a* (Harris et al. 1995; Villanueva and Hastings 2000). Total chlorophyll *a* is defined here as the sum of the concentrations of chlorophyll *a* and the derivatives (pheophorbide *a* methyl ester, pyropheophorbide *a* methyl ester, pheophytin *a*, pyropheophytin *a*, SCEs-*a*, methyl ether of chlorophyllone *a*, and the epimer of chlorophyllone *a*). The amount of chlorophyllone *a* represents the sum of the methyl ether of chlorophyllone *a* and the epimer of chlorophyllone *a*.

The HPLC SCE fractions were collected and hydrolyzed by the addition of 6 M HCl (Prowse and Maxwell 1991), and the resulting sterols were converted to their respective trimethylsilyl ether forms using bis(trimethylsilyl)trifluoroacetamide (BSTFA) and then analyzed by gas chromatography–mass spectrometry (GC–MS). The derived sterols were identified by comparing their mass spectra with those of the appropriate standards, and by using published data (Mühlebach and Weber 1998).

After the extraction of pigments, the residual sediment of the G-6 core sample was oven-dried and suspended in pure water with ultrasonication, followed by a determination of particle size distribution by wet-sieving (<45 μm, 45–75 μm, 75–212 μm, and >212 μm).

2.3 BGS and TOC

BGS was determined by measuring the concentration of dissolved Si in wet-alkaline extracts (Mortlock and Floelich 1989) using atomic absorption and/or inductively coupled plasma (ICP) atomic emission spectroscopy. TOC was determined using an element analyzer, after treatment with 1–2 M HCl overnight followed by three washings with Milli-Q water.

2.4 AMS ^{14}C Measurement

The acid-pretreated sediment samples containing about 0.4 mg organic carbon were heated at 900°C in a sealed quartz tube to convert organic carbon to CO_2. The CO_2 thus produced was purified and converted to graphite on iron powder catalyst (1 mg) by reducing with H_2 at 470°C for 2 h. An accelerator mass spectrometer at the National Institute for Environmental Studies (NIES-TERRA) (Kume et al. 1997) was used for the determination of ^{14}C contents (Yoneda et al. 1998).

3 Record of Photosynthetic Pigments in a Gravity Core G-6 from the Southern Basin for the Last 24 ka

3.1 AMS ^{14}C Age of the Core

The AMS ^{14}C age for the G-6 core is shown in Fig. 2a. Since the age of the lowest part of the G-6 core corresponded to about 24000 yr BP, the age of the G-6 core sediment spanned the transition of the last glacial and postglacial periods. The water content gradually increased upward (Fig. 2d), and fine particles (<45 μm) dominated (more than 80%) the sediment throughout the G-6 core (Fig. 2e).

3.2 Profiles of Chlorophyll Derivatives, TOC, and BGS in the G-6 Core

Chlorophyll *a* (Fig. 3b) and its derivatives, including pheophytin *a* (Fig. 3d), pyropheophytin *a* (Fig. 3e), pheophorbide *a* methyl ester (Fig. 3c), and steryl chlorin esters (SCEs-*a*) (Fig. 3f), were detected in the G-6 core in accordance with previous studies of Lake Baikal sediments (Soma et al. 1996; Tani et al. 2001). The chlorophyll *b* derivatives, pheophytin *b* and pyropheophytin *b*, were detected together with a smaller content of steryl ester of pyropheophorbide *b* (SCE-*b*), which was eluted after pyropheophytins *a* and *b* (see Fig. 1). Total chlorophyll *b* (pheophytin *b* + pyropheophytin *b* + SCE-*b*) (Fig. 3g) was an order of magnitude less than total chlorophyll *a* (Fig. 3a) throughout the G-6 core.

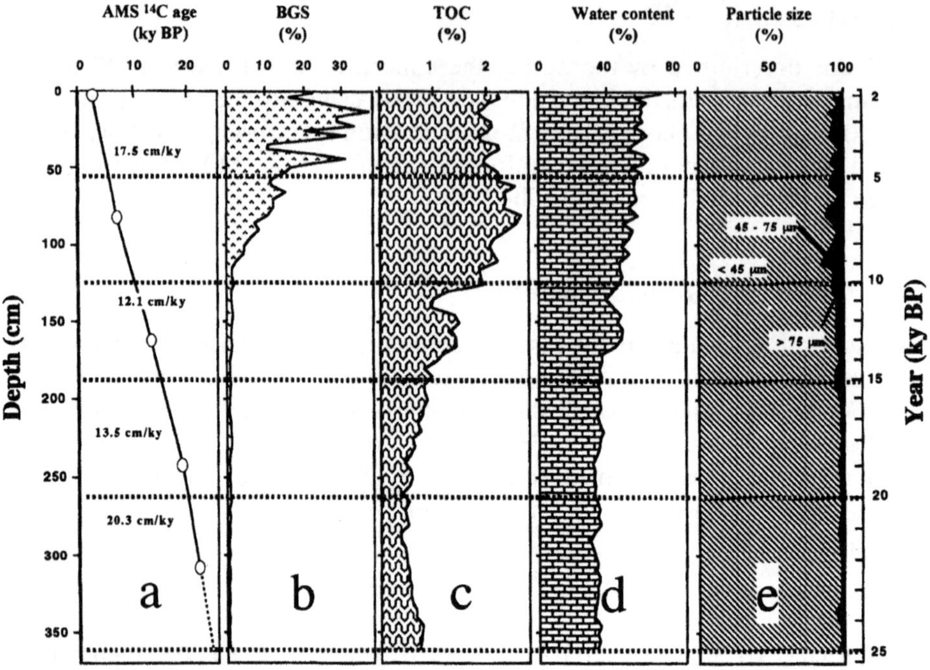

FIG. 2. Depth profiles of the fundamental parameters in the G-6 core. **a** Accelerator mass spectrometer (AMS) ^{14}C age; **b** biogenic silica (BGS); **c** total organic carbon (TOC); **d** water content; **e** particle size distribution (modified from Tani et al. Copyright (2002), with permission from Elsevier Science)

The trends of TOC (Fig. 2c), BGS (Fig. 2b), and chlorophyll derivatives (Fig. 3a–g) from past to present show that all of those indicators exhibit an increase in recent time, starting between 15000 and 10000 yr BP. This is consistent with the increased lake productivity in the postglacial period (Qiu et al. 1993; Carter and Coleman 1994; Horiuchi et al. 2000). Pheophorbide *a* methyl ester was relatively high in the uppermost layer of the sediment (22%–30% at 0–9 cm), and immediately decreased with depth (to less than 10%). Pheophytin *a* exhibited a gradual decrease in relative abundance with depth. In contrast, pyropheophytin *a* and SCEs-*a* increased in relative abundance with depth, and their relative abundance in total chlorophyll *a* exceeded 70%. These changes in the relative concentrations of individual chlorophyll *a* derivatives may reflect the long-term stabilities of these compounds in the sediments of Lake Baikal, i.e., SCEs-*a* $\geq$ pyropheophytin *a* > pheophytin *a* $\gg$ pheophorbide *a* methyl ester. This order of stability corresponds to the lipophilicity of these derivatives, i.e., the inverse order of reverse-phase HPLC elution. Notably, SCEs-*a* and pyropheophytin *a* showed positive correlations with TOC ($R = 0.84$ and 0.91, respectively) throughout the G-6 core, as was anticipated from their profiles in Figs. 2c and 3e–f. Thus, SCEs-*a* and pyropheophytin *a* may be more appropriate biogenic indicators for reconstructing the long-term history of

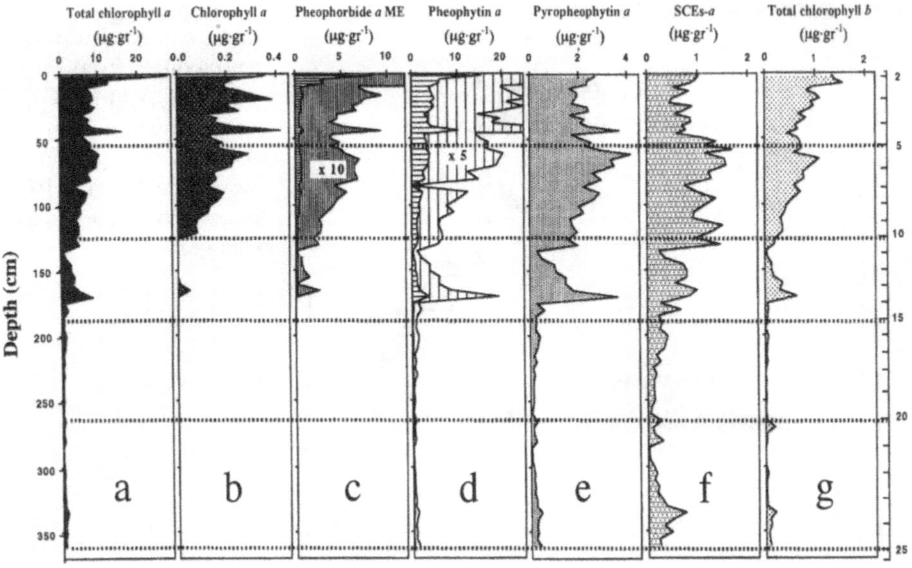

FIG. 3. Depth profiles of chlorophyll derivatives in the G-6 core. **a** Total chlorophyll *a*; **b** intact chlorophyll *a*; **c** pheophorbide *a* methyl ester; **d** pheophytin *a*; **e** pyropheophytin *a*; **f** SCEs-*a*; **g** total chlorophyll *b* (modified from Tani et al. Copyright (2002), with permission from Elsevier Science)

phytoplankton productivity in Lake Baikal. Since SCEs-*a* and pyropheophytin *a* are the main products formed from chlorophyll *a* by the grazing of phytoplankton (Harradine et al. 1996; Talbot et al. 1999a), these results suggest that the grazing of phytoplankton by zooplankton contributes much to the accumulation of TOC in the sediment, or that the amount grazed depends on the primary productivity as reflected in TOC. Fecal transportation is considered to be an important process for the accumulation of chlorophyll derivatives, since rapidly sinking fecal pellets carry the pigments out of the photic zone where otherwise photooxidation effectively degrades chlorophyll derivatives to colorless compounds within a few days (Carpenter et al. 1986; Nelson 1993; Sun et al. 1993).

The BGS was plotted against TOC in Fig. 4a. The data after 6000 yr BP, where BGS accumulation was high, and those before 6000 yr BP were distinguished by different symbols. After 6000 yr BP, when both TOC and BGS were relatively high, TOC showed a negative dependence on BGS, with a correlation coefficient $R = -0.78$. BGS also showed a negative correlation with SCEs-*a* ($R = -0.70$) (Fig. 4b) and pyropheophytin *a* ($R = -0.43$). While BGS and TOC have been inferred to indicate increased biological activity during the Holocene in Lake Baikal (Qiu et al. 1993; Carter and Coleman 1994; Horiuchi et al. 2000), the negative correlation between BGS and TOC indicates that the contribution of diatom residues to TOC was small, but the dilution of TOC by BGS was significant. A negative correlation between BGS and TOC in the surficial sediments in Lake Baikal has been reported (Williams et al. 1993).

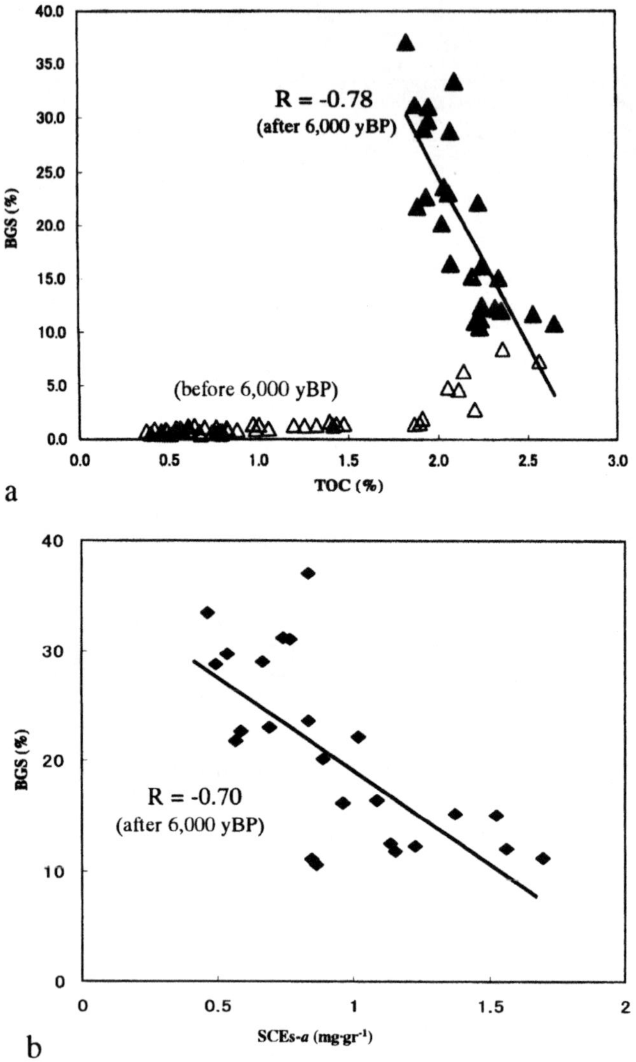

FIG. 4. Relations between TOC and BGS (**a**) (modified from Tani et al. Copyright (2002), with permission from Elsevier Science), and between SCE-*a* and BGS (**b**) in the G-6 core after 6000 years

BGS started to increase after about 10 000 yr BP, while the chlorophyll *a* derivatives and TOC started to increase significantly as early as 15 000 yr BP (Figs. 2b, c and 3b–f). The latter indicators decreased temporarily once between 12 000 and 11 000 yr BP and then increased again (Figs. 2c and 3b–f).

3.3 Carotenoid Distribution in the G-6 Core and Primary Productivity after 15 000 yr BP during Postglaciation

The major carotenoids detected in the G-6 core, which are indicative of various phytoplankton taxa (Young 1993; van den Hock et al. 1995; Jeffrey and Vesk

1997), were diatoxanthin (diatoms, dinoflagellates, and Chrysophyceae), lutein (green algae), canthaxanthin (cyanobacteria), zeaxanthin (cyanobacteria and green algae), alloxanthin (Cryptophyceae), and β-carotene (ubiquitous) (see Fig. 1), which is in agreement with previous results for Lake Baikal sediments (Soma et al. 1996; Tani et al. 2001). Figure 5 shows the depth profiles of these carotenoids in the G-6 core. All the carotenoids except for alloxanthin increased in the sediment above the detection level after 15 000 yr BP (Fig. 5), demonstrating directly the contribution of the relevant algae to the increase in primary productivity.

However, the detailed distributions of the carotenoids were different. Diatoxanthin, which is contained in diatoms, dinoflagellates, and Chrysophyceae, showed a maximum after 4000 yr BP, the period when BGS exhibited a remarkable increase (see Fig. 2b). On the other hand, the other carotenoids showed a maximum before that time. Notably, while the concentration of canthaxanthin, an indicator of cyanobacteria, was high at a depth of around 90 cm, it was low above 50 cm as compared with the other carotenoids.

The amounts and relative compositions of residual photosynthetic pigments in lake sediment should depend not only on their production in water, but also on their whole fate during transportation to the sediment (Carpenter et al. 1986; Nelson 1989; Leavitt and Brown 1988; Leavitt and Carpenter 1990; Nelson 1993; Wetzel 2001) and during diagenetic modification in the sediment (Swain 1985; Flurog and Carpenter 1988; Villanueva et al. 1994). Carotenoids with 5,6-epoxide structure, such as fucoxanthin and peridinin, which are major carotenoids in

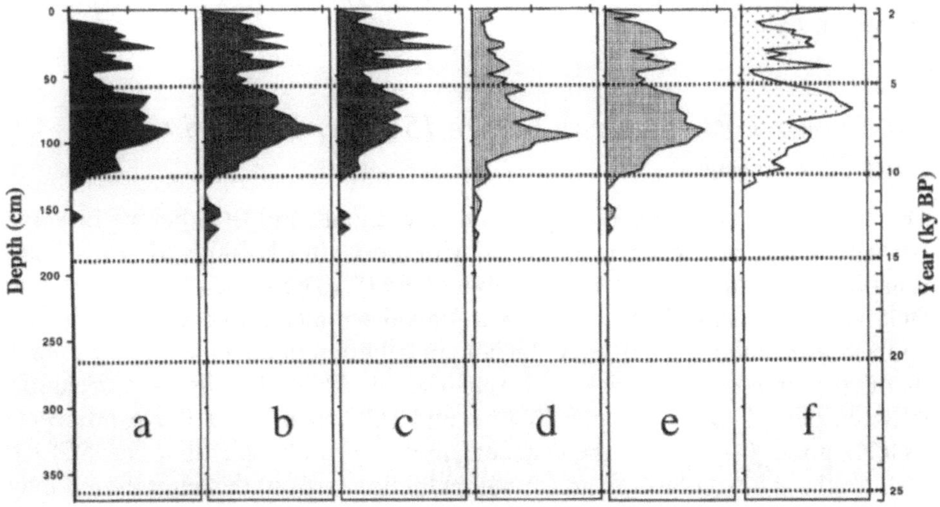

FIG. 5. Depth profiles of carotenoids in the G-6 core. **a** β-carotene; **b** lutein; **c** diatoxanthin; **d** canthaxanthin; **e** zeaxanthin; **f** alloxanthin (modified from Tani et al. Copyright (2002), with permission from Elsevier Science)

diatoms and dinoflagellates, respectively, are known to be susceptible to degradation. However, since the residual carotenoids listed above do not have 5,6-epoxide structure but similar structures, we presume that their relative stabilities have not been so variable, and that their changes reflect the changes in phytoplankton composition in the lake. Then, also assuming that all the residual photosynthetic pigments are of autochthonous phytoplankton origin, we can summarize the changes in phytoplankton population in Lake Baikal in the postglacial period from the variation of TOC, BGS, and photosynthetic pigments in the depths of the core, as described below.

After 15000yr BP, the lake productivity, which had previously been low, began to increase. During the period between 15000 and 5000yr BP, algal groups such as green algae, cyanobacteria, dinoflagellates, and Cryptophyceae, rather than diatoms, dominantly contributed to the primary productivity. The contribution reached a maximum around 8000 to 6000yr BP, and then gradually diminished. In turn, diatoms gradually increased to reach a maximum after 4000yr BP, and became the major primary producers. The period during which the carotenoids, except for diatoxanthin, showed maximum concentrations coincides with the so-called climate optimum of the postglacial period at 6000–7000yr BP (Parker 2000). The increase in diatoms in Lake Baikal, as shown by the increase in BGS and diatoxanthin (although several other algal groups also contain diatoxanthin), after the climate optimum has been reported previously (Qiu et al. 1993; Carter and Coleman 1994; Horiuchi et al. 2000). Horiuchi (1996) reported that in sedimentary records of Lake Baikal, there was a period between 16000 and 6000yr BP with high TOC and relatively low BGS. As the C/N ratio in this period was compatible with an autochthonous phytoplankton origin, algae other than diatoms were considered to have contributed significantly to the primary production in Lake Baikal.

3.4 Primary Productivity before 15000yr BP during the Last Glaciation

The amounts of total chlorophyll *a*, total carotenoids, and BGS before 15000yr BP were almost negligible as compared with those after 15000yr BP, while TOC remained at an appreciable level of about 0.6wt%. The association of very low BGS and appreciable TOC in the Lake Baikal sediments was reported earlier, and was attributed to a relatively higher contribution of an allochthonous input of organic matter to the lake (Horiuchi et al. 2000). The linear correlation between pyropheophytin *a* plus SCEs-*a* and TOC, mentioned above, intersects TOC at about 0.4%, when the concentration of pyropheophytin *a* plus SCEs-*a* is 0. As the latter indicators are of autochthonous origin, this intersection may allow an estimate of allochthonous TOC, possibly during the last glacial period. This amounts to about 70% of TOC, indicating that autochthonous phytoplankton productivity in the last glacial period was still significant.

An almost monotonous increase was observed in the ratio of SCEs-*a* to total chlorophyll *a* with depth in the sediment core (Fig. 6), regardless of the glacial/postglacial transition. Therefore the formation of SCEs-*a* through the predation of phytoplankton by zooplankton would have worked essentially similarly during the glacial period. While carotenoids were not detectable in the sediment before 15 000 yr BP, the composition of detectable SCEs-*a* should also depend on phytoplankton composition and be reflected in the HPLC pattern. Significant differences in the HPLC SCEs-*a* patterns for the glacial and postglacial periods were found (Fig. 7), suggesting a change in the phytoplankton composition through the last glacial/postglacial transition (see the BDP section below). On the other hand, the changes in SCE patterns within the postglacial period were apparently not large, as was the case with SCE patterns in one interglacial period for the BDP-98 core (see below).

At present, it is difficult to show a close link between the composition of carotenoid and the SCE pattern. Figure 7 also shows the depth profile of SCE-*b*, designated as peak 1 in the SCE pattern and determined by the absorbance at 440 nm, along with the profile of SCEs-*a*. Its general features are similar to those of lutein (Fig. 5b), as would be expected from their presumably common origin in green algae. The peak positions, however, are not always coincident. This might be due either to the variation of lutein/chlorophyll *b* ratio in green algae (Yokohama 1982) or to the different fates of lutein and SCE-*b*.

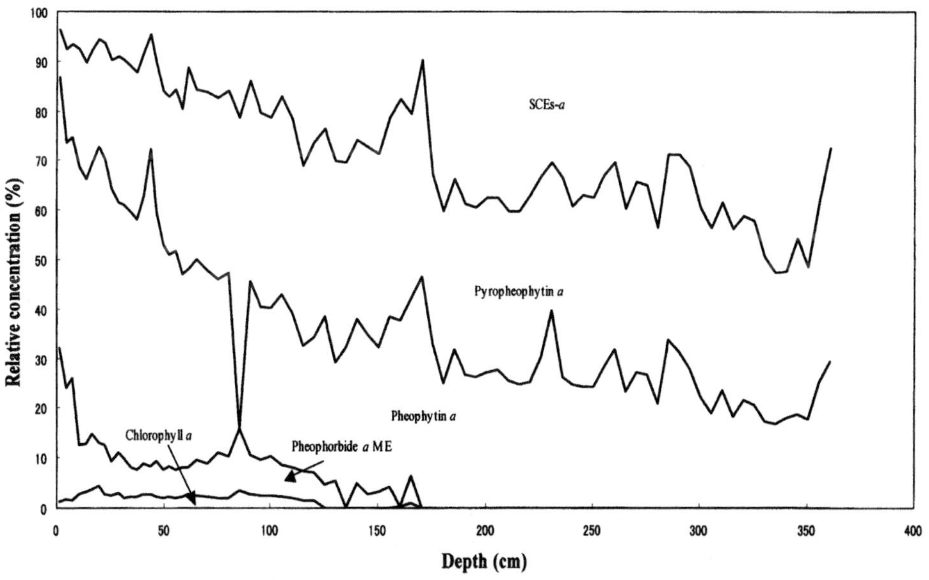

FIG. 6. Relative concentrations of chlorophyll *a* derivatives in the G-6 core (modified from Tani et al. Copyright (2002), with permission from Elsevier Science)

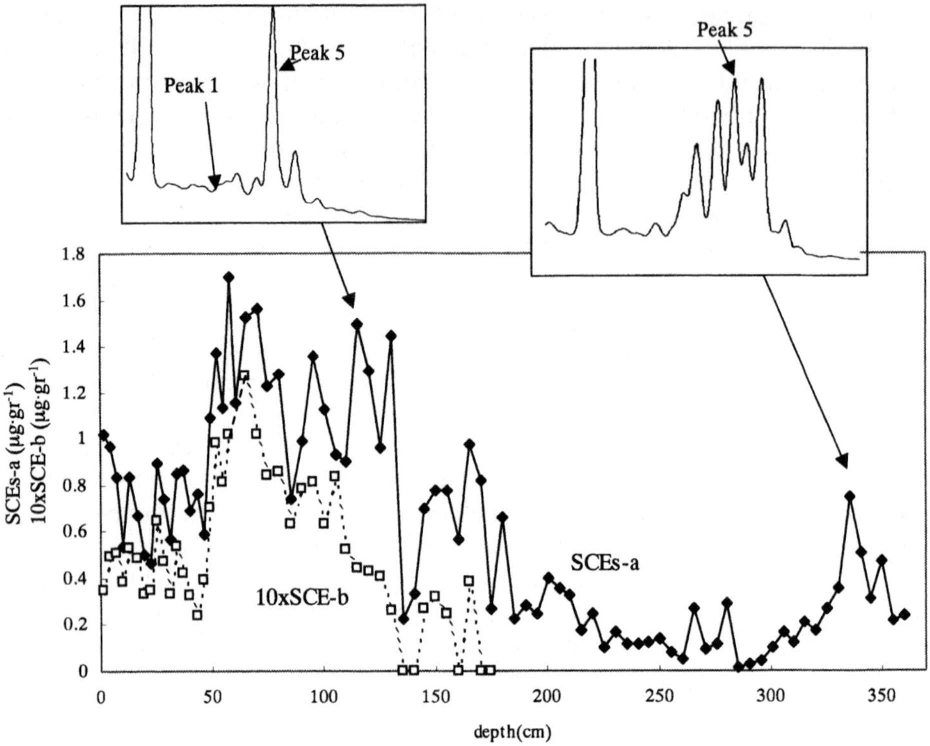

FIG. 7. Depth profiles of SCEs-*a* and SCE-*b* in the G-6 core, and examples of HPLC patterns

3.5 Manifestation of the Younger Dryas Period (11000–12000yr BP) in the Distribution of Photosynthetic Pigments

It has been recognized that, in the Northern Hemisphere, the temperature started to increase after 15000–16000yr BP, but decreased temporarily between 12000–11000yr BP, the period called the Younger Dryas (Cronin 1999). Sakai et al. (2001) noted the Younger Dryas in the magnetic susceptibility during the period 12000–11000yr BP in the same sediment core sample from Lake Baikal described here. In previous studies using BGS as an indicator of primary productivity (Qiu et al. 1993; Carter and Coleman 1994; Horiuchi et al. 2000), the presence of the Younger Dryas was not recognized, probably because the increase in BGS (diatoms) started late in the Holocene, while Colman et al. (1999) reported a temporary decrease in BGS in the BDP-93 sediment core corresponding to the Younger Dryas cooling event. In the present case, the period was discernible not only for TOC, but also for total chlorophyll *a* and carotenoids (Figs. 2c, 3a, and 5). Most of the individual carotenoids detected showed the same trends. Thus, it is likely that the cold climate of the Younger Dryas period was imprinted in the record of sedimentary photosynthetic pigments in Lake Baikal.

4 Record of Photosynthetic Pigments in the BDP-98 Sediment Core (0–100m) for the Last 2.8Ma

4.1 Chlorophyll Derivatives in the BDP-98 Sediment Core

The amounts of photosynthetic pigments found in the surface sediments of the Academician Ridge were very small compared with those of the southern basin (Soma et al. 2001a), although the amount of BGS was similar or greater on the Academician Ridge (Williams et al. 1993; Soma et al. 2001a). This would be due to the fact that the surface sediments were more oxic on the Academician Ridge, where photosynthetic pigments are easily degraded. The amount of carotenoids in algae is generally less than the amount of chlorophylls (Wright and Jeffery 1997). Thus, the limited amounts of BDP-98 sediment samples were insufficient to use carotenoids as markers. Pheophytin *a*, pyropheophytin *a*, and SCEs were observed throughout the core from 0 to 100m, and showed wide variations in concentrations. Unaltered chlorophyll *a* and *b* were detected only in the topmost layers. It should be noted that in the portion of the G-6 core corresponding to the last glacial period, SCEs-*a* and pyropheophytin *a* were detected, while carotenoids were undetectable.

Among the chlorophyll derivatives, SCEs were the most abundant pigments in the sediment layers deeper than 1m. In Fig. 8a, the SCEs/total chlorophyll *a* ratio is plotted against depth where it could be quantified. Except for a few outliers, SCEs account for 60%–90% of the total chlorophyll derivatives in the sediment layers deeper than 1m. This indicates that SCEs are stable in the subsurface sediments, and can persist for a longer period than chlorophyll *a*, pheophytin *a*, or pyropheophytin *a*. In the oligotrophic Lake Baikal, a significant part of the phytoplankton is grazed by zooplankton such as the copepod *Episcura* and by amphipod gammarids (Morris 1984; Watanabe 1994). The SCEs formed during this process would be egested in fecal pellets and sedimented.

A gradual decrease in the SCEs in the layers deeper than 30m is discernible (Fig. 8a). This gradual decrease is due to an increase in the relative amount of chlorophyllone *a* in the deep layers, as shown in Fig. 8b. Chlorophyllone *a* is enzymatically formed from pyropheophorbide *a* in marine bivalves and some diatoms (Watanabe et al. 1993). Chlorophyllone *a* methyl ether and the epimer of chlorophyllone *a* found in the surface sediment of Lake Baikal (Harris et al. 1995) might be directly formed in diatoms, because the major phytoplankton in the recent Baikal sediment is diatoms. However, the gradual increase in the chlorophyllone *a*/SCEs in the section of the core between 40 and 100m suggests that SCEs may diagenetically transform to chlorophyllone *a*. This is possible if SCEs loose steryl moiety, and the remaining propionate of the pyropheophorbide *a* structure is incorporated into the 7-membered ring. Chlorophyllone *a* has been considered to be an intermediate compound for the formation of bicycloalkylporphyrins (Chiller et al. 1993; Keely et al. 1994). The suggested diagenetic processes of chlorophyll *a* observed in Lake Baikal water and sediment are shown in Fig. 9.

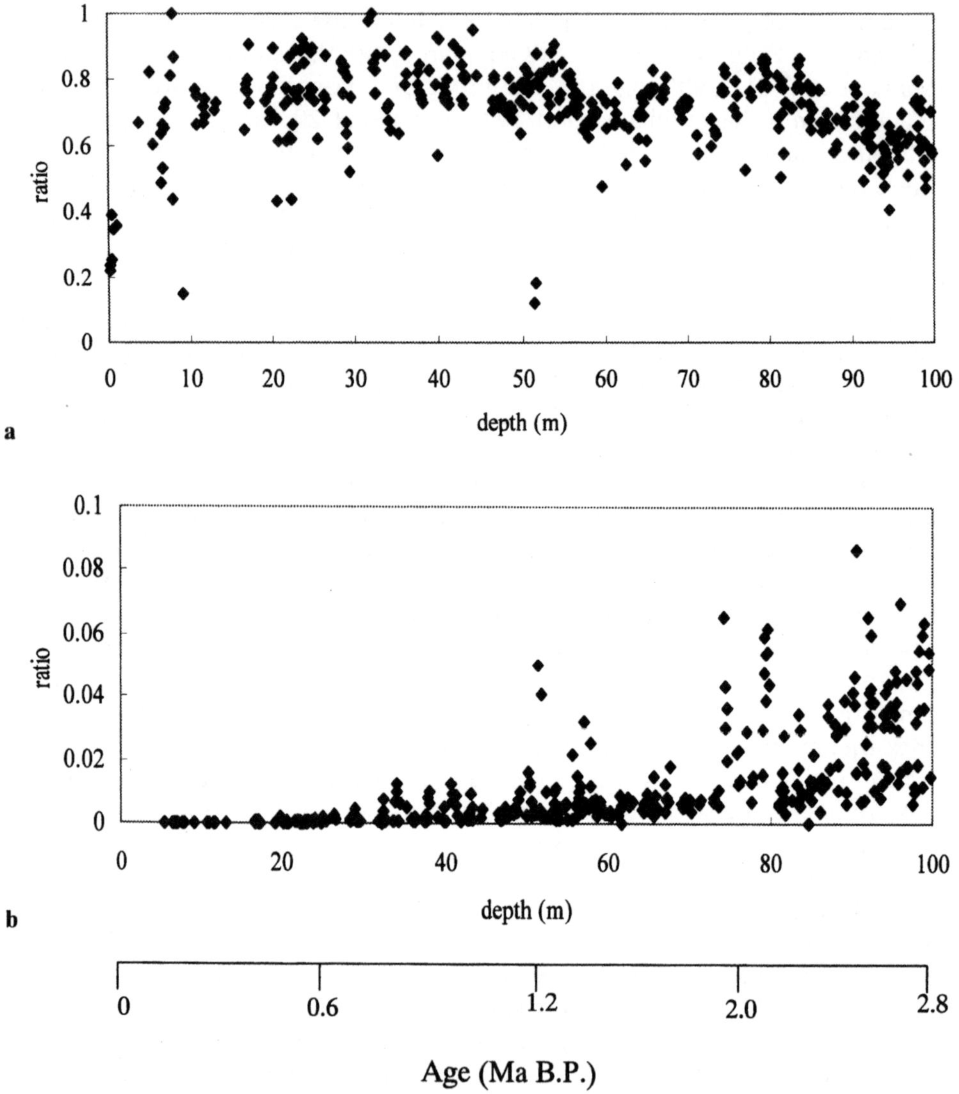

FIG. 8. Ratio of **a** steryl chlorin esters (SCEs) and **b** chlorophyllone *a* to total chlorophyll *a* extracted from BDP-98 sediment core samples

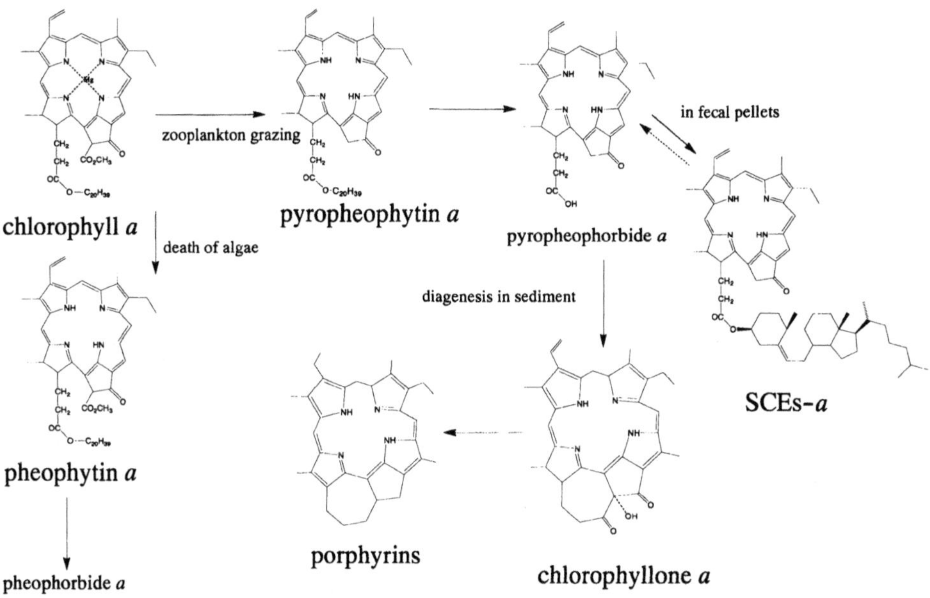

FIG. 9. Diagenesis of chlorophyll *a* in Lake Baikal

4.2 Composition of SCEs

A depth profile of SCEs is shown in Fig. 10. Each peak has a corresponding increase in the BGS profile (Fig. 11). Some examples of typical HPLC chromatograms for the retention time-range of pheophytin *a*, pyropheophytin *a*, and SCEs are shown on the left-hand side of Fig. 10. The pattern of SCEs in the chromatogram varied with depth, although the pattern showed little change within one interglacial period characterized as having relatively abundant TOC, BGS, and SCEs. The various SCE patterns could be brought about by the difference in the composition of sterols in SCEs, and thus is ascribable to the difference in phytoplankton composition. We found four basic patterns (types I–IV, as shown in Fig. 10). The assignment of some SCEs by reference to synthesized authentic samples and comparisons with patterns in the literature can explain these four patterns.

The SCE pattern at the top (type I) was mostly observed in layers less than 50 m deep, which corresponded to 1.2 Ma. Considering that diatoms have been the main phytoplankton in Lake Baikal (Kozhova and Izmest'eva 1998; Khursevich et al. 2000), SCEs of type I might be formed from diatom grazing. The major sterols in SCEs of type I are shown in Table 1. 24-Methylcholesta-5-en-3β-ol and 24-ethylcholesta-5,24-dien-3β-ol have been described as common sterols in diatoms (Patterson 1991; Barrett et al. 1995). However, the type I pattern has some differences from the SCEs pattern which Talbot et al. (1999b) observed in fecal pellets from the copepod *Calanus hegolandicus* grazing on the

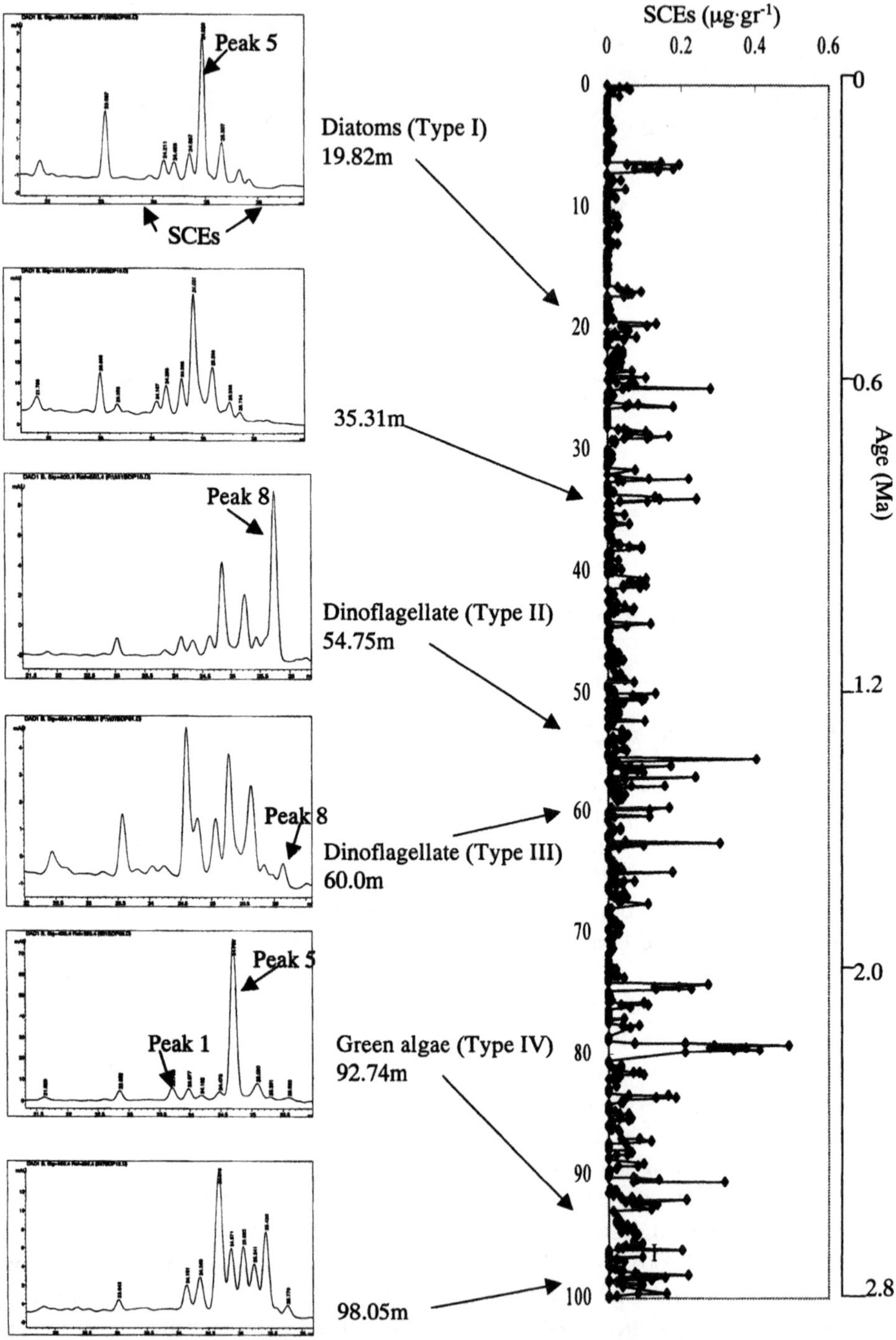

FIG. 10. Depth profile of SCEs in the BDP-98 sediment core from 0 to 100 m, and typical HPLC chromatograms of SCEs

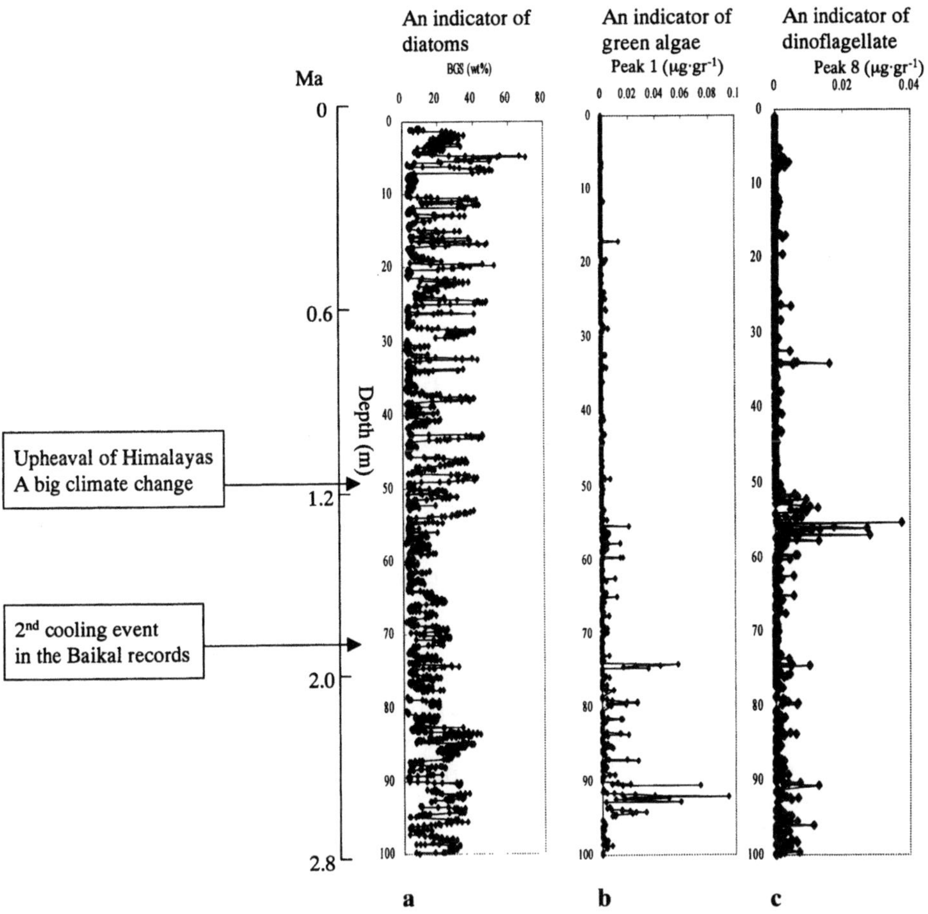

FIG. 11. Depth profiles of phytoplankton indicators in a BDP-98 sediment core from 0 to 100 m. **a** Profile of BGS (wt%); **b** profile of an indicator of green algae (SCE-*b* corresponding to peak 1 in Fig. 8); **c** profile of an indicator of dinoflagellates (C_{30}–SCE corresponding to peak 8 in Fig. 8)

diatom *Thalassiosira weissflogii*. Sterol in diatoms shows variations in composition according to families or species (Volkman 1986; Patterson 1991). Various taxa of diatoms, almost centric, have appeared in Lake Baikal during the last 2.8 Ma, such as *Stephanodiscus*, *Cyclotella*, and *Aulacoseisa* species (Khursevich et al. 2000). This would cause the difference in SCEs.

The third and forth SCE chromatogram patterns (types II and III) were considered to have originated from dinoflagellates, where C_{30} sterols characteristic of dinoflagellates were the major sterols, especially in type II. The type III SCE chromatogram pattern is similar to that of SCEs observed in fecal pellets of copepods grazing on a dinoflagellate (Talbot et al. 2000). In the series of layers

TABLE 1. Major SCE sterols found in the representative HPLC patterns shown in Fig. 10

		Type I	Types II, III	Type IV
1	24-norcholesta-5,22-dien-3β-ol	*		
2	Cholest-5-en-3β-ol	**	*	**
3	5α-cholestan-3β-ol	**	*	
4	5α-cholest-7-en-3β-ol		*	*
5	24-methylcholestan-3β-ol		*	
6	24-methylcholest-5-en-3β-ol (campesterol)	*	*	*
7	4-methylcholestan-3β-ol		*	
8	24-ethylcholesta-5,22-dien-3β-ol (stigmasterol)			**
9	24-methylcholest-7-en-3β-ol			*
10	24-ethylcholest-22-en-3β-ol		*	
11	24-ethylcholest-5-en-3β-ol (β-sitosterol)	**	*	**
12	4,24-dimethylcholestan-3β-ol		**	**
13	24-ethylcholesta-5,24 (28)-dien-3β-ol	**		
14	24-ethylcholestan-3β-ol		*	
15	4,23,24-trimethylcholest-22-en-3β-ol (dinosterol)		**	
16	4,23,24-trimethylcholestan-3β-ol (dinostanol)		**	
17	C_{30}-trimethylcholestanol		**	

*, major sterols detected; **major sterols detected in larger amounts

among which type II and/or type III SCE patterns have been observed, a mixture of the two types were commonly found, and sterols from C_{26} to C_{30} were detected in the SCEs by GC–MS measurements (see Table 1). Among the major SCE sterols were 4α,23,24-trimethyl-5α-cholestan-3β-ol (dinostanol), 4α,23, 24-trimethyl-5α-cholest-22E-en-3β-ol (dinosterol), and 4α,24-dimethyl-5α-cholestan-3β-ol (see Table 1). These compounds are presumed to have derived from dinoflagellates (Volkman 1986; Patterson 1991). The GC–MS total ion chromatograms are consistently similar to those of the sterols from a freshwater dinoflagellate, *Peridinium lomnickii* (Robinson et al. 1984). Therefore, SCEs of both type II and type III could be produced from dinoflagellate grazing. From a comparison with a liquid chromatography–MS analysis of SCEs (Eckardt et al. 1991), we infer that peak 8 was due to C_{30} stanol chlorin esters, including dinostanol SCE. The difference in SCE patterns between types II and III may be due to the difference in phytoplankton or in predators. The endemic copepod *Epischura* and amphipod gammarids are the main predators of phytoplankton in Lake Baikal (Morris 1984; Watanabe 1994).

Type IV is characterized by the existence of peak 1, which was assigned to a steryl ester of pyropheophorbide *b* from the absorption spectrum, which was apparently identical with pheophytin *b*. The intensities of peak 1 (SCE-*b* peak) correlated well with those of a SCE-*a* peak designated as peak 5 in Fig. 10. In the 90–95-m sections where the type IV pattern prevails, the correlation coefficient (R) between the two peaks was 0.98 ($n = 38$), indicating that peak 5 (SCE-*a*) of the type IV pattern is strongly representative of green algae. For the G-6 core, this correlation was lower ($R = 0.82$, $n = 40$), which suggests a contribution of algae other than green algae to peak 5. The type IV chromatogram pattern of SCEs shows some similarity to that of SCEs in fecal pellets produced during the grazing of a copepod on a prasinophyte (chlorophyll *b* bearing green algae, Talbot et al. 1999a). The major sterols in SCEs of type IV are campesterol, β-sitosterol, and stigmasterol, which are characteristic of green algae (Volkman 1986; Patterson 1991). Table 1 shows a list of sterols detected.

The variations in HPLC patterns of SCEs throughout the core (Fig. 10) suggest that the structure of the phytoplankton populations in Lake Baikal has changed significantly during the past 2.8 million years.

4.3 Vertical Distribution of Phytoplankton Indicators

It was reported that the sediment layers of Lake Baikal with high TOC and BGS concentrations corresponded to interglacial periods (Qiu et al. 1993; Horiuchi et al. 2000; Khursevich et al. 2000). The interglacial period is estimated to have lasted from 5000 to 20000 years from the width of individual peaks. Expressed on the basis of oven-dry sediment, the concentration of TOC was in the range from 0.3 to 2.3 wt%, and that of BGS from 3 to 70 wt%. A depth profile of BGS (an indicator of diatoms), an indicator of green algae (SCE-*b*, peak 1 in the SCE chromatogram in Fig. 10), and an indicator of dinoflagellates (C_{30} stanol esters of pyropheophytin *a*, peak 8 in the SCE chromatogram in Fig. 10) are shown in Fig. 11.

Peaks 1 and 8 are prominent in the sediment core below 70 m and 50 m, respectively, while the amount of BGS ranges from 20 to 70 wt% in sediment from warm periods throughout the core. This suggests that phytoplankton other than diatoms, and notably dinoflagellates, proliferated until 1.2 Ma, while green algae were abundant until 1.95 Ma. The uplift of the Himalayas is considered to have induced a significant climate change around 1.2 Ma (Yasunari and Seki 1992). The second cooling event in the Baikal record occurred 1.75–1.45 Ma, which would correspond to the Plio-Pleistocene boundary 1.796 Ma (Karabanov et al. 2000). An analysis of diatoms (Khursevich et al. 2000) showed that the dominant diatom flora changed from *Cyclotella* to such genera as *Stephanodiscus* over a period between 1.25 and 1.12 Ma BP, and new *Cyclotella* species appeared around 1.8–1.7 Ma, which also seems to correspond to these climate changes. A significant climate change around 1.2 Ma also corresponds to the appearance of an outlier of SCEs/total chlorophyll *a* in Fig. 8, where the amount of pyropheophytin *a* predominates over the amounts of SCEs. Therefore, it appears that critical events causing the replacement of, or diminution in, diatom flora also accompanied those in other phytoplankton groups.

5 Conclusion

We showed that the sedimentary record of photosynthetic pigments could provide information about the paleoenvironmental conditions of a large, deep, oxic, oligotrophic old lake such as Lake Baikal. After the glaciation, photosynthetic pigments, including carotenoids, of clearly phytoplankton origin showed the onset of an increase in phytoplankton activity, which generally started as early as 15000 yr BP at the beginning of the postglacial period, i.e., much earlier than the onset of the diatom record (BGS in this study). BGS increased appreciably after 10000 yr BP, and strongly after 5000 yr BP, while the maximum total carotenoids appeared around the climate optimum of 6000–7000 yr BP. The record of photosynthetic pigments revealed a temporary decrease in algal activity during the Younger Dryas period.

In the BDP-98 sediment core, we have shown a significant variation in the structure of the phytoplankton population in the past 2.8 Ma in Lake Baikal by the changes in the compositions of SCEs. It is fortunate that SCEs, which acquire taxonomic specificity through the predation of phytoplankton by zooplankton, are well preserved in the sediment for millions of years, while carotenoids are rarely preserved in the sediments which correspond to the period before the termination of the last glaciation. Each interglacial (warm climate) period may be characterized by the composition of SCEs in the corresponding sediment layer.

Evidence of phytoplankton populations preserved in long cores of lake sediments, spanning millions of years, have so far been confined to fossil diatoms. The results of our investigation of sediments from Lake Baikal suggest that SCEs are useful indicators of phytoplankton communities as a whole, including diatoms.

Acknowledgments. This work was a part of BICER (the Baikal International Center for Ecological Research) project, and of the project Paleo-Environmental Reconstruction in Northern Asia using Baikal Sediment Cores, subsidized by Special Coordination Funds for Promoting Science and Technology. The collaboration of both Russian and Japanese participants in the 1999 summer expedition to Lake Baikal are gratefully acknowledged. We thank Drs. M. Yoneda, M. Hirota, and Y. Shibata for AMS ^{14}C measurements, and Y. Amano for the TOC analysis.

References

Baikal Drilling Project Group (2000) Paleoclimatic record in the late Cenozoic sediments of Lake Baikal (in Russian). Russ Geol Geophys 41:3–32

Barrett SM, Volkman JK, Dunsta GA (1995) Sterols of 14 species of marine diatoms (Bacillariophyta). J Phycol 31:360–369

Carpenter SR, Elser MM, Elser JJ (1986) Chlorophyll production, degradation, and sedimentation: implications for paleolimnology. Limnol Oceanogr 31:112–124

Carter SJ, Coleman SM (1994) Biogenic silica in Lake Baikal sediments: results from 1990–1992 American cores. J Great Lakes Res 20:751–760

Chiller XFrD, Gülaçar FO, Buchs A (1993) A novel sedimentary lacustrine chlorin: characterization and geochemical significance. Chemosphere 27:2103–2110

Colman SM, Peck JA, Karabanov EB, Carter SJ, Bradbury JP, King JW, Williams DF (1995) Continental climate response to orbital forcing from biogenic silica record in Lake Baikal. Nature 378:769–771.

Colman SM, Peck JA, Hatton J, Karabanov EB, King JW (1999) Biogenic silica records from the BDP-93 drill site and adjacent areas of the Selenga Delta, Lake Baikal, Siberia. J Paleolimnol 21:9–17

Cronin TM (1999) Principles of paleoclimatology. Columbia University. Press, New York, p 194–252

Eckardt CB, Keely BJ, Maxwell JR (1991) Identification of chlorophyll transformation products in a lake sediment by combined liquid chromatography–mass spectrometry. J Chromatogr 557:271–288

Eckardt CB, Pearce GES, Keely BJ, Kowalewska G, Jaffe R, Maxwell JR (1992) A widespread chlorophyll transformation pathway in the aquatic environment. Org Geochem 19:217–227

Flurog ET, Carpenter R (1988) Pigment preservation and remineralization in oxic coastal marine sediments. Geochim Cosmochim Acta 52:87–99

Goodwin TW, Britton G (1988) Distribution and analysis of carotenoids. In: Goodwin TW (ed) Plant pigments. Academic Press, New York, p 62–132

Harradine PJ, Harris PG, Head RN, Harris RP, Maxwell JR (1996) Steryl chlorin esters are formed by zooplankton herbivory. Geochim Cosmochim Acta 60:2265–2270

Harris PG, Pearce GS, Peakman TM, Maxwell JR (1995) A widespread and abundant chlorophyll transformation product in aquatic environments. Org Geochem 23:183–187

Horie S (ed) (1992) International project on paleolimnology and late Cenzoic climate. IPPCC Newsletter No. 6

Horiuchi K (1996) Late Quaternary environmental changes in Lake Baikal and surrounding area studied by the lake sediments. PhD thesis, Tohoku University

Horiuchi K, Minoura K, Hoshino K, Oda T, Nakamura TT, Kawai T (2000) Paleoenvironmental history of Lake Baikal during the last 23000 years. Paleogeogr Paleoclim Paleoecol 157:95–108

Jeffrey SW, Vesk M (1997) Introduction to marine phytoplankton and their pigment signatures. In: Jeffery SW, Mantoura RFC, Wright SW (eds) Phytoplankton pigments in oceanography. UNESCO, Paris, p 37–84

Karabanov EB, Kuzmin MI, Prokopenko AA, Williams DF, Khursevich GK, Bezrukova EV, Kerber EV, Gvozdkov AN, Gelety VF, Wel D, Schwab N (2000) Glaciation of central Asia in the late Cenozoic according to the sedimentary record from Lake Baikal. In: Minoura K (ed) Lake Baikal—a mirror in time and space for understanding global change processes. Elsevier, Amsterdam, p 71–84

Kashiwaya K, Ochiai S, Tsukahara H, Sakai H, Kawai T (2002) Some issues to be considered in establishing age models for the long Lake Baikal sediment records. Quaternary International 205–207:95–96

Keely BJ, Harris PG, Popp BN, Hayes JM, Meischner D, Maxwell JR (1994) Porphyrin and chlorins in a late Pliocene lacustrine sediment. Org Geochem 58:3691–3701

Khursevich GK, Karabanov EB, Williams DF, Kuzmin MG, Prokopenko AA (2000) Evolution of freshwater centric diatoms within the Baikal rift zone during the late

Cenozoic. In: Minoura K (ed) Lake Baikal—a mirror in time and space for understanding global change processes. Elsevier, Amsterdam, p 146–154

King LL, Repeta DJ (1991) Novel pyropheophorbide steryl chlorin esters in Black Sea sediments. Geochim Cosmochim Acta 55:2067–2074

King LL, Repeta DJ (1994) Phorbin steryl esters in Black Sea sediment traps and sediments: a preliminary evaluation of their paleooceanographic potential. Geochim Cosmochim Acta 58:4389–4399

King LL, Wakeham SG (1996) Phorbin steryl ester formation by macrozooplankton in the Sargasso Sea. Org Geochem 24:581–585

Kozhova OM, Izmest'eva LR (eds) (1998) Lake Baikal—evolution and biodiversity. Backhuys, Leiden

Kume K, Shibata Y, Tanaka A, Yoneda M, Kumamoto Y, Uehiro T, Morita M (1997) The AMS facility at the National Institute for Environmental Studies (NIES). Jpn Nucl Instrum Methods Phys Res B 123:31–33

Leavitt PR, Brown SR (1988) Effect of grazing by *Daphnia* on algal carotenoids: implications for paleolimnology. J Paleolimnol 1:201–213

Leavitt PR, Carpenter SR (1990) Aphotic pigment degradation in the hypolimnion: implications for sedimentation studies and paleolimnology. Limnol Oceanogr 35:520–534

Mantoura RFC, Llewellyn CA (1983) The rapid determination of algal chlorophyll and carotenoid pigments and their breakdown products in natural waters by reverse-phase high-performance liquid chromatography. Anal Chim Acta 151:297–314

Minoura K (ed) (2000) Lake Baikal—a mirror in time and space for understanding global change processes. Elsevier, Amsterdam

Morris RJ (1984) The endemic faunae of Lake Baikal: their general biochemistry and detailed lipid composition. Proc R Soc Lond B 222:51–78

Mortlock RA, Floelich PN (1989) A simple method for the rapid determination of biogenic opal in pelagic marine sediments. Deep Sea Res A 36:1415–1426

Mühlebach A, Weber K (1998) Origins and fate of dissolved sterols in the Weddell Sea, Antarctica. Org Geochem 29:1595–1607

Naylor CC, Keely BJ (1998) Sedimentary purpurins: oxidative transformation products of chlorophylls. Org Geochem 28:417–422

Nelson JR (1989) Phytoplankton pigments in microzooplankton feces: variability in carotenoid alterations. Mar Ecol Prog Ser 52:129–144

Nelson JR (1993) Rate and possible mechanism of light-dependent degradation of pigments in detritus derived from phytoplankton. J Mar Res 51:155–179

Parker AG (2000) Biotic response to late Quaternary global change—the pollen record. A case study from the upper Thames valley, England. In: Culver SJ, Rawson PF (eds) Biotic response to global change. Cambridge University Press, Cambridge, p 265–287

Patterson GW (1991) Sterols of algae. In: Patterson GW, Nes WD (eds) Physiology and biochemistry of sterols. American Oil Chemists Society, Champaign, p 118–157

Pearce GES, Harradine PJ, Talbot HM, Maxwell JR (1998) Sedimentary sterols and steryl esters: distribution differences and significance. Org Geochem 28:3–10

Prowse WG, Maxwell JR (1991) High molecular weight chlorins in a lacustrine shale. Org Geochem 17:877–886

Qiu L, Williams DF, Gvorzdkov A, Karabanov E, Shimaraeva M (1993) Biogenic silica accumulation and paleoproductivity in the northern basin of Lake Baikal during the Holocene. Geology 21:25–28

Robinson N, Eglinton G, Brassell SC, Cranwell PA (1984) Dinoflagellate origin for sedimentary 4α-methylsteroids and 5α(H)-stanols. Nature 308:439–442

Sakai H, Nomura S, Araki S, Kashiwaya K, Tani Y, Shibata Y, Kawai T, Kravchinskii VA, Peck J, King J (2001) Magnetic susceptibility of the surface sediments of Lake Baikal and Lake Biwa (in Russian). Russ Geol Geophys 42:339–347

Soma Y, Tanaka A, Soma M, Kawai T (1996) Photosynthetic pigments and perylene in the sediments of the southern basin of Lake Baikal. Org Geochem 24:553–561

Soma Y, Soma M, Tani Y, Tanaka A, Kawai T (2001a) Distribution of photosynthetic pigments in the surface sediments of Lake Baikal (in Russian). Russ Geol Geophys 42:213–219

Soma Y, Soma M, Tanaka A, Kawai T (2001b) 2.8 million years of phytoplankton history in Lake Baikal recorded by the residual photosynthetic pigments in its sediment core. Geochem J 35:377–383

Sun M-Y, Lee C, Aller RC (1993) Laboratory studies of oxic and anoxic degradation of chlorophyll-*a* in Long Island Sound sediments. Geochim Cosmochim Acta 57:147–157

Swain EB (1985) Measurement and interpretation of sedimentary pigments. Freshwater Biol 15:53–76

Talbot HM, Head RN, Harris RP, Maxwell JR (1999a) Steryl esters of pyrophaeophorbide *b*: a sedimentary sink for chlorophyll *b*. Org Geochem 30:1403–1410

Talbot HM, Head RN, Harris RP, Maxwell JR (1999b) Distribution and stability of steryl chlorin esters in copepod faecal pellets from diatom grazing. Org Geochem 30:1163–1174

Talbot HM, Head RN, Harris RP, Maxwell JR (2000) Discrimination against 4-methyl sterol uptake during steryl chlorin ester production by copepods. Org Geochem 31:871–880

Tani Y, Yoshii K, Itoh N, Nara F, Soma M, Tanaka A, Soma Y, Yoneda M, Hirota M, Shibata Y (2001) Distribution of photosynthetic pigments and other biogenic compounds in the sediments from the southern basin of Lake Baikal (in Russian). Russ Geol Geophys 42:339–347

Tani Y, Kurihara K, Nara F, Itoh N, Soma M, Soma Y, Tanaka A, Yoneda M, Hirota M, Shibata Y (2002) Temporal changes in the phytoplankton community of the southern basin of Lake Baikal over the last 24000 years recorded by photosynthetic pigments in a sediment core. Org Geochem 33:1621–1634

Van den Hock C, Mann DG, Jahns HM (1995) Algae: an introduction to phycology. Cambridge University Press, Cambridge, p 1–16

Villanueva J, Hastings DW (2000) A century-scale record of the preservation of chlorophyll and its transformation products in anoxic sediments. Org Geochem 64:2281–2294

Villanueva J, Grimalt JO, Wit RD, Keely BJ, Maxwell JR (1994) Sources and transformations of chlorophylls and carotenoids in a monomictic sulfate-rich karstic lake environment. Org Geochem 22:739–757

Volkman JK (1986) A review of sterol markers for marine and terrigenous organic matter. Org Geochem 9:83–99

Watanabe N, Yamamoto K, Ihshikawa H, Yagi A (1993) New chlorophyll-*a*-related compounds isolated as antioxidants from marine bivalves. J Nat Prod 56:305–317

Watanabe Y (1994) Plankton in Lake Baikal. In: Morino H, Miyazaki N (eds) Lake Baikal (in Japanese). University of Tokyo Press, Tokyo, p 113–135

Wetzel RY (2001) Limnology (3rd edn). Academic Press, San Diego, p 798–802

Williams DF, Qui L, Karavanov E, Gvozdkov A (1993) Geochemical indicators of productivity and sources of organic matter in surficial sediments of Lake Baikal. Russ Geol Geophys 33:111–125

Wright SW (1984) Rapid extraction and high-performance liquid chromatography of chlorophylls and carotenoids from marine phytoplankton. J Chromatogr 294:281–295

Wright SW, Jeffery SW (1997) High-resolution HPLC system for chlorophylls and carotenoids of marine phytoplankton. In: Jeffery SW, Mantoura RFC, Wright SW (eds) Phytoplankton pigments in oceanography. UNESCO, Paris, p 327–350

Yasunari T, Seki Y (1992) Role of the Asian monsoon on the interannual variability of the global climate system. J Meteorol Soc Jpn 70:177–189

Yokohama Y (1982) Distribution of lutein and its derivatives in marine green algae. Jpn J Phycol 30:311–317

Yoneda M, Shibata Y, Kume H, Kumamoto Y, Morita M (1998) AMS radiocarbon dating at NIES-TERRA. Anthropol Sci 106:173

Young AJ (1993) Occurrence and distribution of carotenoids in photosynthetic systems. In: Young AJ, Britton G (eds) Carotenoids in photosynthesis. Chapman & Hall, London, p 16–71

Late Pleistocene and Holocene Environmental Changes in the Lake Baikal Catchment

Glacial/Interglacial Changes in the Carbon Cycle of Lake Baikal

ALEXANDER A. PROKOPENKO[1,2] and DOUGLAS F. WILLIAMS[1]

Summary. This chapter presents a conceptual framework for understanding the impact of glacial/interglacial climate changes on the carbon cycle of Lake Baikal. To do so, we bring together evidence from mass-balance estimates, from the carbon and nitrogen compositions of plankton, particulate organic matter and the surface layer of bottom sediments, and watershed soils and riverine sediments, as well as from Lake Baikal organic sedimentary records of the last glacial/interglacial transition and of the last climatic cycle to 130 Ka BP. The goal of this synthesis is to offer new ways of examining organic carbon paleoproductivity records.

Autochthonous primary production is the dominant source of organic matter (OM) in Lake Baikal sediments. The OM composition in pelagic sediments of Lake Baikal lies within the range of the primary produced particulate organic matter which sinks beyond the trophogenic layer. At the same time, the replenishment of the dissolved carbon pool of Lake Baikal is largely dependent on the input of terrigenous organic matter, which is assimilated and actively recycled in the lake. Therefore, allochthonous input does contribute to the observed present-day rate of carbon burial in Lake Baikal. However, by itself, terrigenous OM does not appear to constitute a major portion of OM in sediments, as suggested by mass-balance estimates and by a comparison of the C/N and $\delta^{13}C$ ratios of various carbon sources with those of Lake Baikal bottom sediments.

The sedimentary total organic carbon content and the $\delta^{13}C$ and C/N ratios indicate that past glacial/interglacial changes in OM accumulation were basin-wide, and mostly reflect internal changes in the lake's carbon cycle driven by the regional climate. These changes included the varying efficiency of carbon export

[1] Department of Geological Sciences, University of South Carolina, Columbia, SC 29208, USA
[2] United Institute of Geology, Geophysics and Mineralogy, Russian Academy of Sciences, Novosibirsk 630090, Russia

from the photic zone, and dramatic fluctuations in the carbon pool of the lake, as reflected in the carbon isotope records during the last deglaciation.

Key words. Organic carbon, Stable isotopes, Lake Baikal, Siberia, Paleolimnology

1 Introduction

In this chapter, we offer a synthesis combining new 130-Ka records of biogenic silica, organic carbon content, and C/N ratios from Lake Baikal with unpublished results on the composition of carbon sources in the modern Baikal system, and with some published results on Baikal $\delta^{13}C$ and C/N proxy records during the last glacial/interglacial transition. The goal of our study is to arrive at a new understanding of climatically driven carbon cycle changes in Lake Baikal. In this synthesis, we attempt to integrate modern processes in the carbon cycle (mostly based on Russian literature sources which are not readily available) with new data from the sediments and from the watershed in order to discuss evidence from sediment cores of how the Lake Baikal carbon cycle has functioned in the past.

Given the interest in more paleoclimate-oriented lake drilling and the ubiquitous use of $\delta^{13}C$ and C/N indices in sediments to interpret paleoclimate–paleoproductivity records, this synthesis is of wider interest than for Lake Baikal only, because it reveals distinct climate-dependent modes of carbon cycling. In addition, it demonstrates that instead of merely being a sign of a mixture of organic matter (OM) from different sources, the $\delta^{13}C$ and C/N proxies reflect systematic glacial–interglacial changes in productive efficiency and in the carbon pool of the lake. Because of the low alkalinity of lake waters and the high role of respiratory carbon in the CO_2 pool of Lake Baikal, the link between watershed processes and the internal lacustrine cycling of carbon is especially pronounced in the Baikal records of the composition of sedimentary OM during the last climatic cycle.

2 Paleoproductivity Indices in the Lake Baikal Sedimentary Record

Past changes in productivity in response to glacial/interglacial climate changes have dominated the sedimentation processes in Lake Baikal, affecting sediment geochemistry, physical and rock-magnetic properties, grain size, etc. (BDP Members 1997). The biogenic silica content (BioSi), the most widely used paleoclimate proxy in the Baikal record, is produced by diatoms which account for the major biogenic flux to bottom sediments (BDP Members 1997; Williams et al. 1997). This BioSi paleoproductivity index has been used as the basis for developing orbitally tuned time-scales for the Baikal paleoclimate record, and for exploring variances in Milankovitch frequency bands (Williams et al. 1997; Prokopenko et al. 2001).

Another paleoproductivity index in Baikal sediments is total organic carbon content (TOC). TOC in sediments has been shown to correlate strongly with BioSi content and diatom abundance since the first work on Baikal cores (Ishiwatari et al. 1992; Qiu et al. 1993; Toyoda et al. 1993). Moreover, when BioSi fails to produce a signal during glacials due to the absence of preserved diatoms, a clear TOC signal is still present in the sedimentary record, thus offering evidence of productivity changes during glacials. However, initial interpretations of the Baikal organic carbon records in the context of productivity and regional paleoclimate changes were complicated by uncertainty about the origin of OM in bottom sediments. This uncertainty was largely fed by early interpretations of total organic carbon to nitrogen ratios (C/N), suggesting that the TOC signal reflected a complicated mixture of autochthonous and allochthonous organic matter. This uncertainty ultimately led to a lack of interest in further studies of total organic matter preserved in Baikal sediments. This work is an attempt to change this attitude, and to extract more information from the composition of the total sedimentary OM.

3 Materials and Methods

For this study we used new records from the Baikal Drilling Project drill core BDP-96-2, along with published records from Baikal piston cores and the upper sections of drill core BDP-93-2 (Fig. 1). All these cores share a similar stratigraphy: the interglacial sediments are composed of diatomaceous ooze or mud (semiquantitative diatom abundance from 20% to 80% grains), and the glacial sediments consist of diatom-barren silty clay. The soil samples from the adjacent Baikal and Angara watersheds (Koval et al. 1993) and riverine fine fraction sediments were collected during geochemical surveys by the Institute of Geochemistry, Irkutsk. Plankton was collected by net towing, and particulate organic matter (POM) was collected by filtering water sampled with 10-l Niskin bottles on precombusted glass fiber 0.47 μm filters during the Baikal spring cruise in 1994. Samples of phytoplankton and zooplankton were hand-picked off the filters and screens on which they were collected using tweezers. The surface sediment layer was sampled using box cores during BDP cruises in 1989–1991 (Williams et al. 1993) and during the 1994 cruise.

For isotopic, as well as TOC and nitrogen analyses, we used a standard sample preparation procedure which included drying, grinding, and decalcifying in 1 M H_3PO_4 in an ultrasonic bath. Plankton and POM samples were dried at 50°C. Samples of 0.5–15 mg weight were analyzed on an Optima isotopic ratio mass spectrometer (VG Isotech, England) in line with a Fisons NC 1500 CHNS elemental analyzer (Fisons, Italy). The analyses on most samples were replicated. The standard error for isotopic analyses was from 0.6% for pure phytoplankton to 2% for some of the POM samples. For TOC and total nitrogen analyses (as

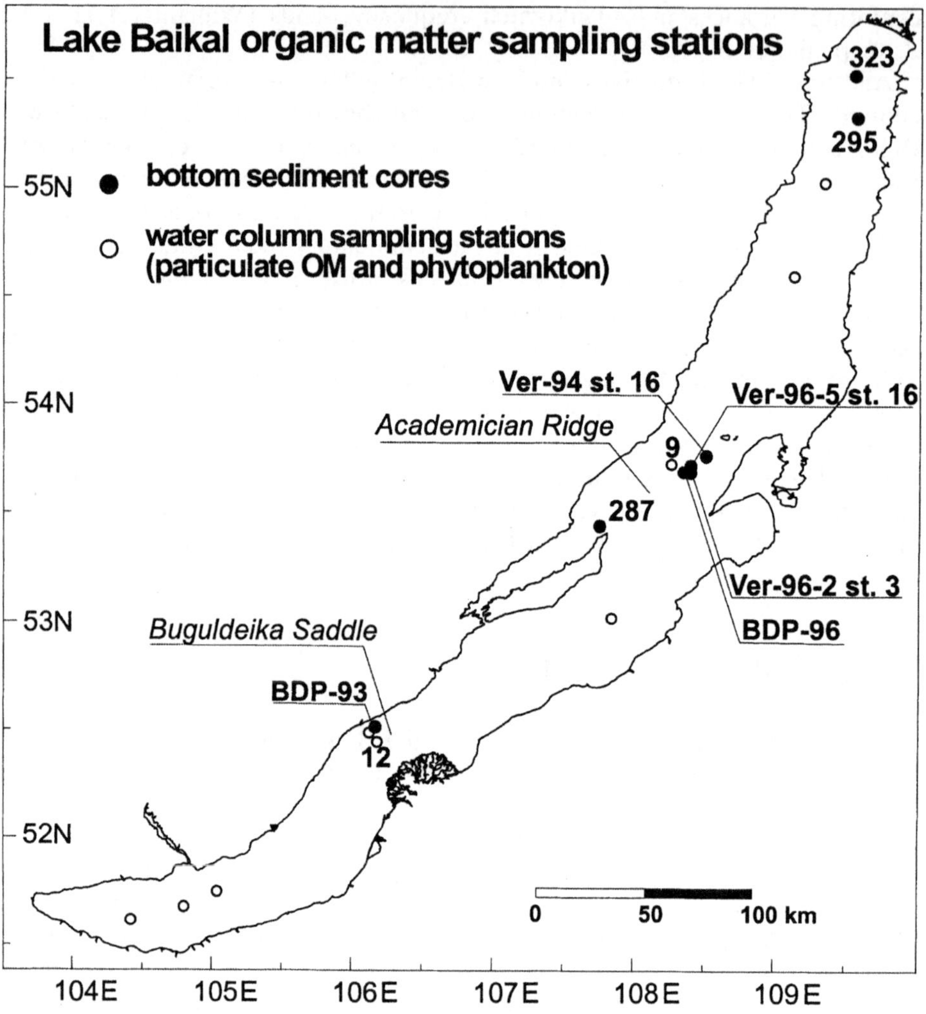

FIG. 1. Lake Baikal sampling sites. *OM*, organic matter

percentage dry sediment weight), the standard errors were 0.5%–1.5% and 1%–4%, respectively, depending upon sample type.

4 Deglacial Organic Carbon Records in Lake Baikal

Based on published stratigraphic and radiocarbon age models (Ishiwatari et al. 1992; Bradbury et al. 1994; Colman et al. 1996; Prokopenko et al. 1999; Minoura et al. 1999; Horiuchi et al. 2000), we can now investigate the last deglacial change in paleoproductivity and C/N indices throughout Lake Baikal. A comparison of sedimentary records from the Buguldeika Saddle (Fig. 2a), Academician Ridge

(Fig. 2b), and northern Baikal (Fig. 2c) reveals a uniform basin-wide increase in both TOC content and C/N ratios from the last glacial to the Holocene. This correlation indicates that past changes in OM accumulation encompassed the entire lake and, were largely climate-dependent. More evident in the higher-resolution records of cores BDP-93-2 (Fig. 2a) and Ver-94 St.16 (Fig. 2b) are the TOC maximum in the mid-Holocene interval, and the progressive trend of increasing C/N ratios during the Holocene. The interpretation of these trends in terms of

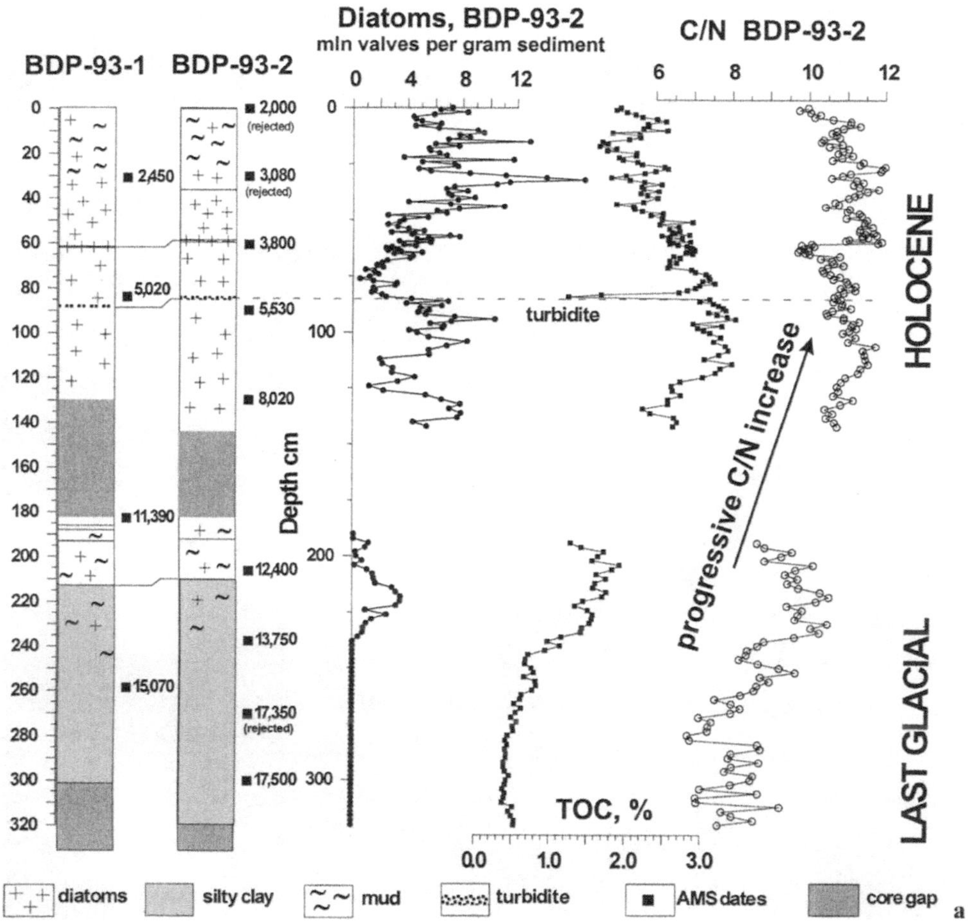

FIG. 2. **a** Diatom abundance, total organic carbon (*TOC*), and *C/N* ratio records in core BDP-93-2 (Prokopenko et al. 1999) from the Buguldeika Saddle of Lake Baikal. Note the distinct trend of increasing C/N from the last glacial to the Holocene. **b** Biogenic silica, TOC, and C/N records in sediment cores 287 K3 (Williams and Jenkins 1993; Prokopenko 1997) and Ver-94 St-16 (Horiuchi et al. 2000) from Academician Ridge in Lake Baikal. **c** Biogenic silica, TOC, and C/N records in sediment cores 295 K3 (Qiu et al. 1993; Prokopenko et al. 1999) and 323 PC1 (Toyoda et al. 1993) from northern Baikal

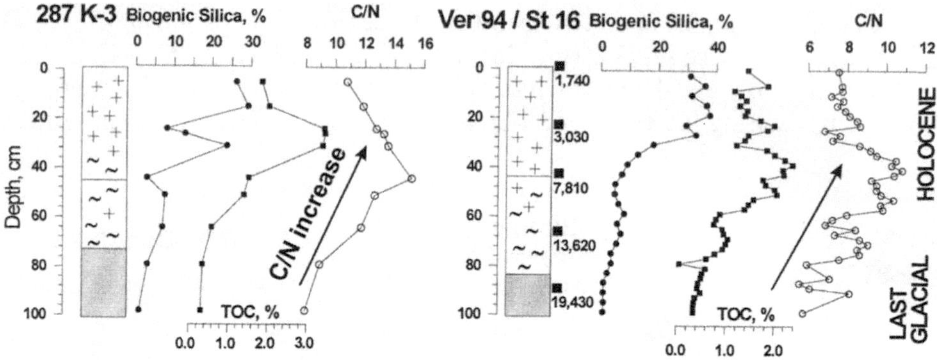

b

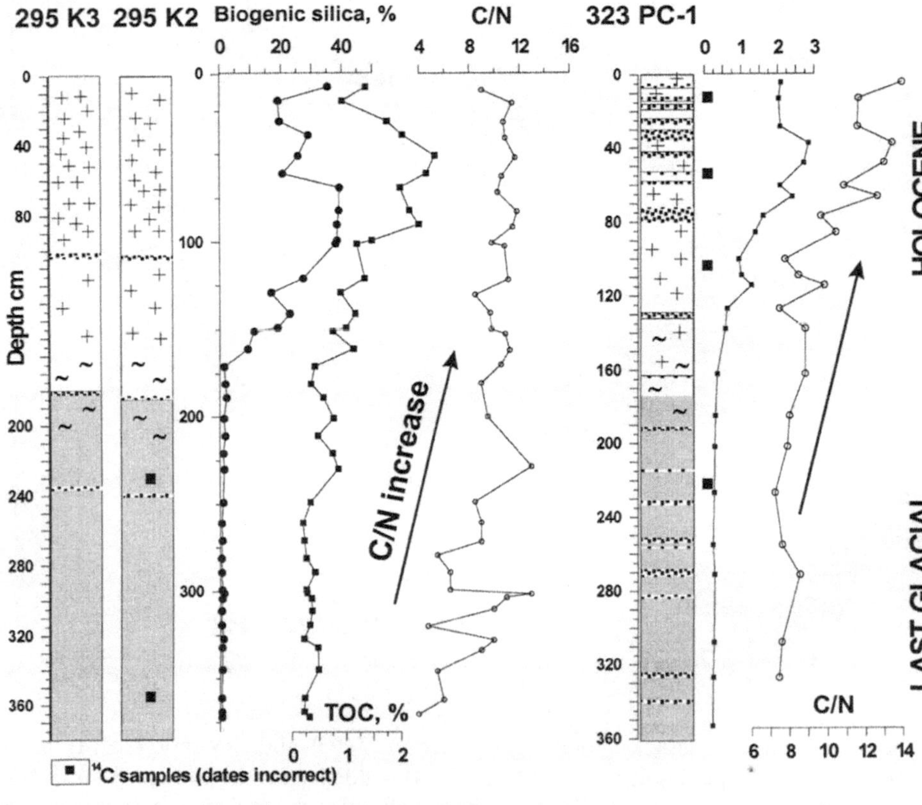

c

Fig. 2. *Continued*

paleoproductivity and paleoclimate signals was previously complicated, because C/N values above 6–7 in sediments were usually regarded as a significant presence of allochthonous OM (Qiu et al. 1993; Toyoda et al. 1993; Colman et al. 1996). However, we use the data reviewed in following sections to argue against such a straightforward interpretation, and to provide new insights into the nature of OM signals in Baikal sediments.

5 Features of the Modern Carbon Cycle in Lake Baikal

Lake Baikal is often referred to as oligotrophic or strongly oligotrophic (Weiss et al. 1991; Colman et al. 1996), thus implying very low primary productivity. The trophic state of lakes may be reflected by a number of indices, such as the volumetric ratios of epilimnion and hypolimnion, the oxygen saturation of deep waters, the TOC content in sediments, nutrient concentrations, the transparency of waters, etc. (Jorgensen 1979; Cole 1988). All these listed characteristics undoubtedly suggest that Lake Baikal is oligotrophic. However, the phytoplankton species which dominate in the biomass as Baikal's primary producers, such as the diatoms of *Aulacoseira*, *Synedra*, and *Cyclotella* genera, are more characteristic of meso- and eutrophic lakes (Jorgensen 1979; Cole 1988). The biomass of Baikal's phytoplankton (Popovskaya 1987), when converted into mg C/m^3 using the coefficient 0.1 (Kozhova 1967), may reach into the mesotrophic–eutrophic range (Jorgensen 1979). This becomes evident if the estimates are made for the trophogenic layer (0–25 m) instead of being divided by the entire volume of this deep lake (Fig. 3), especially in the more productive southern Baikal (Popovskaya 1987). The diatom cell abundance values in the photic zone during productive years may reach values which are characteristic of highly eutrophic lakes (Popovskaya 1987). The Lake Baikal production estimates suggest that it is most likely mesotrophic (Likens 1975). These "mesotrophic" characteristics of the modern Baikal carbon cycle suggest a productive efficiency much higher than that of an "ultraoligotrophic" system, as considered in several paleolimnologic reconstructions. This perspective has important implications for studies of sedimentary organic carbon because it implies that the system is much more dynamic.

General assumptions based on mass-balance estimates support the contention that autochthonous production is the main source of carbon in Lake Baikal. For instance, according to Votintzev (1985), autochthonous production accounts for 90% of the carbon input to the lake. During spring, autochtonous production exceeds the total allochthonous OM input by more than six times (Popovskaya 1967). Another estimate leaves only 7% of the annual OM balance to allochthonous sources (Votintzev and Popovskaya 1973). Among autochthonous sources, the dominant role is that of diatoms, which account for as much as 90% of phytoplankton biomass in productive years (Vykhristyuk 1980). Although the dominant part of primary produced OM undergoes grazing and respiration in the upper layers (Kozhov 1963; Votintzev and Popovskaya 1967), a substantial

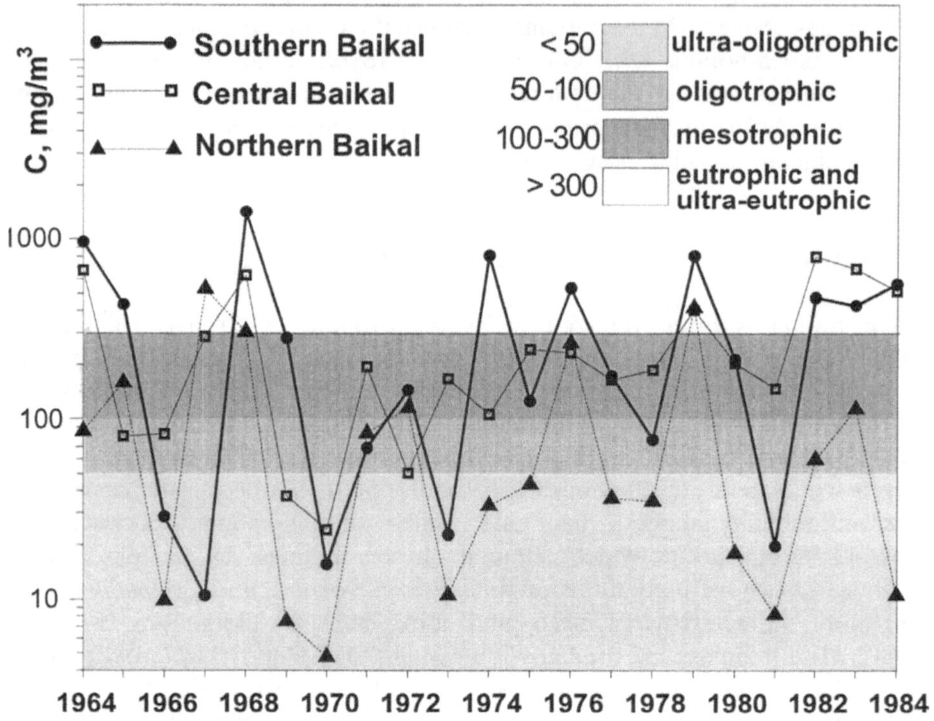

FIG. 3. Long-term changes in the biomass of phytoplankton (Popovskaya 1987) in Lake Baikal in the upper 0 to 25-m layer during spring period (C, mg/m^3) give a perspective on estimating the Lake Baikal trophic state based on trophic level criteria from Jorgensen (1979). Note the irregular pattern of alternating productive and nonproductive years in Lake Baikal (Popovskaya 1987)

amount of carbon (from 2–3 g/m^2 in nonproductive years to 8–11 g/m^2 in productive years) still reaches the lake bottom (Votintzev and Popovskaya 1967). In general, the oxidation of OM takes up 95.5% of the total OM in the Baikal water column (Votintzev and Popovskaya 1973). Based on this estimate, Vykhristyuk (1980) suggested an average value of 4 g/m^2 for the annual TOC flux to the bottom. Using this value, an approximate estimate of OM decomposition at the sediment/water interface can be made. For example, average values for TOC content in surface sediments of about 2.3% (our data) or 1.5% (Vykhristyuk 1980) imply that 50%–60% of TOC is further decomposed at the sediment/water interface. Votintzev et al. proposed a temporal scale for OM decomposition by assuming that 70%–75% of the input is decomposed annually, whereas 1.9% of the annual TOC input is sequestered in sediments (Votintzev et al. 1975).

Concentrations of major components in Baikal waters are lower than in tributaries. For instance, nutrients, silica, and OM are in deficit in Lake Baikal by a factor of 4 to 10 (Votintzev 1961a). Estimates of the annual chemical balance have demonstrated that the outflow of major components is much lower than the input. For instance, 78% of silica and 76% of OM is retained in Lake Baikal, and bottom sediments represent a large sink for these components (Votintzev 1961a). For carbon, another important sink is CO_2 release to the atmosphere (Mizandrontzev and Mizandrontzeva 1994). A comparison of annual cycles of input of dissolved components by major tributaries and output from Lake Baikal through the Angara River (Fig. 4) demonstrates that OM input (shading in Fig. 4) significantly increases with spring runoff, whereas the output is shifted toward autumn months, indirectly supporting its autochthonous origin.

From the above balance estimates, it may be concluded that allochthonous OM does not constitute a major part of the annual balance, and that it is successfully utilized and assimilated in Lake Baikal. In the following sections we compare the isotope and C/N compositions of modern living plankton, POM, and soil and riverine OM with those of the surface layer of bottom sediments in order to estimate the relative contributions of different sources to sedimentary OM in the present-day Lake Baikal.

6 Autochthonous Organic Matter within the Lake Baikal Water Column

Phytoplankton was sampled in spring 1994, which was a highly productive year in Lake Baikal. The samples were almost entirely comprised of *Aulacoseira baicalensis*, a typical dominant species of the Baikal spring phytoplankton assemblage. This species produces most of the annual biomass during the winter–spring period, and accounts for the major BioSi and TOC flux to the bottom (Popovskaya 1987). The isotopic composition of this characteristic Baikal phytoplankton significantly varied spatially (Fig. 5). On average, however, the composition of the Baikal pelagic phytoplankton in the northern basin ($\delta^{13}C = -32.3‰$; C/N = 6.4), the southern basin ($\delta^{13}C = -31.1‰$; C/N = 7.5), and the Selenga Delta area ($\delta^{13}C = -29.7‰$; C/N = 6.9) corresponded to that of lakes in temperate latitudes (McKenzie 1985), and is close to the average composition of phytoplankton in high-latitude oceans (Rau et al. 1989). C/N ratios for Baikal phytoplankton also lie within normal lacustrine (Meyers and Ishiwatari 1995) and marine (Redfield ratio ~6.3) C/N ranges (Romankevich 1977).

Comparisons of the phytoplankton composition with those of POM and zooplankton (Fig. 5) show that the latter cover the same wide $\delta^{13}C$ range as the phytoplankton. However, C/N values for each of these groups of samples are distinctly different (Fig. 5). Zooplankton, as expected, contains more nitrogen, whereas POM contains less owing to the partial decomposition of primary produced OM. Even in the upper 50- to 100-m water layer above the thermocline, plankton-derived POM shows evidence of decomposition based on elevated C/N ratios (see Fig. 5, shading). This is in agreement with observations of the high

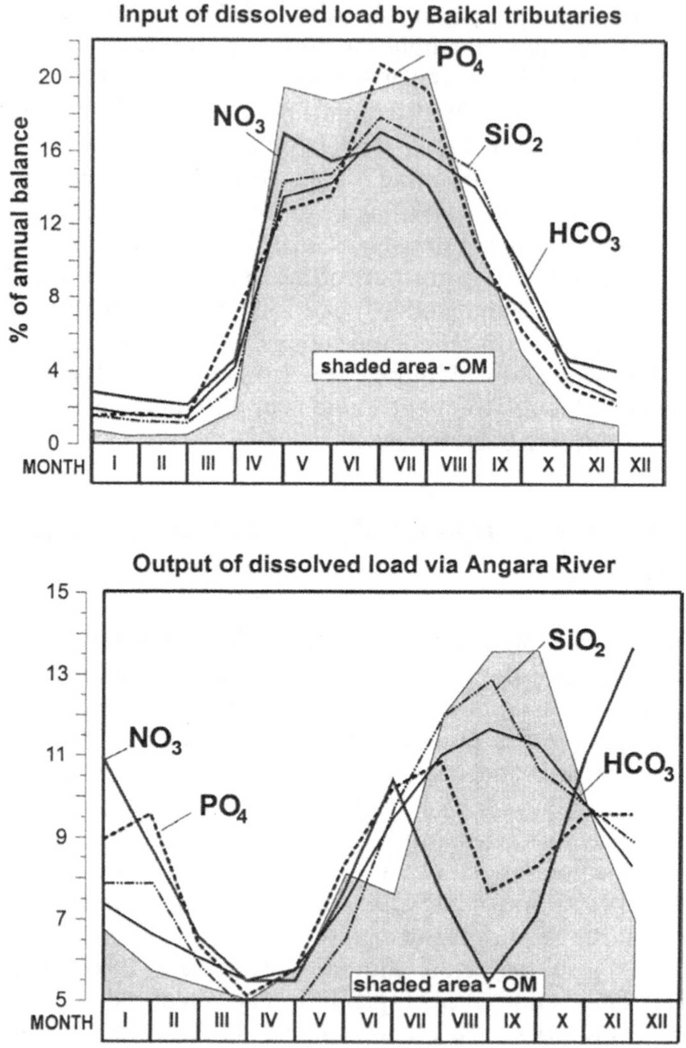

FIG. 4. Balance estimates of the input and output of dissolved substances in Lake Baikal as percentages of the annual balance (modified from Votintzev et al. 1965). The lack of a close temporal relation between the maximum input and output of dissolved organic carbon indirectly suggests that allochthonous OM is retained and assimilated in Lake Baikal

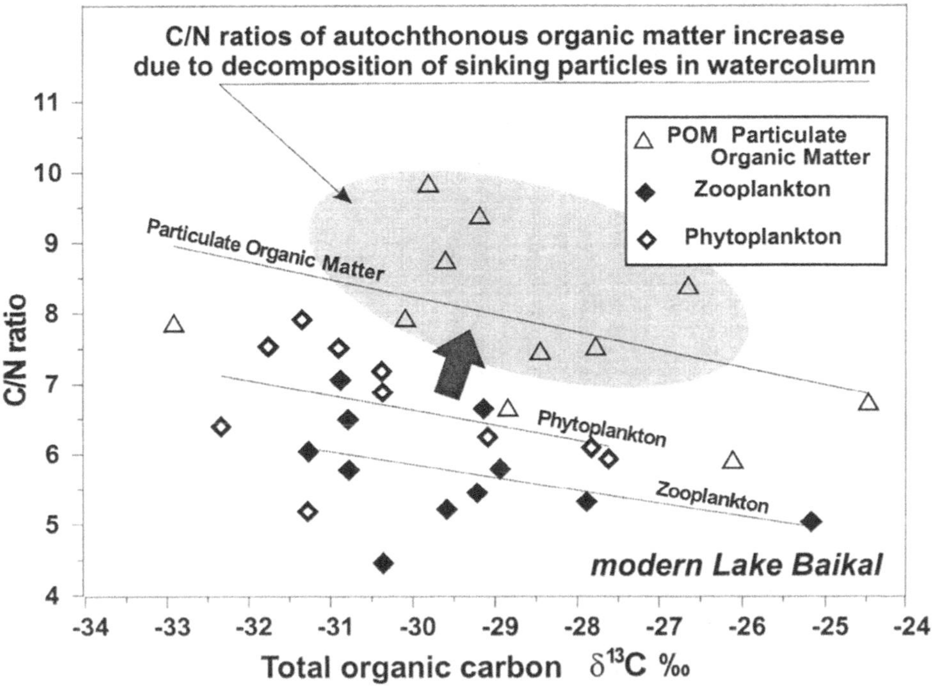

FIG. 5. Carbon stable isotope composition ($\delta^{13}C$) versus C/N ratios in phytoplankton, zooplankton, and particulate organic matter (*POM*). Note that for the entire range of autochthonous $\delta^{13}C$ values, the C/N ratios of POM (*shading*) are systematically higher as a result of degradation of autochthonous OM. *Lines* show linear best fits for each of the datasets

efficiency of carbon cycling in Lake Baikal. For instance, Votintzev (1961b) observed that the OM content in the photic zone waters unexpectedly decreases drastically with the onset of phytoplankton bloom (Fig. 6). He attributed this observation to "photosynthetic aeration," i.e., to high oxygen production, which literally "burns out" organic matter (Votintzev 1961b).

The high oxygen content of Baikal waters is indeed a favorable factor for OM decomposition. Water-column profiles reveal evidence for a deep degradation of primary produced organic matter as it sinks into deeper layers, as shown by $\delta^{13}C$ and C/N values of POM. Below approximately 100 m (the thermocline depth), the loss of nitrogen-rich labile compounds results in a C/N increase of 3–4 units, and the preferential loss of ^{12}C results in a corresponding increase in $\delta^{13}C$ by 3‰–4‰ towards heavier values (Fig. 7). This is observed throughout Lake Baikal, as shown by sampling stations in both the Selenga area and northern Baikal.

It may therefore be concluded that although the initial composition of the autochthonous source of Baikal OM is fairly typical, it changes significantly due

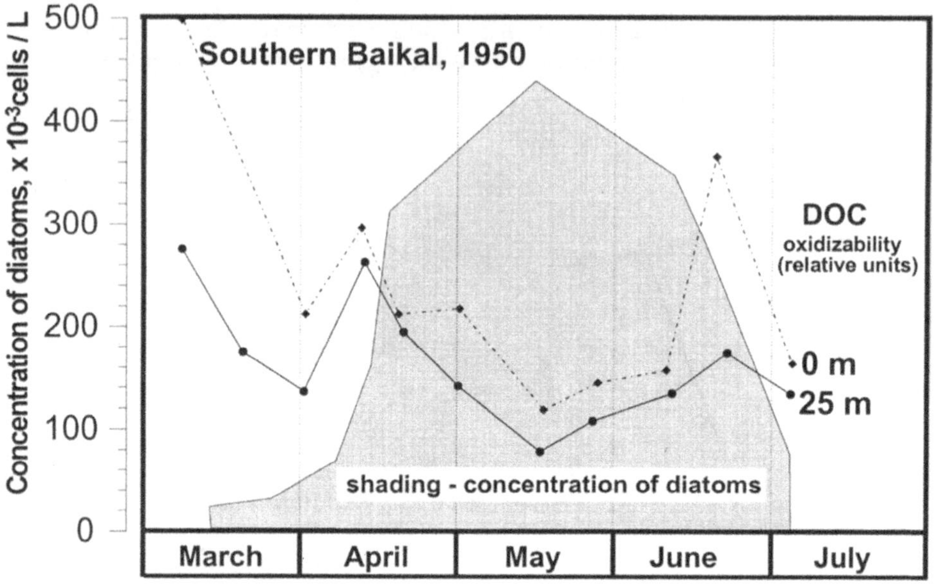

FIG. 6. The phenomenon of "photosynthetic aeration" in Lake Baikal, reflecting increased rates of OM degradation in the photic zone due to high oxygen production by phytoplankton (modified from Votintzev 1961b). The oxidizability as a function of the concentration of dissolved organic carbon (*dashed line*, 0 m depth; *solid line*, 25 m depth) is inversely related to the concentration of diatoms (*shading*)

to decomposition within the water column. By the time the autochthonous flux reaches the lake bottom, the composition of primary produced OM is likely to be significantly shifted towards higher C/N and heavier $\delta^{13}C$. In the following section we compare the compositions of primary produced POM and allochthonous OM from different sources in order to determine whether the latter comprise a significant flux of carbon to bottom sediments.

7 Organic Matter in Soils, Stream Sediments, and Bottom Sediments of Lake Baikal

The composition of soil OM in the Baikal region (Fig. 8) is representative of a wide range of landscapes, including steppes, tundra, taiga, and alpine meadows. For every soil section, samples from each of the soil horizons were taken. Soil profiles were sampled along the following lateral profile: soil of the watershed topographic high, slope soil, alluvial plain soil, and the fine fraction of stream sediments (Koval et al. 1993). Despite the diversity of landscapes, the soil OM analyzed has a rather narrow range of $\delta^{13}C$ values both in terms of soils of different types and within each of the soil sections. The C/N values are scattered

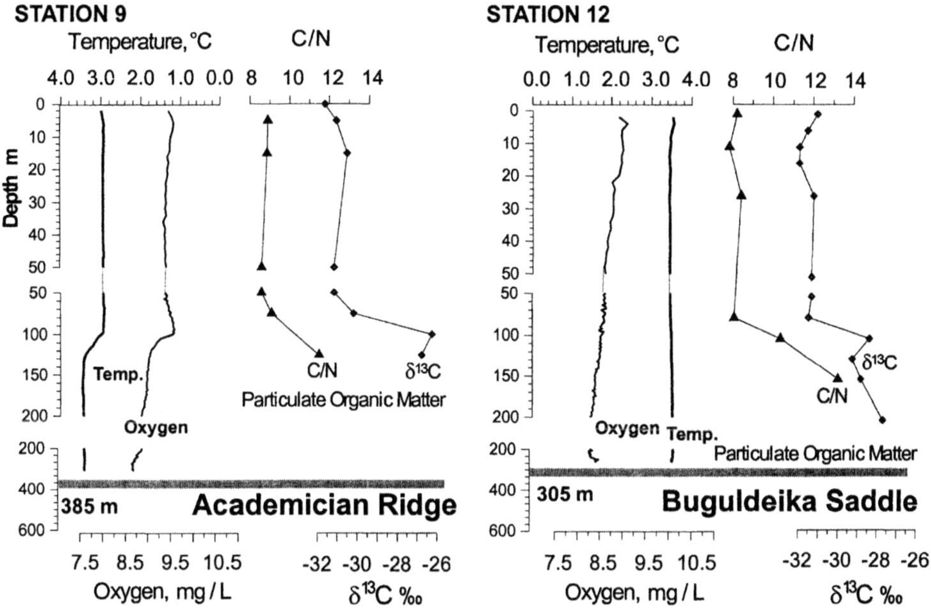

FIG. 7. Water column profiles of $\delta^{13}C$ and C/N at Academician Ridge (station 9) and in the Buguldeika area (station 12) of Lake Baikal (Prokopenko 1997). The dramatic shifts in both organic indices below the thermocline result from the decomposition of sinking organic matter

more widely (Fig. 8). The average values for soil OM are $\delta^{13}C = -25.8‰$ (SD 0.4‰) and C/N = 15.5 (SD 1.82).

The relatively uniform isotopic composition of soil OM presents an argument regarding possible changes in the paleoabundance of C3 and C4 plants in the Lake Baikal area. It has been suggested that during glacial periods, cold steppes would have supplied isotopically heavy C4 plant material into Lake Baikal (Qiu et al. 1993; Meyers and Ishiwatari 1995). However, the soil profiles studied did not reveal the presence of isotopically heavy OM either in steppe or in alpine or forest soils, suggesting that a C3/C4 change in the Baikal watershed might not have taken place. This conclusion is supported by the modern distribution of C4 plants, which shows that they are limited by low temperatures, and do not extend north further than Gobi in Mongolia. Because of a dramatic depression of temperatures during glacial periods, it is highly unlikely that C4 plants could have become more abundant around Baikal than they are today.

The $\delta^{13}C$ and C/N composition in fine sediments from Lake Baikal tributaries, sampled at their discharge sites, may be used to characterize the average composition of the riverine OM source. It is of interest to compare the riverine OM with that of soils and bottom sediments in order to assess to what extent OM in Baikal sediments is representative of these two end members of allochthonous flux. As seen in Fig. 8, the range of C/N values is essentially the same as in soils

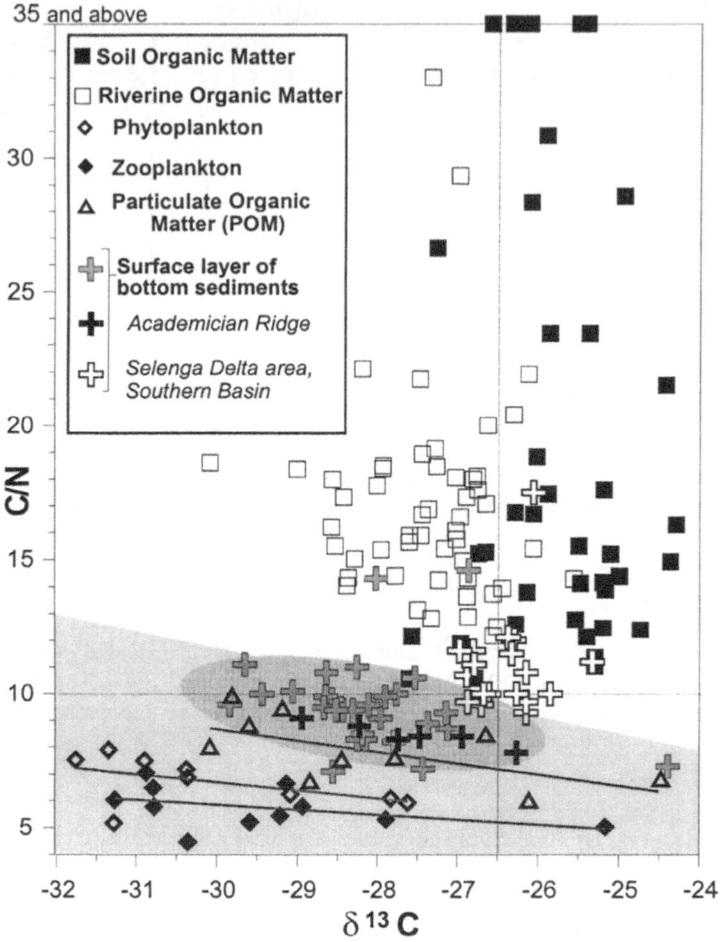

FIG. 8. A comparison of the $\delta^{13}C$ and C/N ratios of OM from different sources in Lake Baikal with those in surface layer of bottom sediments (Williams et al. 1993; Prokopenko 1997). The composition of sediment samples, especially those from Academician Ridge, corresponds to the typical range of modern autochthonous POM except for a cluster of points from the Selenga area and the southern basin. Note the clear distinction in $\delta^{13}C$ composition between soil and riverine OM

(C/N = 17.3; SD 4). The riverine $\delta^{13}C$ values are on average about 1‰–2 ‰ lower, suggesting higher contents of labile, isotopically light fresh OM as compared with the degraded OM in soils. The average isotopic value for riverine OM ($\delta^{13}C$ = –27.15‰; SD 1.38) is, in fact, identical to the average isotopic composition for terrestrial C3 plants (Deines 1980). This observation further supports the conclusion about the presence of relatively undegraded OM in the riverine sediments analyzed.

The assumption that the average composition of riverine OM adequately represents the allochthonous organic matter flux into the lake implies that deep oxidation of terrigenous OM takes place within the water column and at the sediment/water interface in Lake Baikal (Prokopenko 1997). This explanation is consistent with the balance estimates, indicating that allochthonous OM is largely assimilated in Baikal. As with the soil dataset, the tributary dataset does not reveal isotopically heavy organic carbon, thus arguing against a discernible contribution of C4 plants to soil and stream TOC.

The distributions of TOC content, C/N ratio, and $\delta^{13}C$ values for the surface layer of Lake Baikal sediments, based on a set of over 50 box and gravity core samples, have been discussed previously (Williams et al. 1993). We use this dataset here (see Fig. 8) to discuss the allochthonous versus autochthonous nature of sedimentary OM. Most data points for the surface layer of bottom sediments in Fig. 8 fall within the shaded $\delta^{13}C$—C/N range of modern POM within the water column. The exception is the cluster of data points with higher C/N and $\delta^{13}C$ from the Selenga Delta area, which apparently have a terrigenous imprint (Prokopenko 1997).

The important conclusion to be drawn from this distribution is that there is no need to involve a significant contribution of allochthonous OM in order to explain the composition of modern bottom sediments, because in general, the latter corresponds to the presently observed composition of the autochthonous POM. For instance, when discussing the composition of Baikal sediments, Colman et al. (1996) interpreted C/N values equal to 13.7 as indicating the presence of 30% allochthonous OM. In another study, C/N values of 13.32 were interpreted as showing 45% of terrestrial OM in sediments (Ishiwatari et al. 1992). Such interpretations are unlikely, mainly because the end-members used in these calculations are not representative of Lake Baikal OM sources. The C/N ratios of 7.4 (Ishiwatari et al. 1992) or 6.4 (Toyoda et al. 1993) used to characterize autochthonous OM in sediments are excessively low because they do not account for any degree of degradation of buried OM, which cannot be true in the oxygen-saturated, deep Lake Baikal.

When applied to sedimentary sections, the conclusion regarding the dominance of autochthonous OM in sediments implies that sedimentary $\delta^{13}C$, TOC, and C/N records reflect mostly internal changes in the lake's carbon cycle driven by climate change. This is a useful vantage point for looking at these signals.

8 Lake Baikal Organic Carbon Record of the Past 130Ka

The conclusion that Lake Baikal sedimentary records from Academician Ridge are dominated by the flux of autochthonous OM is clearly supported by the close correlation of TOC and total nitrogen (TN) with BioSi and diatom abundance profiles in core BDP-96-2 during the last climatic cycle (Fig. 9). The TOC profile shows all the climatic–stratigraphic intervals which are known for Lake Baikal from the correlation of BioSi profiles with marine oxygen isotope stages (MIS):

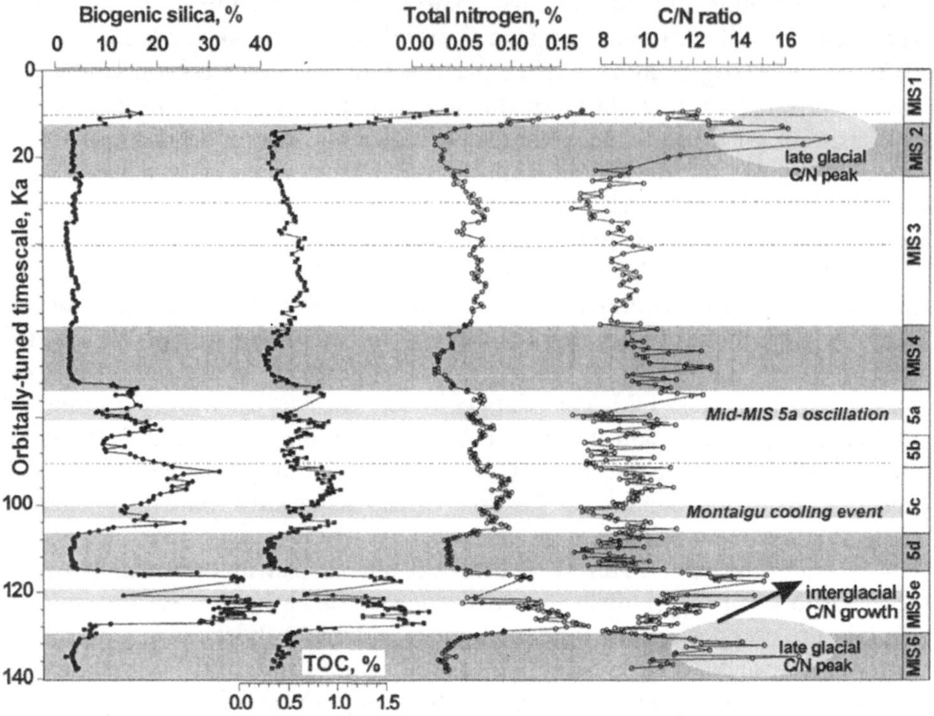

FIG. 9. Biogenic silica, total organic carbon, total nitrogen, and C/N records in the upper section of the BDP-96-2 drill core, encompassing the past 130 Ka. *Shading* marks glacial intervals, *white areas* are millennial cooling events. The distinct interglacial pattern of C/N increase during the last interglacial is a likely indication of the increasing efficiency of carbon export from the photic zone via the Baikal "biological pump." Dramatic glacial/interglacial variations in the C/N index reflect the switching between different modes of carbon cycling in Lake Baikal. A distinct feature of the C/N record is the repetition of deglacial C/N peaks at both MIS 6/5 and MIS 2/1 major glacial–interglacial transitions

(1) the TOC maximum during the last interglacial MIS 5e; (2) the abrupt decrease in TOC during the regional MIS 5d glaciation; (3) the increased TOC accumulation during warm MIS 5c and MIS 5a intervals, punctuated by short-term millennial cooling events; (4) the long period of low Baikal productivity during MIS 4-2; (5) the TOC increase during the last deglaciation, the MIS 2/1 transition (Fig. 9). The close correlation of the productivity indices implies that the bulk of OM preserved in the sediments was exported from the photic zone with the biogenic diatom flux.

The exception to the close correlation of the OM and BioSi records is the MIS 3 interstadial, in which diatom abundance and BioSi are very low, whereas TOC and TN form a well-defined peak (see Fig. 9). This divergence might be due to the fact that primary producers other than diatoms could be developing during glacials. For instance, even in the present-day Baikal during years which are unproductive for the spring assemblage of large diatoms, a development of abundant picoplankton occurs, especially in northern Baikal, where the phytoplankton production is generally lower (see Fig. 3) (Popovskaya 1987). The picoplankton consists of the typical Baikal summer assemblage (Nagata et al. 1994), and accounts for the apparent discrepancy between high productivity and low biomass in summer (Popovskaya 1987).

The profile of C/N ratios during the past 130 Ka reveals an interesting pattern. On the one hand, C/N appears to reflect a general climate-driven cyclicity with lower ratios during glacials and higher ratios during interglacial and interstadial periods (see Fig. 9). On the other hand, however, the C/N index behaves differently from the TOC and TN profiles, which seem to essentially parallel each other. The distinct feature of the C/N profile is the progressive increase during the last interglacial (MIS 5e) (see Fig. 9). This C/N growth is similar to that observed during the Holocene (see Fig. 2a–c). Instead of interpreting these C/N increases as merely a mechanical mixture of OM from different sources, we hypothesize that this consistent interglacial pattern of progressive C/N growth reflects a dynamic process of an increasing efficiency of carbon export from the photic zone via Baikal's "biological pump." This interpretation is in agreement with the observation that the composition of modern Baikal sediments corresponds to that of autochthonous POM (see Fig. 8). It is quite likely that a long-term effect of accumulating nutrients and increasing the overall carbon pool of the lake, as a result of assimilation and recycling of allocthonous OM during interglacials, leads to increasingly higher productivity and to more efficient carbon pumping to the bottom sediments. In this scenario, the repetitive pattern of progressive basin-wide C/N growth during the Holocene and during the last interglacial (see Fig. 2a–c, Fig. 9) gives an insight into the different operational modes of the Baikal lacustrine system under different climatic regimes.

Another distinct feature in the C/N profiles of both the MIS 2/1 and the MIS 6/5 transitions is the late glacial C/N peak (see Fig. 9). The recurrence of this C/N signal suggests that this peak is not likely to be an artifact, but the origin of this feature is unclear at the moment. These peaks do not appear to signal an allochthonous input because the corresponding TOC contents remain at low glacial values (see Fig. 9). The deglacial C/N peaks might reflect changes in OM cycling or preservation, and thus warrant further investigation. Interestingly, at both climatic transitions, C/N values returned to a "normal" lower range by the beginning of both the last and the present interglacial diatomaceous intervals (see Fig. 9).

9 Glacial and Interglacial Carbon Cycles in Lake Baikal

A comparison with the carbon cycle of the world ocean allows us to take a broad view of Baikal's carbon cycle. In the world ocean, the roles of primary production of phytoplankton, the phytobenthos, and the input of allochthonous

OM are 94.7%, 0.5%, and 4.8%, respectively, of the average annual OM balance (Romankevich 1977). For Baikal, these values are 91%, 0.6%, and 6.7%, respectively (Votintzev and Popovskaya 1973). The average dissolved organic carbon content in the world oceans is 1.3 mg/l (Romankevich 1977), and it is 1.02–1.5 mg/l in Lake Baikal (Votintzev 1985). The ratio of 100:1.7:0.16 in the pyramid of carbon masses in the world ocean was suggested as a planetary constant reflecting the dynamic equilibrium of dissolved, particulate and living OM (Romankevich 1977). In Baikal, this ratio may be very roughly estimated as ca. 100:6:1 based on different sources (Popovskaya 1967; Vykhristyuk 1980; Tarasova and Mesheryakova 1992). As can bee seen from these relationships, the role of benthos and the relative parts of POM and living OM in the modern carbon balance of Lake Baikal are currently shifted toward higher values. It is quite likely that this "excess" should be attributed to a progressive long-term trend in the accumulation of assimilated OM and nutrients originating from the watershed. The behavior of the C/N index during interglacials in sedimentary record appears to support this notion. Kozhov (1963) and Votintzev (1985) also noted that such long-term cumulative effects of allochthonous input of nutrients could affect the Lake Baikal ecosystem.

In general, the rate of OM decomposition within the Baikal water column (95.5% of the total, according to Votintzev and Popovskaya 1973) is the same as in the oceans (92%–97%, according to Romankevich 1977), and demonstrates the high efficiency of carbon cycling. The available data on decomposition at the sediment/water interface suggest a relative "excess" accumulation of OM in Baikal sediments. For instance, decomposition at the bottom of the oceans is as efficient as within the photic zone (92%–97% of POM flux), whereas in Baikal it is lower (about 60% according to Vykhristyuk (1980) and Votintzev (1985)). As with the "excess" in particulate and living OM, excess carbon accumulation in Baikal must be supported by the assimilation of allochthonous OM imported from the watershed. This conclusion is consistent with calculations of the carbon dioxide balance and CO_2 release from present day Lake Baikal (Mizandrontzev and Mizandrontzeva 1994, personal communication, 2000), and may also explain the higher Baikal fossilization coefficient as compared with oceanic values of 0.1%–0.7% (Romankevich 1977). The Baikal organic carbon fossilization coefficient, calculated as

$$F = \frac{\text{TOC}\,(\%) \times \text{SedRate (tons/km}^2)}{100 \times \text{Production (autochthomous} + \text{allochthomous) (tons/km}^2)}$$

equals 1.2% of the total annual OM input (Vykhristyuk 1980). Although elevated relative to oceanic ratios, this value is still very low for inland water bodies. For example, this coefficient for the Caspian Sea is 5% (Vykhristyuk 1980). Using more recent data of an average TOC of 2% (Qiu et al. 1993; Prokopenko 1997), dry bulk density of 0.3 g/cm^3 (Colman et al. 1994), and average sedimentation rate of 25 cm/Ka during the Holocene (Colman et al. 1996), we arrived at a similar estimate of 1.1%.

It is difficult to estimate the Baikal fossilization coefficient for the last glacial period without a reliable estimate of production changes. Nevertheless, using a dry bulk density of 0.9 g/cm^3 for glacial clays at Academician Ridge, an average sedimentation rate of 4 cm/Ka, and TOC of 0.3% (from the same sources) in the equation above yields a very low F of 0.25%. This estimate for Academician Ridge, although highly tentative, compares with $F = 0.1\%$ in red clay of the abyssal Pacific (Romankevich 1977). This exercise illustrates that the Baikal carbon cycle during glacials was very different from the modern one. Climate-driven changes appear to have resulted in a complete restructuring of the lake's carbon cycle and trophic relationships, which was potentially catastrophic for pelagic biota of Baikal, as indicated by numerous extinctions of dominant diatom species during the past 800 Ka (Khursevich et al. 2001). During glacials, the terrigenous OM source was suppressed owing to widespread permafrost and the low productivity of land vegetation. The lack of Baikal carbon replenishment from external sources served to further amplify the rapid switching of carbon cycling in Lake Baikal from interglacial to glacial modes. Such dramatic and rapid switching is reflected in productivity proxy records, for instance, at MIS 5e/5d and MIS 5/4 transitions (see Fig. 9).

Evidence supporting these deduced dramatic changes is found in $\delta^{13}C$ records (Fig. 10). Under glacial low-carbon conditions, Baikal's carbon cycle apparently becomes very sensitive to changes in the carbon input from external sources, as indicated by strong deglacial carbon isotope shifts in sedimentary records throughout Baikal (Fig. 10). These dramatic negative $\delta^{13}C$ shifts, paralleled by growing atmospheric pCO_2 levels (Prokopenko et al. 1999), indicate significant changes in the Baikal carbon isotope inventory, and signal the higher availability of isotopically light carbon to Baikal primary producers. During glacials,

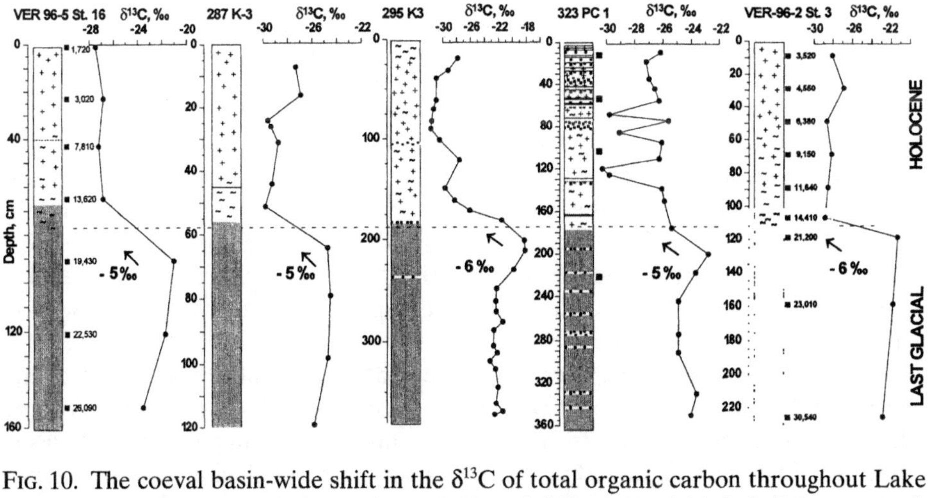

FIG. 10. The coeval basin-wide shift in the $\delta^{13}C$ of total organic carbon throughout Lake Baikal during the last deglaciation in cores Ver 96-5 St 16, Ver 96-2 St 3 (Minoura et al. 1999), 287 K3 (Prokopenko 1997), and 323 PC1 (Toyoda et al. 1993)

distinctly positive $\delta^{13}C$ compositions indicate a dramatic reduction in the exchangeable carbon pool of Lake Baikal.

10 Conclusions

The Lake Baikal sedimentary records of TOC reflect past paleoproductivity changes. These records closely correlate with biogenic silica and diatom abundance, suggesting that during interglacial periods, organic matter is produced primarily by diatoms. Balance estimates and a comparative study of the Baikal OM from different sources, including terrestrial soil and riverine OM, and lacustrine planktonic and particulate organic matter, suggest that sedimentary OM is supplied primarily from an autochthonous planktonic source. The C/N ratio of autochthonous OM exported from the photic zone in Lake Baikal progressively increases with depth owing to degradation. By the time OM reaches the bottom sediments, the C/N ratios may significantly exceed typical planktonic ratios and reach values of 10–12. These observations constrain previous estimates of the fraction of terrestrial OM in sediments, and offer new ways of interpreting the nature of the systematic glacial/interglacial C/N changes observed in sedimentary records.

The TOC content and C/N ratios of sedimentary OM progressively increase during the last deglaciation and during the Holocene, reflecting the overall increase in carbon supply on the one hand, and the more effective export of carbon from the trophogenic layer on the other, most likely owing to the increasing efficiency of Baikal's biological pump. This pattern is not only characteristic of the last glacial/interglacial transition; it is systematically repeated with every glacial/interglacial transition during the past 130 Ka. This repetitive pattern implies that the carbon cycle of Lake Baikal has distinctly different operational modes under different climatic regimes.

Organic carbon records imply dramatic glacial/interglacial shifts in the trophic state of Lake Baikal from nearly present-day interglacial conditions with several mesotrophic characteristics to a glacial ultraoligotrophic state. These changes result in a drastic decrease in carbon export from the trophogenic layer by an order of magnitude. Sedimentary records reflect a complete restructuring of the Baikal carbon cycle and of the lake's trophic relationships, which was potentially catastrophic for the pelagic biota of Baikal. The lack of Baikal carbon replenishment from external sources during glacial periods further amplified the overall impact of climatic changes on the rapid switching between modes of carbon cycling in Lake Baikal.

Under glacial conditions, the carbon pool of Lake Baikal was reduced dramatically. This carbon deficit apparently made the Baikal system sensitive to changes in the carbon input from external sources. The transition to the present interglacial was characterized by a strong, basin-wide, deglacial, negative carbon isotope shift, which appears to parallel the atmospheric pCO_2 increase as recorded in polar ice core records. This isotopic shift indicates significant changes in the Baikal isotope inventory, and signals the higher availability of isotopically light carbon to Baikal primary producers.

The observations presented in this chapter provide a general framework, albeit incomplete, for understanding glacial/interglacial changes in the carbon cycle of Lake Baikal, and will be useful for future studies aimed at linking lacustrine paleoproductivity proxy records with regional climate change.

Acknowledgments. This work was implemented as part of the Baikal Drilling Project, and was supported by US National Science Foundation grants EAR-9119537, EAR-93-1720401, and EAR-9614770, by the Siberian Branch of the Russian Academy of Sciences, and by the Department of Geological Sciences, University of South Carolina. We thank the scientific drilling team of Nedra Enterprise, the crew of R/V *Ulan-Ude*, and the scientists of the Institute of Geochemistry, the Limnological Institute, and the Institute of the Earth Crust (Irkutsk) who participated in primary descriptions and sampling of the BDP drill cores. We thank Dr. Kashiwaya for his leadership and foresight in organizing this special volume.

References

BDP Members (1997) A continuous record of climate changes for the last five million years from the bottom sediment of Lake Baikal. Russ Geol Geophys 39:135–154

Bradbury JP, Bezrukova EV, Chernyaeva GP, Colman SM, Khursevich GK, King J, Likhoshway Y (1994) A synthesis of post-glacial diatom records from Lake Baikal. J Paleolimnol 10:231–251

Cole GC (1988) Textbook of limnology. Waveland Press, Prospect Heights, 401 p

Colman SM, Carter SJ, Hatton J, Haskell BJ (1994) Cores collected in Lake Baikal, Siberia, by USGS, 1990–1992: visual descriptions, photographs, X-radiographs, bulk-density measurements, and grain size analyses. USGS Open-File Report 94–445

Colman SM, Jones GA, Rubin M, King JW, Peck JA, Orem WH (1996) AMS radiocarbon analyses from Lake Baikal, Siberia: challenges of dating sediments from a large, oligotrophic lake. Q Sci Rev 15:669–684

Deines P (1980) The isotopic composition of reduced organic carbon. In: Fritz P, Fontes J (eds) Handbook of environmental isotope geochemistry. Elsevier, Amsterdam, p 329–406

Horiuchi K, Minoura K, Hoshino K, Oda T, Nakamura T, Kawai T (2000) Paleoenvironmental history of Lake Baikal during the last 23000 years. Palaeogeogr Palaeoclimatal Palaeoecol 157:95–108

Ishiwatari R, Uzaki M, Yamada K, Ogura K (1992) Organic matter records of environmental changes in Lake Baikal sediments. Carbon isotopes, organic carbon and nitrogen. In: Horie S (ed) IPPCE Newsletter 6. Universitätsverlag Wagner, Innsbruck, p 80–88

Jorgensen SE (1979) Lake management. Pergamon, New York, 167 p

Khursevich GK, Karabanov EB, Prokopenko AA, Williams DF, Kuzmin MI, Fedenya SA, Gvozdkov AN (2001) Insolation regime in Siberia as a major factor controlling diatom production in Lake Baikal during the past 800000 years. Q Int 80–81:47–58

Koval PV, Kitaev NA, Williams DF, Grebenschikova VI, Romanov VA (1993) Stable isotope composition of carbon in soils and bottom sediments of river valleys in Prebaikalia. Russ Geol Geophys 34:186–194

Kozhov MM (1963) Lake Baikal and its life. Junk, The Hague, 344 p

Kozhova OM (1967) Primary production of plankton and the conditions of its formation in the system Baikal–Bratsk–Irkutsk reservoirs. In: Galazyi GI (ed) Cycling of matter and energy in lakes (in Russian). Nauka, Moscow, p 223–228

Likens GE (1975) Primary production of inland aquatic ecosystems. In: Lieth H, Whittaker RH (eds) Primary productivity of the biosphere, vol 14. Springer, New York, p 185–202

McKenzie JA (1985) Carbon isotopes and productivity in the lacustrine and marine environment. In: Stumm W (ed) Chemical processes in lakes. Wiley, New York, p 99–118

Meyers PA, Ishiwatari R (1995) Organic matter accumulation records in lake sediments. In: Lerman A, Imboden DM, Gat JR (eds) Physics and chemistry of lakes. Springer, Berlin, p 279–328

Minoura K, Oda T, Nakamura T, Tanchi C, Sakai H, Kawai T (1999) Estimation of the loss from the upper layer of sediments in the cores from boreholes BDP-96-1 and BDP-96-2 (Lake Baikal). Russ Geol Geophys 40:1262–1268

Mizandrontzev IB, Mizandrontzeva KN (1994) Partial pressure and the exchange of free CO_2 in the system water–atmosphere in Lake Baikal (in Russian). Vodnye Resursy 21:72–79

Nagata T, Takai K, Kawanobe K, Kim DS, Nakazato R, Guselnikova N, Bondarenko N, Mologawaya O, Kostrnova T, Drucker V, Satoh Y, Watanabe Y (1994) Autotrophic picoplankton in southern Lake Baikal: abundance, growth and grazing mortality during summer. J Plankton Res 16:945–959

Popovskaya GI (1967) Baikal phytoplankton and its role in the production of autochthonous organic matter. In: Galazyi GI (ed) Cycling of matter and energy in lakes (in Russian). Nauka, Moscow, p 216–222

Popovskaya GI (1987) Phytoplankton of the deepest lake in the world (in Russian). Proceedings of the Zoology Institute, 172. AN SSSR, Leningrad, p 107–116

Prokopenko AA (1997) Stable isotopic composition of total organic carbon in bottom sediments of Lake Baikal as a proxy for paleoclimatology. PhD thesis, University of South Carolina, 279 p

Prokopenko AA, Williams DF, Karabanov EB, Khursevich GK (1999) Response of Lake Baikal ecosystem to climate forcing and pCO_2 change over the last glacial/interglacial transition. Earth Planet Sci Lett 172:239–253

Prokopenko AA, Karabanov EB, Williams DF, Kuzmin MI, Shackleton NJ, Crowhurst SJ, Peck JA, Gvozdkov AN, King JW (2001) Biogenic silica record of the Lake Baikal response to climatic forcing during the Brunhes chron. Q Res 55:123–132

Qiu L, Williams D, Gvozdkov A, Karabanov E, Shimaraeva M (1993) Biogenic silica accumulation and paleoproductivity in the northern basin of Lake Baikal during the Holocene. Geology 12:25–28

Rau GH, Takahashi T, Des Marais DJ (1989) Latitudinal variations in plankton $\delta^{13}C$: implications for CO_2 and productivity in past oceans. Nature 341:516–518

Romankevich EA (1977) Geochemistry of organic matter in the ocean (in Russian). nauka, Moscow, 256 p

Tarasova EN, Mesheryakova AI (1992) Present-day state of the hydrochemical regime of Lake Baikal (in Russian). Nauka, Novosibirsk, 141 p

Toyoda K, Takemura K, Ishiwatari R (1993) Comprehensive investigation of Lake Baikal core 323 PC-1. Russ Geol Geophys 34:101–110

Votintzev KK (1961a) Hydrochemistry of Lake Baikal (in Russian). Nauka, Moscow, 311 p

Votintzev KK (1961b) On the role of photosynthetic aeration in organic matter cycling in Lake Baikal (in Russian). Dokl AN SSSR 139:211–213

Votintzev KK (1985) Main features of the hydrochemistry of Lake Baikal (in Russian). Vodnye Resursy 2:43–55

Votintzev KK, Popovskaya GI (1967) Annual autotrophic production of phytoplankton of Lake Baikal (in Russian). Dokl AN SSSR 174:419–422

Votintzev KK, Popovskaya GI (1973) On the cycle of organic matter in Lake Baikal. Cycle of matter and energy in lakes and reservoirs (in Russian). Nauka, Novosibirsk, p 75–77

Votintzev KK, Glazunov IV, Tolmachyova AP (1965) Hydrochemistry of rivers of the Lake Baikal basin (in Russian). Nauka, Moscow, 495 p

Votintzev KK, Mesheryakova AI, Popovskaya GI (1975) Cycle of organic matter in Lake Baikal (in Russian). Nauka, Novosibirsk, 189 p

Vykhristyuk LA (1980) Organic matter of bottom sediments of Lake Baikal (in Russian). Nauka, Novosibirsk, 79 p

Weiss RF, Carmack EC, Koropalov VM (1991) Deep-water renewal and biological production in Lake Baikal. Nature 349:665–669

Williams DF, Jenkins PB (1993) The Holocene–late Quaternary record of sedimentary organic matter in cores from Academician Ridge, Lake Baikal. Russ Geol Geophys 34:76–83

Williams DF, Qiu L, Karabanov E, Gvozdkov A (1993) Geochemical indicators of productivity and sources of organic matter in surficial sediments of Lake Baikal. Russ Geol Geophys 34:111–125

Williams DF, Peck JA, Karabanov EB, Prokopenko AA, Kravchinsky V, King J, Kuzmin MI (1997) Lake Baikal record of continental climate response to orbital insolation during the past 5 million years. Science 278:1114–1117

Lake Hovsgol in the Late Pleistocene and Holocene: On-Land Geological Evidence for a Change in Its Level

SERGEY K. KRIVONOGOV[1], ELENA V. BEZRUKOVA[2], HIKARU TAKAHARA[3], and FRANK RIEDEL[4]

Summary. We present geomorphological and sedimentological evidence for a change in the level of Lake Hovsgol during the second half of the late Pleistocene and Holocene. Both transgressive and regressive stages of lake evolution have been deduced from a study of lacustrine, river, slope, and glacial sediments and landforms. The events which caused the lake level to rise took place in the Karginian,* late Sartanian, early Holocene, and recent time, and were a result of warming periods. The lowering of the level resulted from periods of glaciation and the middle Holocene arid stage.

The maximum water level in the lake was controlled by the top of the Egiin-gol river spillway, and in the latest Pleistocene and Holocene it was 6 and 2 m higher than the present-day level, respectively.

We suggest that all the changes in the lake level are related to one geomorphological stage. The stair-like on-land terraces possibly indicate the older stages of the lake's evolution, which took place at other hypsometric levels.

Key words. Lake Hovsgol, Lake levels, Geomorphology, Sedimentology, Remote sensing

---

[1] United Institute of Geology, Geophysics and Mineralogy Siberian Branch of the Russian Academy of Sciences, Koptyug Ave., 3, 630090 Novosibirsk, Russia

[2] Limnological Institute Siberian Branch of the Russian Academy of Sciences, Ulanbatorskaya str. 3, Irkutsk 664033, Russia

[3] Kyoto Prefectural University, 1-5 Hangi-cho, Shimogamo, Sakyo-ku, Kyoto 606-8522, Japan

[4] Institute of Geological Sciences, Free University of Berlin, Malteserstrasse, 74-100 Haus D, D-12249 Berlin, Germany

* Members of the Siberian stratigraphic scheme: Kazanian (Kazantsevo) = Eemian; early Zyryanian (Ermakovian) = early Weichselian; middle Zyryanian (Karginian) = middle Weichselian; late Zyryanian (Sartanian or Sartan) = late Weichselian

1 Introduction

The depression of Lake Hovsgol in central–northern Mongolia is at the southwestern extremity of the Baikal rift zone. The evolution of the lake is similar to that of Lake Baikal. Both basins are still tectonically active structures. The results of deep-water drilling showed that Lake Baikal has existed as a deep lake for more than 12 Ma (Kawai et al. 2001). The structure of the deposits in Lake Hovsgol has been studied only by geophysics. A comparison of measured thicknesses of the sediments in the Baikal, Tunka, and Hovsgol depressions led to a suggestion that in Hovsgol they are no younger than the Pliocene (Zorin et al. 1989). The youngest basalts were formed in the prerift stage of the evolution of the Hovsgol area at 8–9 Ma (K–Ar dating) (Ivanenko et al. 1989; Shuvalov and Nikolaeva 1989). It was suggested that the Hovsgol area of the Baikal rift zone is much younger than the main part. Support for this comes from a comparison of their tectonic-topography features (Zolotarev et al. 1981) and shore landforms (Rogozin 1993).

The early stage of the development of the Hovsgol basin is shown by high terraces at elevations of 80–85, 42–45, and 34–36 m. These are older than the middle Pleistocene (Zolotarev et al. 1982), and possibly indicate the formation of the Hovsgol graben. Terraces of the middle Pleistocene age have been discovered at depths of 90–120, 50–80, and 20–40 m (Zolotarev et al. 1982). The authors refer the lowering of the lake level to glaciation. In the late Pleistocene, the level of the lake surface rose, and a stair of terraces at 24–26, 18, 12, and 5 m was formed. The highest terrace (IV) is of Kazanian age (Belova and Kulakov 1982). The Holocene terrace (I) is the closest to the lake shore, and consists of a series of beach ridges up to 3–4 m high (Rogozin 1993).

Unfortunately, the above data do not indicate the authors' opinion about the change in the lake level during the late Pleistocene and Holocene. A descending sequence of terraces, as described by the authors, can be a result of regression. River and lacustrine deposits may occur at different hypsometric levels in the terraces, but no correlations have been made.

In previous publications, the submerged terraces were referred to as from the middle Pleistocene (Zolotarev et al. 1982). This confidently suggests a subsidence of the lake, which could also have taken place during the late Pleistocene glaciation events, i.e., at least a part of the above terraces are younger than the middle Pleistocene.

According to the interpretation by Rogozin (1993), the level of the Hovsgol was higher in the first half of the Holocene than it is at the present time. The data obtained by Dorofeyuk and Tarasov (1998) on the Hatgal Bay are indicative of an essential regression of the lake in the middle Holocene.

Thus, the above-mentioned information does not allow an exact interpretation of the dynamics of the Hovsgol level during the late Pleistocene and Holocene times, as in the Baikal (Mats et al. 2000). One stumbling block is the question of whether the oscillations of the lake level resulted from climatic events, or were they controlled by the tops of the spillway: the Angara river headwaters for the Baikal and the Egiin-gol river headwaters for the Hovsgol.

2 Field Research in 1999–2001 at Lake Hovsgol

In order to carry out a geological investigation of the signatures for the change in Lake Hovsgol level, which are recorded in the outcrops and landforms of the shore, a field mission has started. The research area covers the southwestern, eastern, and northern shores of Lake Hovsgol. Figure 1 shows the location of the study sites, which are mainly accumulative landforms: river and lake terraces, beaches and beach ridges, alluvial cones, slopes, and eolian and permafrost formations at river mouths and shore bays. Moraines and other glacial deposits are found in the mountains around the northern and northwestern parts of the lake.

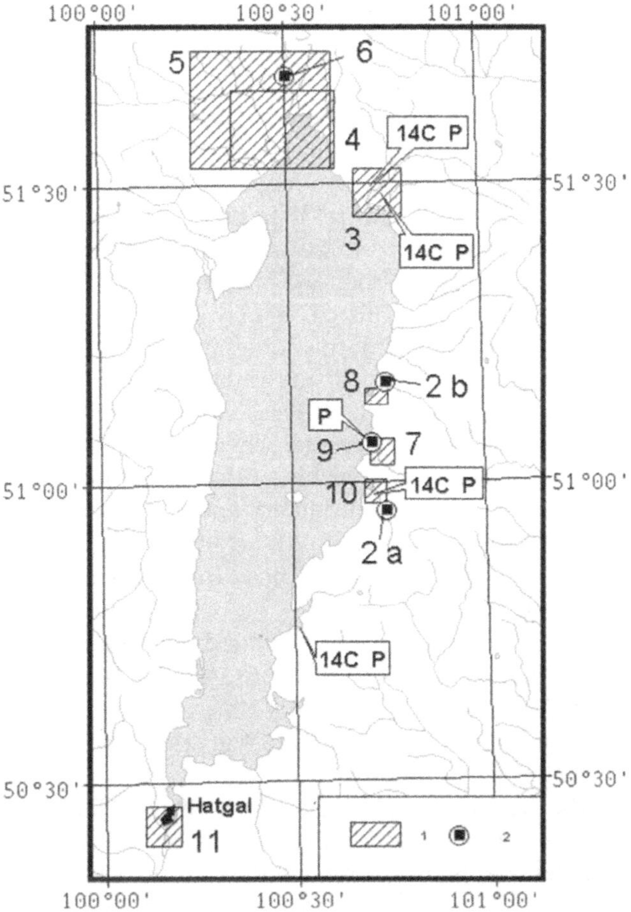

FIG. 1. Map of the area investigated. *Hatched areas* (*1*), frames of the maps; *black squares in a circle* (*2*), photographs shown in later figures. Framed labels show the locations of outcrops studied and their sampling: *14C*, radiocarbon; *P*, palynological. The parameters for all maps are: projection UTM, datum WGS-84, zone 50

These glacial deposits locate in the mouth of the Ih-Horoo-gol and the Munku–Sardyk range.

Five outcrops were sampled for palynological analysis, and four of them were sampled for radiocarbon dating (Fig. 1). Besides direct geological observations, we performed geomorphological interpretations with the help of Landsat-7 satellite images from 1999 and 2000.

3 Geomorphological Features and Sedimentological Evidence

3.1 River and Fluvial Terraces

Most rivers flowing into the Hovsgol from the east are 10–30 km long and have irregular annual runoff and low flowing power. There are only flood-plains in the valleys of these rivers, which meet slope deposits just near their sides (Fig. 2a). In the lower reaches, the flood-plains usually cut the loose slope deposits and forms scarps of variable height (Fig. 2b). Terraces are found only in the valleys of the two biggest rivers in the north, the Ih-Horoo-gol and the Hanh-gol.

Three morphological elements have been distinguished in the lower reaches of the Hanh-gol (Fig. 3): oblique plane, fluvial terrace, and flood-plain. The valley is mostly occupied by the oblique plane. This plain gradually descends to the axis of the valley; there 15- to 30-m-high scarps were formed due to erosion by the river. In the upper part, the scarps are composed of cover units: sands and silty sands with variably rounded clastic material. There are traces of soil and permafrost processes. This cover may possibly overlap the older elements of the valley. The evidence for this comes from the structure of a 19- to 20-m-high exposure, which is located in the right bank of the river, at its junction with the Hanh–Hatgal motor-road (Fig. 3, point *A*). In this exposure, the 7-m-thick cover is underlain by river deposits, which consist of flood-plain and channel sand-size facies of a 6-m visible thickness. Both flood-plain and channel facies contain the lenses of decayed organic matter and wood detritus, i.e., branches and thin trunks of trees. There are few, if any, localities containing abundant tree fragments in Mongolia. The presence of those fragments is indicative of climatic conditions favorable for vegetation.

The radiocarbon age of the flood-plain facies of a 19-m terrace is more than 45 Kyr BP (46710 ± 1030 yr BP SOAN-4303), obtained by the fraction of humic acids. This estimate suggests a Karginian age for this alluvium. This was previously proposed by Belova and Kulakov (1982) based on the results of palynological investigation.

The second morphological unit of the Hanh valley is a 10-m-high pebble terrace extending as a narrow belt along the left bank (Fig. 3, point *B*). As the pebbles are poorly sorted and rounded, we can regard them as alluvial fan deposits, and relate their formation to the periodic outflow of thaw water from

Fig. 2. **a** Morphology of the middle of the Bortsog-gol valley. **b** Erosion slopes of the Sevsuuliyn-gol valley in its lower reaches

a

b

the mountain glacier areas. We call these units fluvial terraces because their genesis is different from that of river terraces. Age of this terrace is of 28070 ± 965 yr BP (SOAN–4743).

The Holocene flood-plain step of the Hanh valley is 1–2 km wide and 1.5 m high. The flood-plain facies are represented by silty sands containing organic matter. The channel facies are sorted pebbles. The total thickness of alluvial deposits is about 5 m. The flood-plain merges into the coastal zone of the Hovsgol. The present-day outline of the river mouth was formed not very long ago: satellite images show a submarine continuation of the riverbed. This suggests a lower lake level in the Holocene stage.

The river valleys of the western shore of the Hovsgol meet extended alluvial cones. The formation of these cones resulted from the activity of much more powerful streams than the present-day ones. The time of their for-

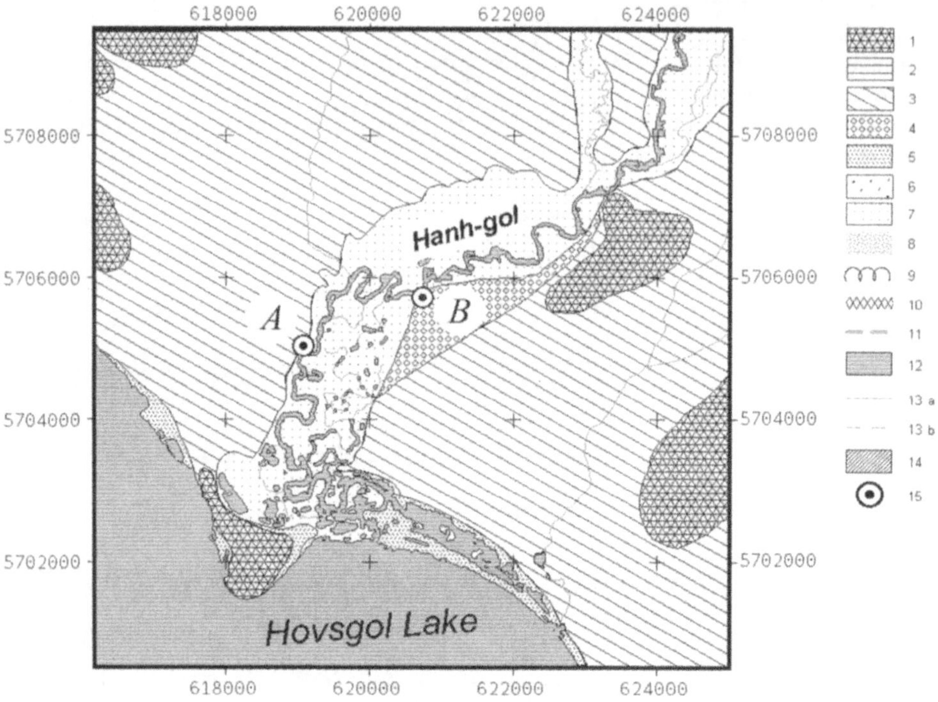

FIG. 3. Geomorphological map of the lower reaches of the Hanh-gol river. The key applies to Figs. 3, 7, 8, 10, and 11. *1*, mountain areas; *2*, ancient terrace-like steps, probably of lake genesis, older than Q_3^4; *3*, oblique plains, i.e., valley bottoms and slopes, Q_3^4; *4*, Proluvial (fan) terraces and cones, Q_3^4; *5*, former bottoms of bays, Q_4; *6*, Proluvial (fan) terraces and cones, Q_4; *7*, flood plains, Q_4; *8*, eolian sands, Q_3^4 and Q_4; *9*, beach ridges, Q_4; *10*, abrasion scarps, Q_4; *11*, reconstructed shorelines; *12*, Lake Hovsgol; *13*, streams, *a*, permanent, *b*, seasonal; *14*, Hatgal village; *15*, outcrops (labeled *A–F*)

mation should coincide with the age of a 10-m terrace of the Hanh-gol. The distal parts of the alluvial cones have been removed by lake abrasion. In such places, the 4- to 8-m-high cliffs show exposed cobblestones and pebbles of the alluvial cones. In the north, a series of inset fluvial or fluvioglacial terraces of the Ih-Horoo-gol valley could also have been formed during the last glaciation. They obviously record the outflow of glacier water and the termination of glaciation.

3.2 Glacial Formations

The late Pleistocene glaciation occupied the northern and northwestern mountain frames of the Hovsgol basin. The glacier source area was restricted by specific orographic features, except for the Ih-Horoo-gol valley.

3.2.1 Ih-Horoo-gol

The Ih-Horoo-gol is the largest river disgorging into the Hovsgol, and it is mainly fed by streams flowing down the Munku–Sardyk range. During glaciation periods, its valley contained large glaciers ending in the Hovsgol depression. A gigantic moraine complex , located in the Ih-Horoo-gol mouth, consists of two end-moraines (Fig. 4). The largest probably belongs to the early Zyryanian glaciation (Kulakov 1981) Only the sides of this moraine ridge, which are about 8-km wide, are located on the plain between the mountains and the Hovsgol shore. The outer part of the moraine ridge is flooded by the lake water. The morphology of the observable part of the moraine ridge suggests that the glacier moved outside the mountainous area for 6–8 km. The thickness of the ice tongue was at least 300 m, judging from the side moraines on the slopes of the 2298-m Hoyd Bayan mountain. Such a giant glacier should have removed a great amount of debris from the mountains, but the moraine ridges are as low as 50 m, and therefore most of the glacial deposits are now below the surface of the piedmont plain and the lake bottom. The smallest end-moraine unit is located in the mouth of the trough, and can be linked to late Zyryanian times. The late Zyryanian

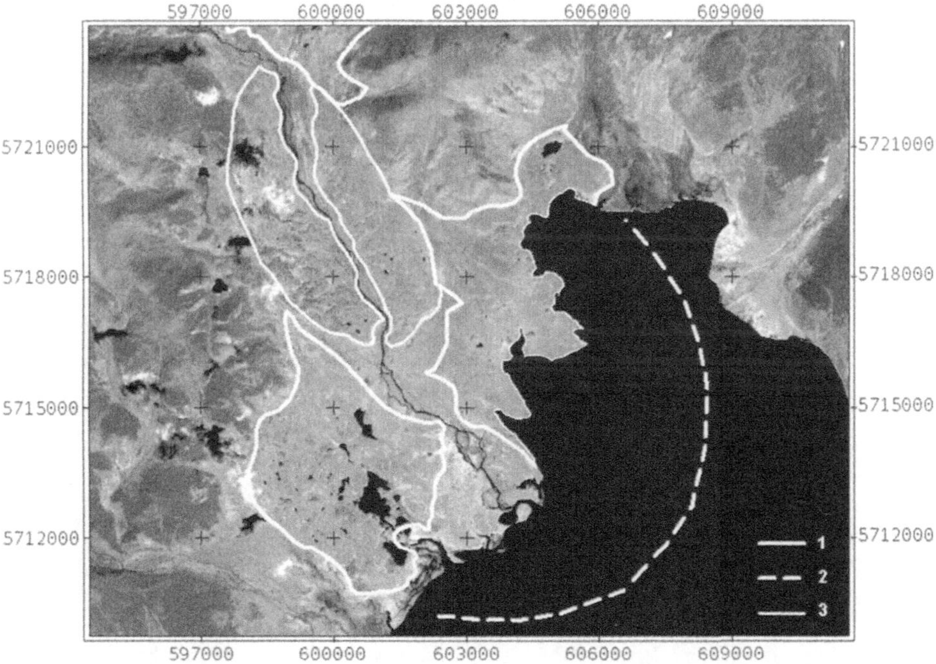

FIG. 4. The Landsat 7 satellite image of the lower reaches of the Ih-Horoo-gol. *1*, early Zyryanian end moraine; *2*, reconstructed front of the end moraine ridge; *3*, late Zyryanian end moraine

glacier was located within the mountain area, and was several times smaller than the early Zyryanian glacier.

3.2.2 Munku–Sardyk Range

The early Zyryanian glaciers of Munku–Sardyk apparently did not reach the Hovsgol depression (Fig. 5). For example, the length of the glaciers moving down the Munku–Sardyk range in the vicinity of the Dzargalan-gol was less than 5 km. The end moraines of those glaciers are located at a piedmont, and form a belt of amalgamated moraines at altitudes of about 1800 m. Only fluvioglacial deposits of this glacial complex could reach the lake basin. The late Zyryanian glaciers were much less widespread there. Well-preserved moraine ridges, probably of that age, have been found in the valley situated to the right of the Dzargalan-gol at a distance of 1.5–2 km from the kars (Fig. 6). The fluvial activity of glacial water was recorded in gigantic canyons transecting older moraines, and in a vast alluvial fan meeting the northern extremity of Lake Hovsgol.

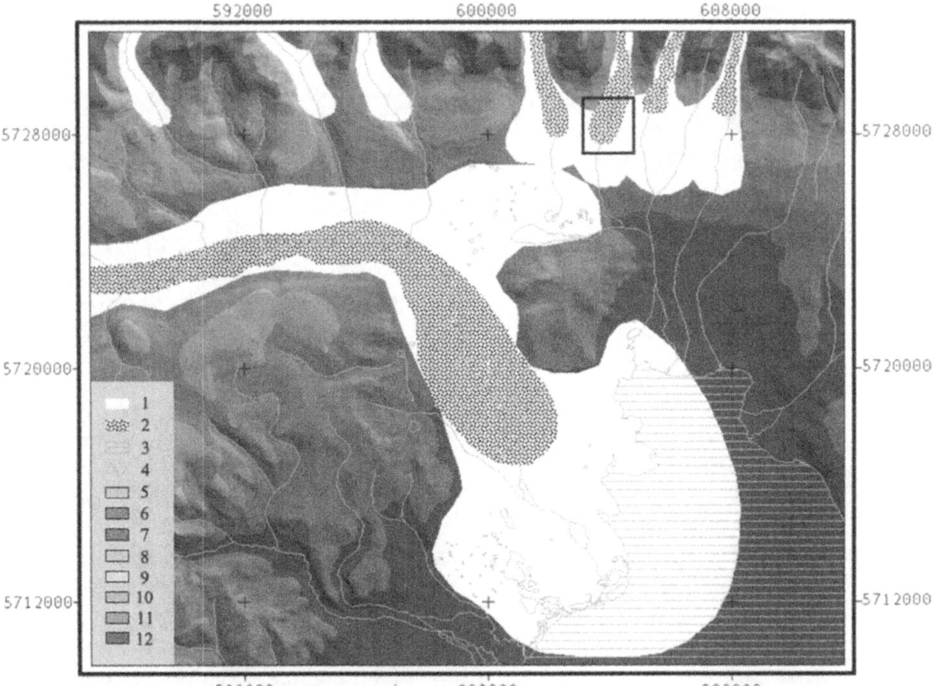

FIG. 5. Extension of late Pleistocene glaciers in the southern Munku–Sardyk range: interpretation of Landsat 7 satellite imagery. ArcView TIN topography is shown in the background. *1*, early Zyryanian glaciers; *2*, late Zyryanian glaciers; *3*, the present Lake Hovsgol area; *4*, modern hydrological objects; *5–12*, elevation

FIG. 6. The late Zyryanian end moraine in the valley located to the right of the Dzargalan-gol (location shown in Fig. 5)

3.3 Cover Sediments in Shore Sections

In many places the lake erodes the oblique plains joining the mountain ranges which form the basin frame. On the western shore, the peripheral areas of those plains consist of unsorted clastic fragments of colluvial and proluvial genesis. The elevation of the shore scarps is 3–5 m, showing that several tens of meters of the piedmont plain have been eroded.

On the eastern shore, the slope sediments are better manifested, and consist of the sediments of variable genesis and age. They are dominated by differently grained sands, the portion of rock debris depending on the distance from the lake to the rock outcrops. Originally, the deposits were colluvial and deluvial. Several sections contain eolian layers, traces of soil formation and permafrost, and the sediments of local streams. The slope sediments can be regarded a single unit of sediments, with a facial structure which depends on local conditions and stratification, i.e., on the duration of sedimentation. Typical examples are the exposures in Ih-Dalbayn-gol Bay (Fig. 7, point *C*) and in the mouth of the Tusgalyn-gol (Fig. 8, point *D*).

The slopes of the eastern shore are largely washed away owing to their high content of sand material. Mainly, only the back areas close to the mountains are

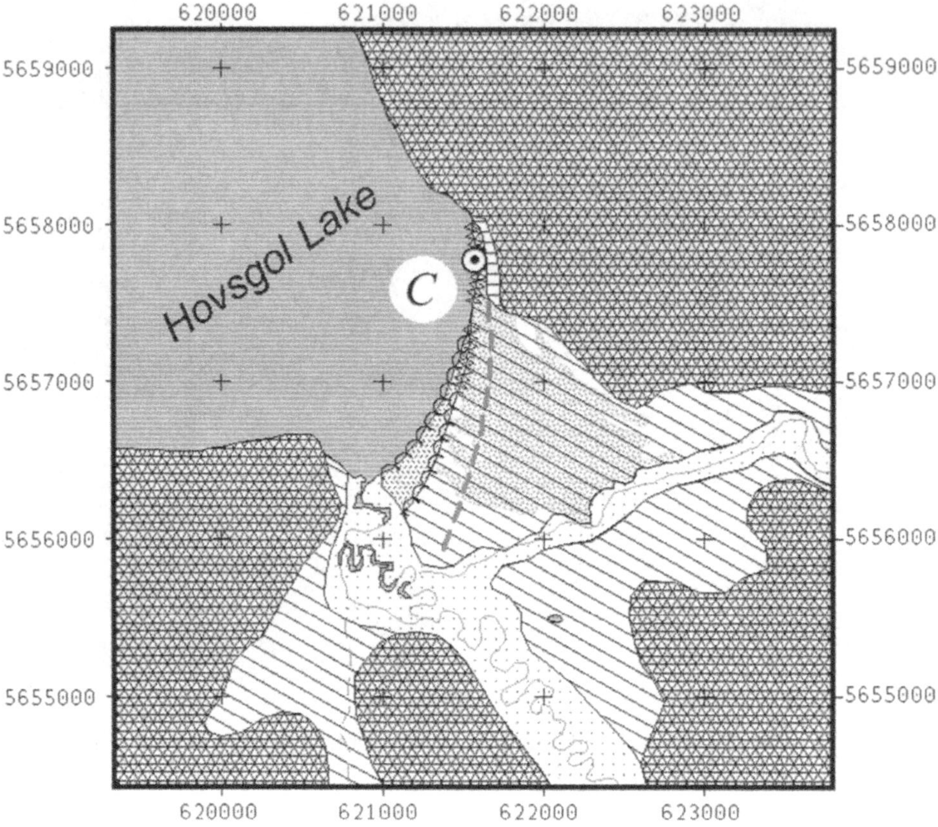

FIG. 7. Geomorphological map of the Ih-Dalbain-gol lower reaches (see legend in Fig. 3)

preserved. The elevation of exposures reaches 15–25 m. If we draw an oblique line which continues the surface of those slopes, it would meet the water surface at a distance of 200–500 m from the shoreline. The slope deposits are almost always submerged under the waterline. Only in two places they are underlain by lacustrine sediments.

3.4 Buried Lacustrine Deposits in Shore Sections

The lacustrine units in the Ih-Dalbayn-gol section consist of beach pebbles and littoral gravel (Fig. 9). The surface of the lacustrine deposits is 2.5 m higher than the present level of the lake; therefore, the water level at the moment of beach facies formation was 1–2 m higher than the present one. Beach facies have been investigated in two trenches of the section, and indicate a regressive stage in the lake's history. We suggest that the lake was 3–6 m higher at the maximum stage.

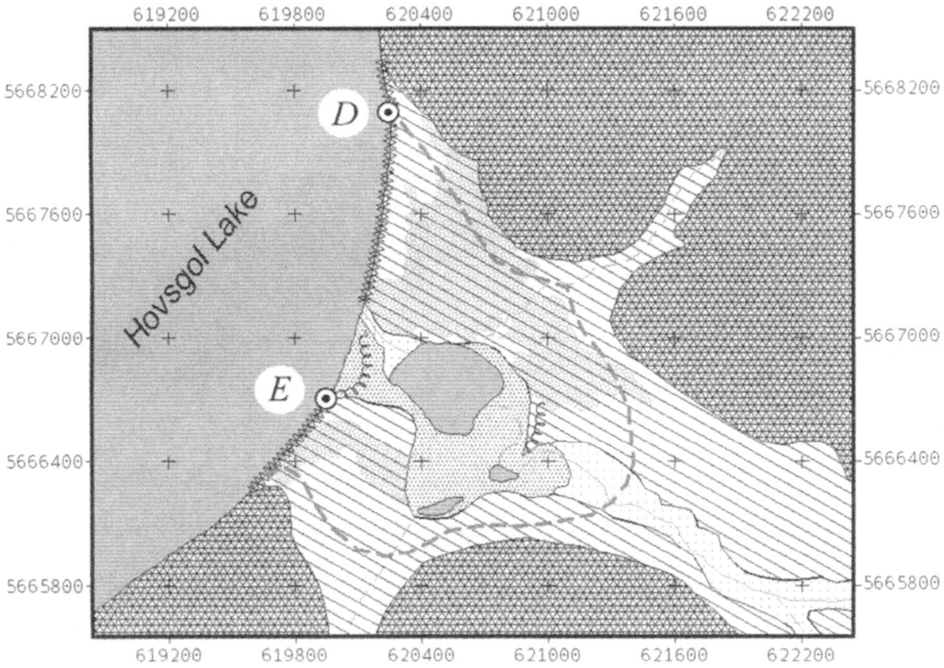

FIG. 8. Geomorphological map of the Tusgalyn-gol lower reaches (see legend in Fig. 3)

At that level, the shoreline would be very close to the shore rocks, and go far into the Ih-Dalbayn-gol valley to form a bay (see Fig. 7).

The age of these lacustrine deposits is conventionally estimated according to the complicated structure of the cover deposits: the abundant traces of soil and permafrost formation are present in the upper and lower parts of the subaerial sequence, respectively. This leads to the suggestion that the cover sequence reflects several climatic stages of sedimentation. For this reason, the cover unit of the Ih-Dalbayn-gol section is similar to that of the Belyi Yar II section in the Tunka basin [description by Ravsky (1972) and Admenko et al. (1975); interpretation by Kulchitsky et al. (1994) and Krivonogov (1995)]. The cover unit is underlain by Karginian sediments. It is not inconceivable that the high level of Lake Hovsgol recorded in the Ih-Dalbayn-gol section took place at the same time, i.e., in the Karginian.

The second location of lacustrine sediments, i.e., north of the mouth of the Tusgalyn-gol river (Fig. 8, point *D*), contains a 1.5-m-thick gravel sequence. This means that the surface of the gravel unit is hypsometrically higher than it is in the Ih-Dalbayn-gol section. A vast bay existed in that part of the valley during the formation of the gravel. The pebble coast of the bay should have been deposited in inland parts of the ancient bottom of this valley. At that stage, the

FIG. 9. Lacustrine units in the Ih-Dalbayn-gol section: *p*, beach pebbles; *g*, sublittoral gravel

transgression was comparable to that recorded in the Ih-Dalbayn-gol section. On the other hand, the regression proceeded very fast.

The sediments overlapping the lacustrine unit in the second location are not very thick, and have a simpler structure than the covering sequence in the Ih-Dalbayn-gol section. Therefore, this unit could have been formed due to a later transgression than the sediments in the first location.

3.5 Limnic Landforms

Younger (Holocene) episodes of lake activity are recorded in the landforms. The records are prints of bays, old beach ridges, and present beach ridges and stra nds. They are located mainly in river mouths and are connected with flood-plain steps. The most representative set of landforms is found on the eastern shore.

3.5.1 Bay Prints

It was noted above that the flood-plains in the river mouths of the eastern shore are more or less deeply inserted into the pre-Holocene bottoms of valleys, which

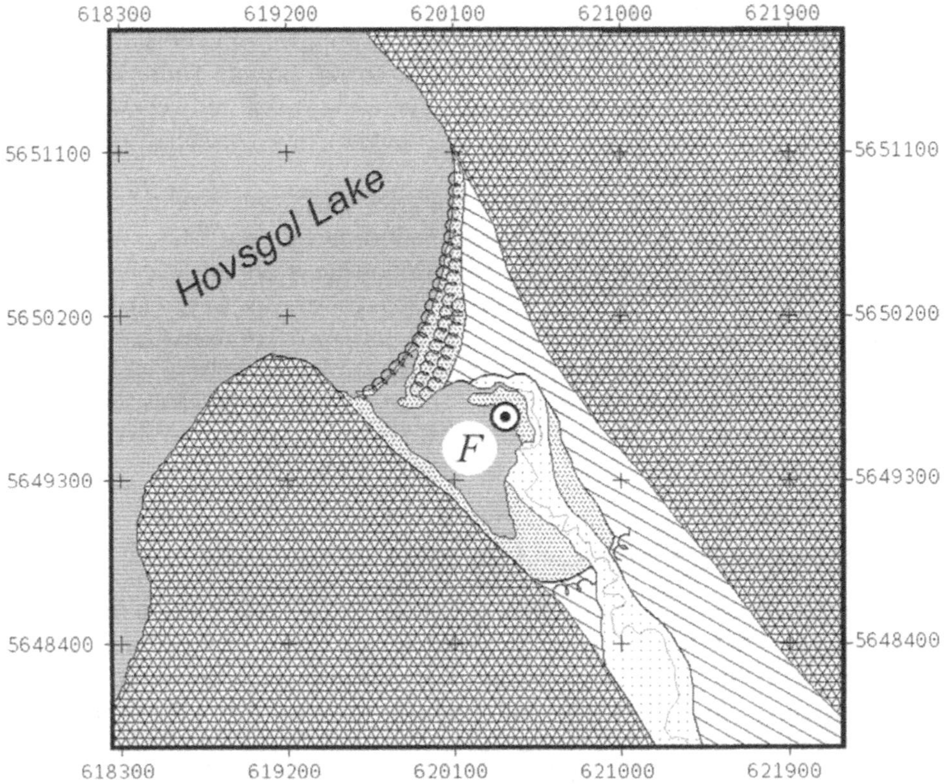

FIG. 10. Geomorphological map of the Bortsog-gol lower reaches (see legend in Fig. 3)

are chiefly composed of slope deposits. These areas of flood-plain contain the remnants of ingression bays, jutting 1–1.5 km into the land, in the southern part of the eastern shore (Figs. 8 and 10). In the northern part, the remnants of Holocene ingression bays occur as 100- to 300-m-wide narrow segments extending parallel to the present-day shoreline. In the field, they are clearly seen owing to their bright green vegetation, large lakes, and vast areas of stagnant water. The shore landforms bounding these bays are ridges which cross the river valleys and the abrasion scarps. These beach ridges are usually strongly eroded. Such flooding could be possible in a 2-m increase in the water level.

3.5.2 Old Beach Ridges

The phases of reduction of Holocene bays are recorded by old beach ridges. They have been found only in the southern part of the eastern shore. Their position and interrelation with other landforms in the valley are variable. One, and frequently two, beach ridges block part of the Holocene bays or extend parallel to the present shore (see Figs. 7, 8, and 10). The elevation of the ridges is 1–2 m above the surrounding surface. The ridges are composed of pebbles, and structurally and compositionally they are similar to the present shore drifts. The surface of those ridges is smoothed and covered with soil and vegetation, and locally with trees. This suggests a time gap between the formation of old and present shore landforms.

Several deep-water bays were crossed by spits. This resulted in the separation of part of a bay and the formation of a small drainage lake. Mineral–biogenic and biogenic sediments were delivered by inflowing rivers and accumulated in the lakes, e.g., in the drainage lake of the Borsog–gol river (see Fig. 10). The depth of this lake changes from 7 m in the central part to 2–3 m near the shore. The bottom sediments consist of turf detritus squeezed by the bottom ice to an elevation of 4 m above the lake level (Fig. 10, point *F*). Thus, the thickness of biogenic sediments in the central part could be 5 m or more. Their age is controlled by radiocarbon date of 4510 ± 40 yr BP (Beta-163953).

3.5.3 Present Beach Ridges and Beaches

The present-day accumulative shores of Lake Hovsgol occupy 6% of its perimeter (Rogozin 1993). They occur mainly in the bays and river mouths, i.e., in the same places as the old beaches. Rogozin described their morphology in detail. The present position of beach landforms is a result of the abrupt rise of the lake level in the end of XIX century (Tomilov and Dashidorz 1965). The water level rose about 3 m, and a small (23 m long, 11 m wide, and 2–3 m high) island opposite to Hanh (Turt) Village was submerged (Marinov 1967). The river mouths were flooded within the Holocene ingression bays. The shore landforms were submerged and washed out. Rogozin (1993) mentioned the presence of beach ridges at depths of about 3 m, which have been recognized in aerial photographs. The level of Lake Hovsgol remains high at the present time. The transgression resulted from a rise at the top of the spillway in the Egiin-gol headwater due to stone runs moving down the right contiguous Ulhen–Sair valley. The stone-run event of 1971–1972 was analogous to that of 1886 (Cherkasov et al. 1972; Rogozin and Yakimov 1977). The maximum offset of the shoreline was 14 m.

4 Control over Lake Level in the Egiin-gol Headwater

Although the Egiin-gol river flowing out from the southern Hovsgol has a well-developed valley, there are no river terraces to allow a reconstruction of erosion and accumulation stages. The expanded flood-plain is bounded by slopes and alluvial cone deposits.

The height of the river headwater is controlled by the stone drifts of the Ulhen–Sair valley. This place, of a fairly simple structure (Fig. 11), is a key to

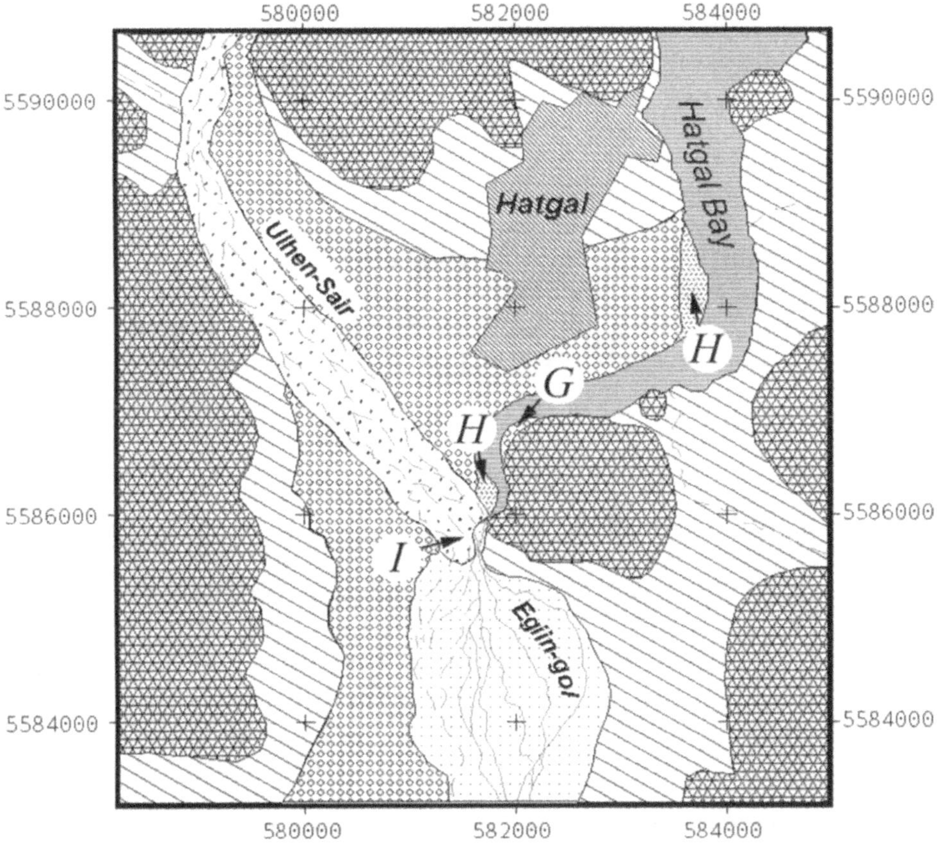

FIG. 11. Geomorphological map of the Egiin-gol river headwater (see legend in Fig. 3). *G*, *H*, *I*, sites described in the text

understanding lake-level variations during the Holocene and late Pleistocene. A vast alluvial cone is composed of boulder–pebble deposits and occupies the whole mouth of the Ulhen–Sair valley. It blocked the Egiin-gol valley, and since then the lake level has been controlled by this alluvial cone. Its margin has been preserved as a terrace-like surface in the left side of the valley, not far from the present Egiin-gol headwater (Fig. 11, site *G*). The alluvial cone formed in subaerial conditions at a low level of the Hovsgol when there was no runoff. The composition of the layers is clearly seen in abundant exposures occurring along the present water streams cutting through this cone. The height of the shore cliffs in the middle and on the margins of the cone is 12m and 6m, respectively. The 6-m height defines the rise in the level of Lake Hovsgol, which was necessary for the cone to start washing out.

This alluvial cone probably formed during the late Zyryanian. The washing out of the cone started after the rise in the lake level, obviously in the Karginian, and continued until the top of the spillway became stable and its height reached the modern level. This stage is recorded by a floodplain-like flat surface, which has been preserved in the right bank of the Egiin-gol valley, opposite Hatgal Village (Fig. 11, sites marked *H*). This surface was flooded during the Holocene transgressions. The scarp bounding the cone cut down by the headwater seems to have been abraded not long ago, because recent river activity is impossible in that place. The recent (Holocene) Ulhen–Sair valley is cut into a pre-Holocene alluvial cone. This valley has its own alluvial cone, which drives the Egiin-gol headwater to the left of the valley, and this defines the elevation of the top of the spillway (Fig. 11, site *I*).

5 Lake Level Changes at Lake Hovsgol in the Late Quaternary

The geomorphological features and sedimentological evidence in coastal areas of Lake Hovsgol indicate the obvious fluctuations during the late Quaternary. The transgressions and regressions described below were presumed.

5.1 Transgressions

The buried river sediments of the Karginian age in the Hanh-gol valley were accumulated in a plain type of river. This fact, and abundant vegetation remnants, lead to the suggestion that these sediments correspond to the sediments of delta and near-mouth river facies. As the surface of Karginian alluvial facies is now 8–9 m above the present lake level, the paleolevel could be 6–7 m higher (if the valley has not experienced neotectonic uplift).

The buried lacustrine sediments of the Ih-Dalbain-gol section are conventionally regarded as being Karginian. They mark the 3- to 6-m rise in the lake level. It would be incorrect to make direct correlations between the events recorded in sections of the Hanh-gol valley and Ih-Dalbay-gol Bay, but we can regard them as evidence for a high lake level during the Karginian warming period in the late Pleistocene.

The +6 m water level is controlled by the elevation of the Ulhen-Sain alluvial cone, and correlates with the shore sedimentary section. In terms of regional paleogeography, a significant rise in the Hovsgol level should have occurred in the phase of active deglaciation, i.e., 12–11 ka ago. The cutting of the Ulhen–Sair alluvial cone we assign to the same time-period.

In the Holocene, the rise of the Hovsgol could have occurred only within the variation of the top level of the spillway. The shore landforms indicate a water rise no more than 2 m above the present water level.

5.2 *Regressions*

Significant lake regressions occurred during glaciation events. The evidence for this comes from the position of an early Zyryanian end-moraine ridge in the Ih-Horoo-gol mouth and Sartanian cover deposits along the shoreline. These regressions are recorded by submerged terraces, which are probably not middle Pleistocene (Zolotarev et al. 1982), but are late Pleistocene. We performed reconstructions of the late Zyryanian slope deposits and showed that as they could have formed only within the shoreline, at that time the lake level was 20 m lower. The indirect evidence for the low water level in the Sartanian is the absence of Sartanian fluvial sediments in the Egiin-gol valley.

The records of Holocene regressions can be recognized only in the near-shore sediments of Lake Hovsgol. According to the data obtained by Dorofeyuk and Tarasov (1998), about 6000 years ago the level of Lake Hovsgol was 5 m below the present level, and it then rose unevenly. These authors reconstructed the variations in lake level, but these variations all seemed to be less than the present level. So we can conclude that transgression, which was recorded in the beach ridges, occurred in the early Holocene.

6 Conclusions

6.1 *Terraces*

1. Most accumulative forms located along the shoreline of Lake Hovsgol are recognized as slope sedimentary units and alluvial cones. The terrace-like appearance of these landforms has no relation to the lake's evolution.

2. Types of terraces: (a) a buried river terrace of the interstadial epoch (the 19-m Hanh-gol terrace, Q_3^3); (b) a fluvial terrace of the interstadial epoch (the 10-m Hanh-gol terrace, Q_3^3); (c) a fluvioglacial terrace of the glaciation zone (the Ih-Horoo-gol terraces, Q_3^2–Q_3^4).

3. The late Pleistocene and Holocene lacustrine sediments created beach ridges and beaches, which had a slight influence on the shape of the slopes. Undoubtedly they formed in one abrasion and accumulation stage. The older stages of abrasion and accumulation are recorded in high terrace-like scarps, which are often erroneously interpreted as young lake terraces.

6.2 *Lake Level*

1. The transgression of the lake took place in the warming stages. In the Karginian, the level was probably 3–7 m higher than the modern level. In the latest Pleistocene (Allerod), it probably corresponded to the top of the spillway of the Egiin-gol headwater but 6 m higher. In the early Holocene the level was not higher than +2 m.

2. The rise of Lake Hovsgol was controlled by the top of the spillway in the Egiin-gol headwater. Since the late Pleistocene, the lake level has not gone higher tha +6 m.

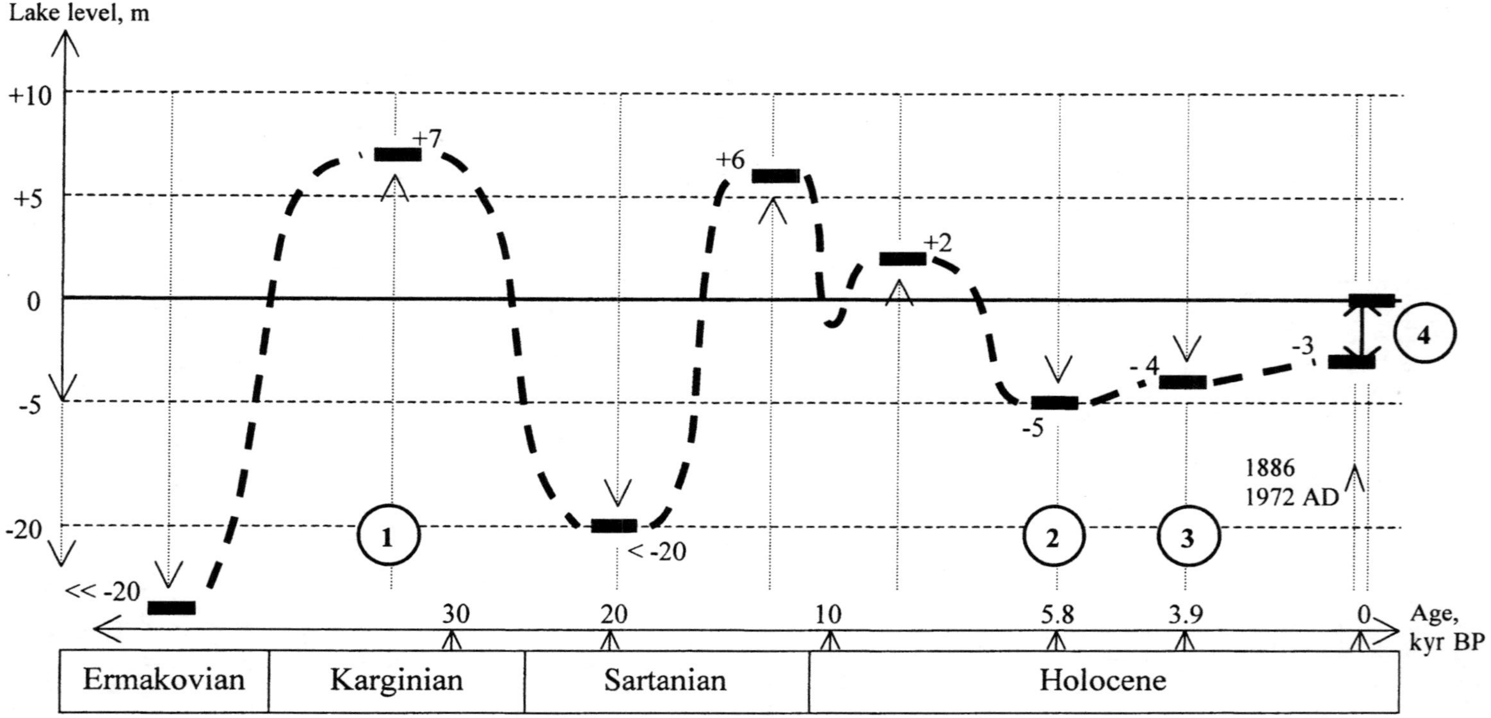

FIG. 12. Variations in the Lake Hovsgol water level in the late Pleistocene and Holocene based on geological data. Numbers in circles: *1–3*, levels dated by radiocarbon (*1*, >45 kyr BP, Hanh-gol; *2*, *3*, 5800 ± 100 kyr BP (TA-671) and 3910 ± 60 kyr BP (TA-670), from Dorofeyuk and Tarasov 1998); *4*, recurrent fluctuations from 1886 and 1972 AD due to the elevation of the Egiin-gol headwater

3. The lake regression is recorded in submarine terraces, glacial deposits, and slope deposits. The submarine terraces studied by echo sounding (Zolotarev et al. 1982) indicate 20, 90, and 120m lake regressions, and require further investigation. Most probably they should be related to the glacial stages of the Lake Hovsgol area evolution. According to bathymetry data, the front of the Ih-Horoo-gol end-moraine unit occurs at depths of about 100m, but the late Sartanian slope deposits were submerged to a depth of 20m.

4. Owing to climatic changes in the Holocene, the water level of the lake fell by at least 5m (Dorofeyuk and Tarasov 1998). After this event, the lake level has fluctuated, but has not exceeded the present water level.

Figure 12 shows the water-level variation in Lake Hovsgol in the late Pleistocene and Holocene based on geological evidence. It should be noted that in the late glacial and Holocene, it coincides with the water-level fluctuations of the lakes in northwestern Mongolia (Walter and Batnasan 2001).

Acknowledgments. This work was carried out within the Integration Program for the Study of Lake Hovsgol by collaborating teams from the Russian Academy of Sciences (Siberian Branch), the Mongolian Academy of Sciences, the Free University of Berlin, and Kyoto Prefectural University. The authors are grateful to Prof. O. Tomurtogoo (Director of the Institute of Geology and Mineral Resources MAS), who helped to arrange our field mission to Lake Hovsgol, Dr. Clyde Goulden (Academy of Natural Science, Philadelphia, USA), and Dr. J. Tsogtbaatar (Director of the Institute of Geoecology MAS) for satellite images and maps of the Lake Hovsgol area.

References

Adamenko OM, Belova VA, Popova SM, Adamenko RS, Ponomareva EA, Efimova LI (1975) Biostratigraphy of the late Pleistocene deposits of the Tunkin depression (in Russian). Russ Geol Geophys 6:78–85

Belova VA, Kulakov VS (1982) History of the Lake Hovsgol basin development. In: Late Cenozoic history of the lakes in the USSR (in Russian). XIth Congress of INQUA in the USSR, Moscow, 1982. Nauka, Siberian Branch, Novosibirsk, p 76–80

Cherkasov AE, Shumeev VP, Batsukh N (1972) Hydrogeologic regime of rivers in the Hovsgol basin. In: Natural conditions and resources of Hovsgol area: Irkutsk–Ulaanbaatar (in Russian). p 74–86

Dorofeyuk NI, Tarasov PE (1998) Vegetation and levels of the lakes in the north of Mongolia during the last 12500 years according to data from palymological and diatom analyses. Stratigraphy (in Russian). Geol Correl 6:73–87

Ivanenko VV, Karpenko MI, Yashina RM, Andreeva ED, Ashikhmina NA (1989) New data on the potassium–argon age of the basalts of the western board of the Hovsgol rift (MPR) (in Russian). Rep USSR Acad Sci 309:915–930

Kawai T, Kashiwaya K, Minoura K, Kuzmin MI (2001) A new object of the Baikal–Khubsugul drilling project following the first stage of the Baikal Drilling Project (BDP). In:

Tomurhuu D, Takahara H (eds) 2001 International Workshop for the Baikal and Hovsgol Drilling Project in Ulaanbaatar. Abstracts, p 1

Krivonogov SK (1995) Origin of relief in the Tunka depression (in Russian). In: Abstracts of the geomorphologic seminar Origin of Relief, Irkutsk, p 87–89

Kulakov VS (1981) Ancient and modern glaciations of the north Mongolia (trans-Khubsugulia). In: Transactions of the Soviet–Mongolian complex Khubsugul expedition (in Russian). Irkutsk, p 10–19

Kulchitsky AA, Osadchyi SS, Misharina VA, Popova SM, Chernayeva GP, Fileva TS, Orlova LA, Krivonogov SK (1994) Results of an investigation of the sandy sediments in the Tunka depression (sections Belyi Yar I and II). In: Baikal and mountains around it (Cenozoic geology, geomorphology, recent tectonics and geologic natural memorials) (in Russian). Abstracts. Earth Crust Institute SB RAS, Irkutsk, p 100–103

Marinov NA (1967) The Khubsugul (Kosogol) Lake depression in the North Mongolia. In: Materials of the commission on the study of the underground water of Siberia and Far East. Issue III (in Russian). Siberian Branch of USSR Acad Sci. USSR Academy of Science Publishing, Moscow, p 186–196

Mats VD, Fujii S, Mashiko K, Osipov EYu, Yefimova IM, Klimansky AV (2000) Changes in Lake Baikal water levels and runoff direction in the Quarternary period. In: Minoura K (ed) Lake Baikal. Elsevier, Amsterdam, p 15–34

Ravsky EI (1972) Sedimentation and climates of internal Asia during the Antropogene (in Russian). Nauka, Moscow, p 335

Rogozin AA (1993) Shore zone of the Baikal and Khubsugul: morphology, dynamics and the development history (in Russian). Nauka, Siberian Branch, Novosibirsk, p 168

Rogozin AA, Yakimov AA (1977) Activization of banks erosion in the Khubsugul. In: Natural conditions and resources of some regions of the MP (in Russian). Irkutsk, p 9–11

Shuvalov VF, Nikolaeva TV (1989) On the age and geomorphologic position of the Cenozoic plate basalts in central, north and southeast Mongolia (in Russian). Bull Leningrad State Univ 7:102–106

Tomilov AA, Dashidorz A (1965) Lake Hovsgol and possibilities of its pisciculture. In: Limnological investigations of the Baikal and some lakes of Mongolia. Nauka, Moscow, p 164–180

Walter M, Batnasan N (2001) Lake level and environmental delopment in north-west Mongolia in the last 15 000 years. In: Tomurhuu D, Takahara H (eds) 2001 International Workshop for the Baikal and Hovsgol Drilling Project in Ulaanbaatar. Abstracts, p 48–50

Zolotarev AG, Suldin VA, Kulakov VS (1981) Structure and present movements of the Khubsugul depression in north Mongolia. In: Natural conditions and resources of trans-Khubsugulia (in Russian). Transactions of the Soviet–Mongolian Complex Khubsugulian Expedition, Irkutsk, p 20–30

Zolotarev AG, Mangazyev VY, Dyemin AA (1982) The main geoniorphological levels of Prikhubsugulye. In: Late Cenozoic history of the lakes in the USSR (in Russian). XIth Congress of INQUA in the USSR. Moscow, 1982. Nauka, Siberian branch, Novosibirsk, p 80–88

Zorin YuA, Tumtanov EKh, Arvisbaatar N (1989) Structure of Cenozoic basins of the Prekhubsugul region from gravity data (in Russian). Russ Geol Geophys 10:130–136

Pollen Record from the Chivyrkui Bay Outcrop on the Eastern Shore of Lake Baikal since the Late Glacial

Hiroko Kataoka[1], Hikaru Takahara[2], Sergey K. Krivonogov[3], Elena V. Bezrukova[4], Lubov Orlova[3], Svetlana Kropivina[4], Norio Miyoshi[1], and Kimiyasu Kawamuro[5]

Summary. Bogs have formed on the shore of Chivyrkui Bay near Svyatoy Nose Peninsula on the eastern shore of Lake Baikal. Two sediment columns were taken from the outcrop of one of bogs (53°39′N, 109°12′E) to investigate the vegetation history of the eastern shore area of Lake Baikal. The forests around the bog are composed mainly of *Pinus sylvestris*, *Pinus sibirica*, *Larix sibirica*, and *Abies sibirica*. The results of a palynological study of the cores with 14 radiocarbon dates revealed the following vegetation history since the late glacial. The late glacial and early Holocene vegetation from 12000 to 9000 ^{14}C years BP was a shrub tundra with spruce and birch trees, thickets of dwarf birch and alder, and herbs such as Gramineae and *Artemisia*. Between 9000 and 8000 ^{14}C years BP, spruce forests expanded in the area, accompanied by some fir and pine, indicating the initial development of the dark taiga. Pine trees gradually increased after 8000 ^{14}C years BP and then started a dramatic increase, to become dominant in 6000 ^{14}C years BP in the dark taiga forests. The vegetation from 6000 ^{14}C years BP to the present was mainly composed of pines (*Pinus sylvestris* and *P. sibirica*) and birch, accompanied by spruce, fir, and larch.

Key words. Lake Baikal, Dark taiga, Late glacial, Holocene, Pollen record

[1] Okayama University of Science, 1-1 Ridai-cho, Okayama 700-0005, Japan
[2] Kyoto Prefectural University, 1-5 Hangi-cho, Shimogamo, Sakyo-ku, Kyoto 606-8522, Japan
[3] United Institute of Geology, Geophysics and Mineralogy, Siberian Branch of the Russian Academy of Sciences, Novosibirsk, Russia
[4] Limnological Institute, Siberian Branch of the Russian Academy of Sciences, Ulanbatorskaya str. 3, Irkutsk 664033, Russia
[5] Kansai Research Center, Forestry and Forest Product Research Institute, Fushimi-ku, Kyoto 612-0855, Japan

1 Introduction

Long-term records of vegetation changes from the Lake Baikal long cores have recently become available (BDP-93 Baikal Drilling Project Members 1997; Bezurkova 1998; Bezrukova et al. 1999; Kawamuro et al. 2000; Maki et al. 2003; Miyoshi et al. 1999, 2002; Oda et al. 2000; Takeuti et al. 1995). These papers have revealed some important facts, such as an extinction of a plant group through the Tertiary and Quaternary. Palynological data from the bottom sediment of Lake Baikal contain long continuous vegetation histories for a huge area because of the lake's large catchment basin. Local vegetation histories of the shore area of Lake Baikal have been revealed from bog sediment cores (Bezrukova et al. 1996; Takahara et al. 2000). The modern vegetation around Lake Baikal contains several biomes, such as deciduous conifer forests of larch, evergreen conifer forests of pine, fir, and spruce, and steps of grasses and mugwort. To understand the dynamics of these biomes, it is important to correlate local vegetation changes in the shore area (Takahara et al. 2000).

Bezrukova et al. (1996) reported the vegetation history of a peat outcrop on the same site as Chivyrkui Bay. New sediment cores from the Chivyrkui Bay outcrop, described here, have given us more precise dating and information about the vegetation history.

2 Materials and Methods

Sediment samples were taken from a peat bog (53°39′N, 109°12′E, 460 m altitude) on the shore of Chivyrkui Bay, which is surrounded by Svyatoy Nose Peninsula and Barguzin Mountain (Fig. 1). The forests around the bog are composed mainly of *Pinus sylvestris*, *Pinus sibirica*, *Larix sibirica*, and *Abies sibirica*.

There is an outcrop on the shore of Chivyrkui Bay. The stratigraphy of this outcrop is shown in Fig. 2. Two sediment columns were taken directly from the outcrop. One is from the upper part of the sediments (site 1), and the other is from the lower (site 2). The core at site 1 is 220 cm in length from the surface of the ground (460 m altitude), and mainly consists of peat with peaty clay from 192 to 196 cm in depth. The core at site 2 is 180 cm in length above the lake level (456 m altitude), and 50 cm below the lake level. The core from site 2 mainly consists of peat, with peaty clay in the upper part of the core. The detailed stratigraphic correlation by sediment features between the cores at sites 1 and 2 is not clear, because the collapse of the uppermost part of the outcrop made it difficult to see the correlation between sites 1 and 2.

Subsamples of 1 cm^3 were taken from these two cores every 10 cm for pollen analysis. These samples were prepared by standard KOH, HF, and acetolysis mixture procedures (Faegri and Iversen 1989, with our modification). To determine the pollen concentrations, a known concentration of 25-μm microspheres in suspension was added to each sample prior to the standard procedure (Ogden

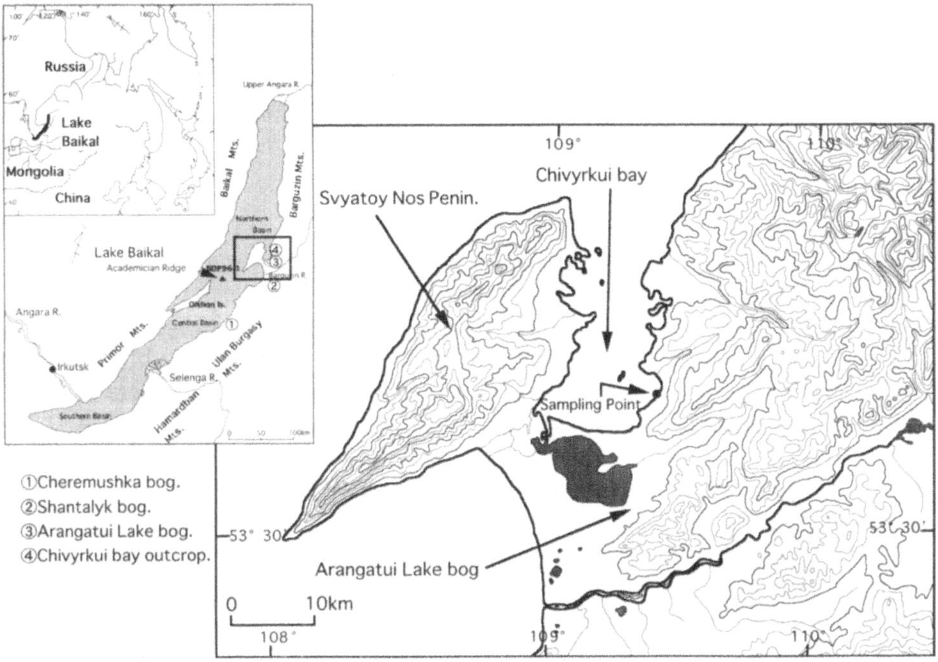

FIG. 1. Map showing the location of the study area. The sampling point was at 53°39′N, 109°12′E. The four bogs indicated on the left-hand map are mentioned in Fig. 7

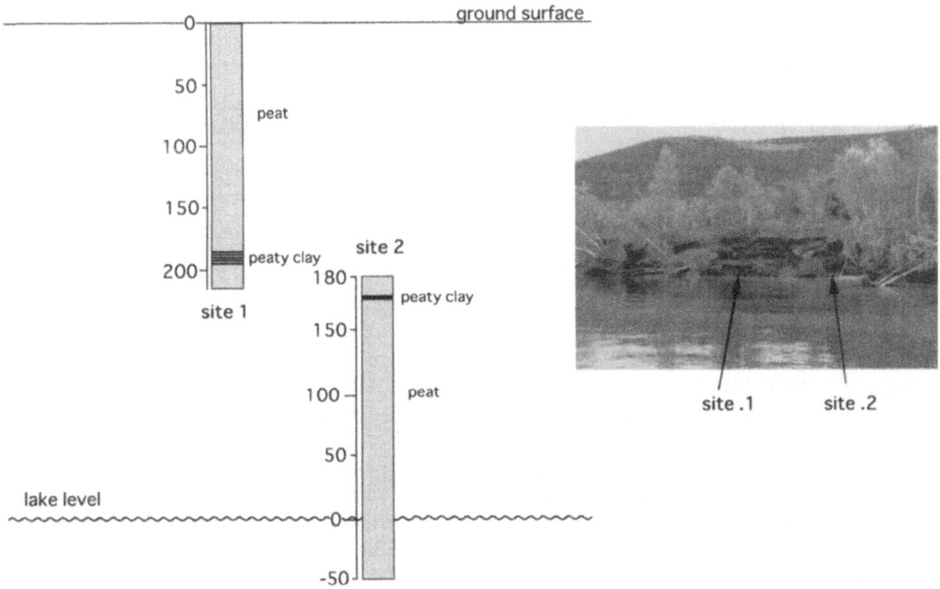

FIG. 2. *Right*, Chivyrkui Bay outcrop, and *left*, column sections

1986). Pollen identification was carried out using an optical microscope with a magnification ×600. A pollen total of at least 500 grains was counted. In the pollen percentage diagrams (see Figs. 4 and 5), the percentage values of each taxon were calculated based on tree pollen totals.

3 Radiocarbon Chronology

Thirteen samples of bulk organic matter were taken from two cores for the radiocarbon dating. The radiometric dating of these samples was carried out in the United Institute of Geology, the Geophysics and Mineralogy, Siberian Branch of the Russian Accademy of Sciences (UIGGM). The $^{13}C/^{12}C$ ratios of these samples were not measured. A wood sample was also dated by the accelerator mass spectrometry (AMS) at Beta Analytic, USA. The results of the dating are shown in Table 1. These radiocarbon dates were calibrated to calendar years using the program CALIB 4.3 (Stuiver and Reimer 1993) with INTCAL98 (Stuiver et al. 1998). Figure 3 shows the age–depth relationships of the cores from sites 1 and 2 for radiocarbon and calibrated ages. The calibrated ages (calendar years) are used to calculate pollen influx (see Fig. 6). The date of 3720 ± 185 ^{14}C years BP at 75–80 cm is older than the date at 115–120 cm at site 1. There is also an inversion between the two deepest samples in the core from site 2. These discrepancies can be seen in the radiocarbon dates.

The radiocarbon chronology described above indicates that the core from site 1 corresponds to the period since the mid-Holocene, and the core from site 2 is from the late glacial to the mid-Holocene. The average sedimentation rate is 0.30 mm/years (site 1: calibrated ages based on the series of dates without the

TABLE 1. Radiocarbon dates from Chivyrkui bay outcrop

Site	Depth(cm)*	Lab number**	Radiocarbon age (^{14}C yr B.P.)	Dated material
site 1	35–40	SOAN-3 803	1 450 ± 75	peat
site 1	75–80	SOAN-3 804	3 720 ± 185	peat
site 1	115–120	SOAN-3 805	3 500 ± 90	peat
site 1	155–160	SOAN-3 806	4 795 ± 80	peat
site 1	195–200	SOAN-3 807	5 400 ± 125	peat
site 1	215–220	SOAN-3 808	6 605 ± 130	peat
site 2	190–185	SOAN-3 809	5 645 ± 85	peat
site 2	170–165	SOAN-3 810	6 105 ± 220	peat
site 2	150–145	SOAN-3 811	7 025 ± 230	peat
site 2	110–105	SOAN-3 812	8 340 ± 300	peat
site 2	70–65	SOAN-3 813	9 165 ± 150	peat
site 2	30–25	SOAN-3 828	9 600 ± 130	peat
site 2	0–5	SOAN-3 829	10 810 ± 150	peat
site 2	−50	Beta-113 969	9 700 ± 70	wood

* Base point (zero) is grand surface for site 1 and lake level for site 2.
** SOAN, UIGGM; Beta, Beta Analytic

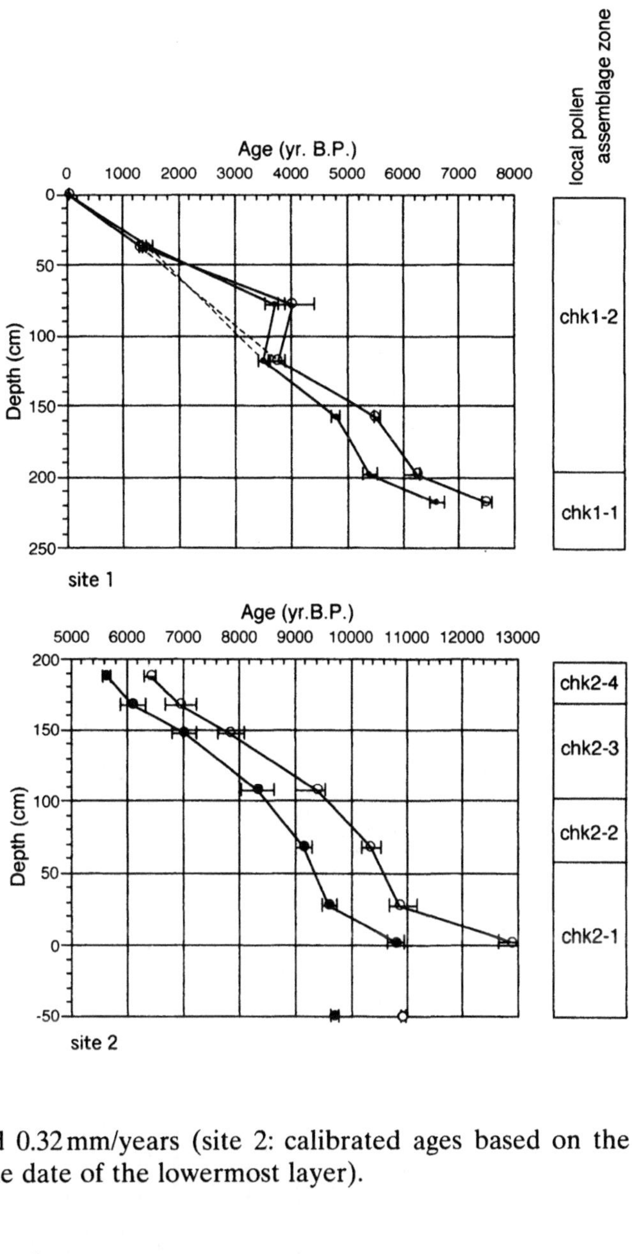

FIG. 3. Age–depth relationships for cores from sites 1 and 2 from the Chivyrkui Bay outcrop. *Solid circles*, radiocarbon ages; *open circles*, calibrated ages (calendar years). *Error bars* are one standard deviation. The *broken line* is inserted for chronological control

date from 75–80 cm) and 0.32 mm/years (site 2: calibrated ages based on the series of dates without the date of the lowermost layer).

4 Palynological Results

Pollen percentage diagrams of sites 1 and 2 are shown in Figs. 4 and 5, respectively, and pollen influx diagrams of sites 1 and 2 are shown in Fig. 6. The zonation of the pollen diagrams was made by visual inspection of the local pollen assemblage zones, chk1-1, and chk1-2 for site 1, and chk2-1, chk2-2, chk2-3, and chk2-4 for site 2.

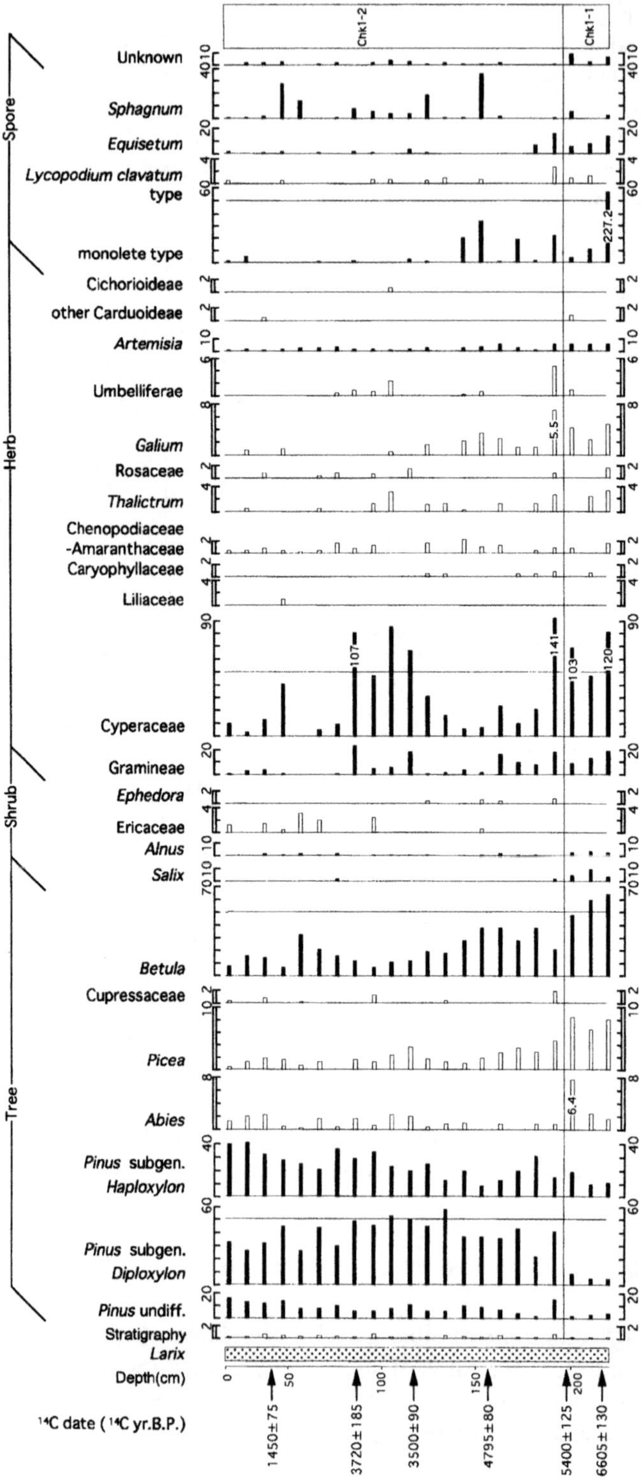

FIG. 4. Pollen percentage diagram of site 1 from the Chivyrkui Bay outcrop. Percentage values for each taxon were calculated based on tree pollen totals. *Dotted area*, peat

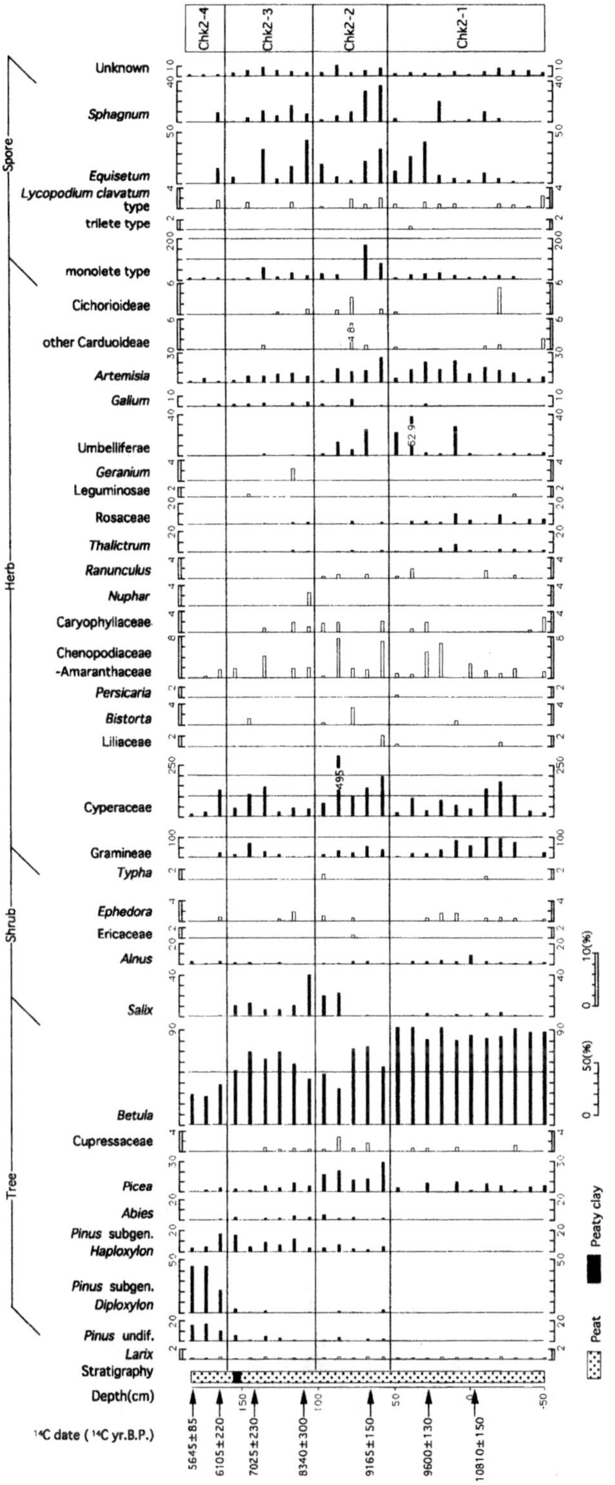

FIG. 5. Pollen percentage diagram of site 2 from the Chivyrkui Bay outcrop. Percentage values for each taxon were calculated based on tree pollen totals

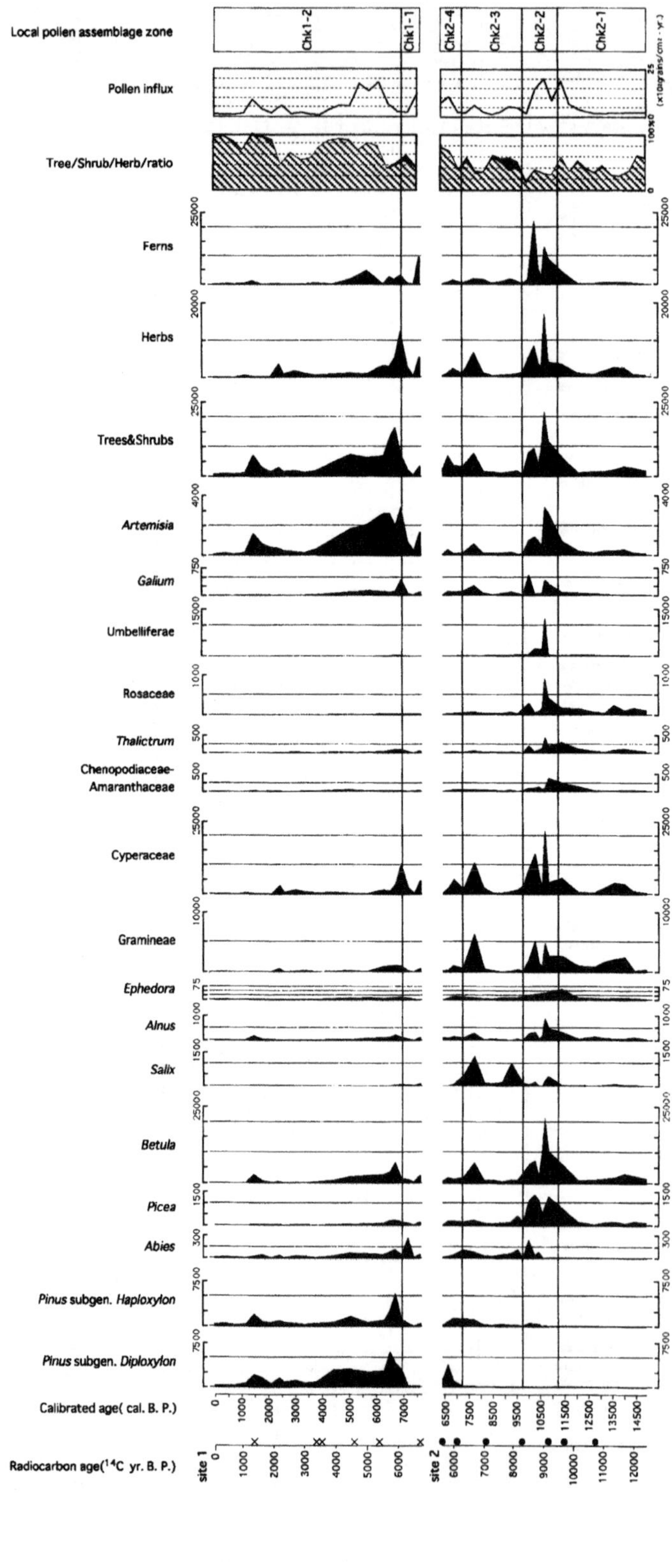

FIG. 6. Pollen influx diagram for selected taxa from the Chivyrkui Bay outcrop. *Crosses*, radiocarbon dates at site 1; *solid circles*, radiocarbon dates at site 2. Gross-hatching, trees; *black*, shrubs; *white*, herbs

4.1 Site 1

In the lowest zone of the core of site 1 (chk1-1), *Betula* pollen is dominant, accompanied by *Pinus* subgenus *Haploxylon*, *Abies*, *Picea*, and *Salix* pollen. Zone chk1-2 is characterized by high percentages of *Pinus* pollen, including *P.* subgenus *Diploxylon* and *Haploxylon* (Fig. 4). *Abies* and *Picea* pollen percentages decline in this zone, and show several different values. Total pollen influx values become high in the lower part of this zone (>20000 grain/cm^2 year), and decline (<500 grain/cm^2 year) in the upper part. The *Pinus* pollen influx increases in the lower part of the zone. Tree pollen ratios (Fig. 6, third diagram from the right) are also high (up to 80%–90%).

4.2 Site 2

The lowest zone (chk2-1) is characterized by high percentages of *Betula* (over 70%), Cyperaceae (20%–170%), and Gramineae pollen (up to 100%), with relative high percentages of *Artemisia* (5%–20%) (Fig. 5). *Betula nana*-type pollen was recognized as about 10% of the total *Betula* pollen and *Picea* pollen reached 10%. Pollen influx values are high for *Betula*, Gramineae, Cyperaceae, and *Artemisia* in this zone (Fig. 6). The total herb pollen ratio reaches 40%-70% (Fig. 6).

In zones chk2-2 and chk2-3, *Betula* pollen percentages are still high, but have decreased to about 50% (Fig. 5). *Picea* pollen shows relatively high percentages of 13%–30% in zone chk2-2. In the upper part of zones chk2-2 and chk2-3, *Salix* pollen reaches 10%–40%. Through zones chk2-2 and chk2-3, *Pinus* pollen gradually increases, and *Abies* pollen is found at low percentages (0.4%–5%). The total pollen influx shows a decline in zone chk2-3 (500–100 grains/cm^2/year). Zone chk2-4 is characterized by an abrupt increase in *Pinus* pollen.

5 Discussion

The following correlations between the local pollen assemblage zones of sites 1 and 2 were recognized on the basis of features of pollen assemblages described above. Zones chk1-1 and chk1-2 at site 1 correspond to zones chk2-3 and chk2-4 at site 2, respectively.

5.1 Late Glacial

Zone chk2-1 corresponds to the late glacial period based on the age–depth curve (see Fig. 3, site 2). The late glacial vegetation around this site was composed of spruce, birch, and alder, with herbs such as Gramineae, Cyperaceae, *Thalictrum*, and *Artemisia*, according to the pollen assemblage. Because of the high ratio of upland herbs, the landscape during this period was a shrub tundra with spruce and birch trees, thickets of dwarf birch and alder, and herbs.

5.2 Early Holocene

During the period from 9200 to 6000 ^{14}C years. BP (corresponding to zone chk2-2), the abundance of birch declined compared with the previous period. The percentage and influx values for *Picea* pollen increased (zone chk2-2 in Figs. 5 and 6), indicating the expansion of spruce forests between 9200 and 8300 ^{14}C years BP. The appearance of *Abies* and *Pinus* pollen at low values indicates the initial development of a dark taiga. In the period between 8300 and 6000 ^{14}C years BP (corresponding to zones chk2-3 and chk1-1), the total pollen influx values declined, suggesting that the taiga forests retreated. *Salix* pollen was more common than in zone chk2-1. *Betula* pollen increased, and included *Betula nana* type (about 10% of total *Betula* pollen). These changes in the pollen assemblages indicate that shrubs of birch and willow developed in this period.

5.3 Mid- and Late Holocene

Pinus pollen percentages and influx dramatically increase in the lowermost part of zones chk2-4 and chk1-2,) suggesting the rapid spread of pine, probably, *Pinus sylvestris* (Diploxylon pine) and *P. sibirica* (Haploxylon pine), which were established in their current habitat around 6000 ^{14}C years BP.

According to the relationship between surface pollen assemblages and the surrounding vegetation (Ogura A. and Takahara H., unpublished data), pine and birch pollen belong to long-distance dispersal types, while *Abies*, *Picea*, and *Larix* do not disperse their pollen over long distances. For example, *Pinus* pollen ranges from 20% to 30% of the total tree pollen in samples of present-day fir forests where *Pinus sylvestris* does not occurr. In addition, *Abies* pollen shows several percentage values even in dense fir forests. *Larix* and *Picea* pollen show similar values to *Abies* pollen. Therefore, the pollen assemblages in zone chk1-2 suggest that forests in the mid- and late Holocene vegetation were mainly composed of pines (*Pinus sylvestris* and *P. sibirica*) and birch, accompanied by spruce, fir, and larch.

The decline in the pollen influx after 3500 ^{14}C years BP does not imply a retreat of vegetation, because it might be caused by the low density of loose sediment in the upper sedimentary conditions. We are extending our study to other sites on the eastern shore of Lake Baikal. Pollen influx data from other sites will provide a more precise description of the late Holocene vegetation.

5.4 Correlations between the Pollen Data in the Eastern Shore Area

This study provided the finest dating control of all the sites we investigated around Lake Baikal. By using this dating control and a new pollen record from the Chivyrkui Bay outcrop, we improved the correlations (Fig. 7) between the local pollen assemblage zones in the eastern shore area given by Takahara et al. (2000). Around the eastern shore of Lake Baikal, the following vegetation changes since the late glacial were recognized.

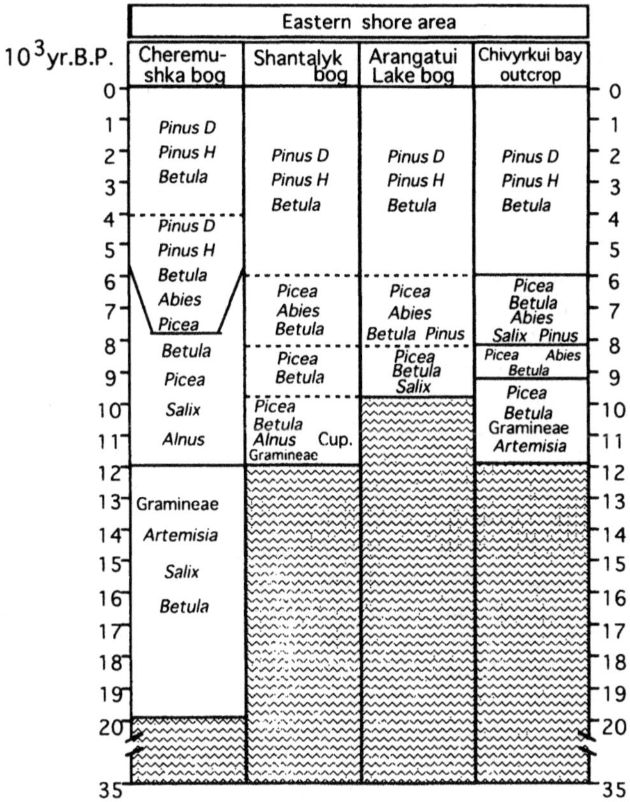

FIG. 7. Correlations between the local pollen assemblage zones in the eastern shore area of Lake Baikal (revision of Takahara et al. 2000). The location of each site is indicated in Fig. 1. *Solid lines*, dated (^{14}C dating) zone boundary; *dotted lines*, undated zone boundary; *shaded area*, sediments not taken

The late glacial and early Holocene vegetation from 12000 to 9000 ^{14}C years BP was a shrub tundra with spruce and birch trees, thickets of dwarf birch and alder, and herbs such as Gramineae and *Artemisia*. Between 9000 and 8000 ^{14}C years BP, spruce forests expanded in the area, accompanied by fir and pine, indicating the initial development of the dark taiga. Pine trees gradually increased after 8000 ^{14}C years BP, and then started to increase dramatically to become dominant in 6000 ^{14}C years BP in the dark taiga forests. The vegetation from 6000 ^{14}C years BP to the present day was mainly composed of pines (*Pinus sylvestris* and *P. sibirica*) and birch, accompanied by spruce, fir, and larch.

Further investigations of mire sediments which have already been taken from Lake Baikal will provide a more detailed history of the taiga.

Acknowledgments. This work was supported by the Science and Technology Agency of Japan and the Russian Foundation for Basic Research. We thank Mikhail Grachev for giving us opportunities for fieldwork, and Nadezhda Cherepanova for arranging it. We also thank Yoshimune Morita, Hideaki Noi, Yoshiki Shinomiya, Akihide Takehara, and Takashi Uchiyama for their cooperation in the fieldwork, and Kyoko Tanida and Naoko Sasaki for their help in the laboratory.

References

BDP-93 Baikal Drilling Project Members (1997) Preliminary results of the first scientific drilling on Lake Baikal, Buguldeika site, southeastern Siberia. Quaternary Int 37:3–17

Bezrukova EV (1998) Palynology of late glacial and Holocene sediments of Lake Baikal (in Russian). Geogr Nat Resource 3:54–61

Bezrukova EV, Mats VD, Letunova PP, Nakamura T, Fuji S (1996) Holocene peat bogs in prebaikalia as an object of paleoclimatic reconstructions (in Russian). Russ Geol Geophys 37:78–92

Bezrukova EV, Kulagina NV, Letunova PP, Shestakova ON (1999) Climatic and vegetation changes in the Baikal region for the last 5 Ma (in Russian). Russ Geol Geophys 40:739–749

Faegri K, Iversen J (1989) Textbook of pollen analysis, 4th edn (revised by Faegri K, Kaland PE, Krzywinski K). Wiley, New York, p 328

Kawamuro K, Shichi K, Hase Y, Iwauchi A, Minoura K, Oda T, Takahara H, Sakai H, Morita Y, Miyoshi N, Kuzmin MI (2000) Forest–desert alternation history revealed by the pollen record in Lake Baikal over the past 5 million years. In: Minoura K (ed) Lake Baikal. Elsevier, Amsterdam, p 101–107

Maki T, Hase Y, Kawamuro K, Shichi K, Minoura K, Oda T, Miyoshi N (2003) Vegetation changes in the Baikal region during the late Miocene based on pollen analysis of the BDP-98-2 core. In: Kashiwaya K (ed) Long continental records from Lake Baikal. Springer, Tokyo, p 123–135

Miyoshi N, Takeuchi T, Kataoka H, Ueda K, Morita Y, Kawamuro K, Takahara H, Hase Y, Inouchi Y, Oda T, Minoura K (1999) Pollen analysis of upper sediment (VER 94/5-St.21) in Lake Baikal (in Japanese, with English summary). Jpn J Palynol 45:27–34

Miyoshi N, Kataoka H, Shichi K, Oda T, Takahara H, Hase Y, Kawamuro K, Bezrukova EV (2002) Quaternary vegetation history based on pollen analysis of bottom sediment (BDP-96-1) taken from Lake Baikal, Russia (in Japanese, with English summary). Quaternary Res (Daiyonki Kenkyu) 41:171–184

Oda T, Minoura K, Fujimura C, Nakamura T, Kawai T (2000) Vegetation history of the Baikal drainage basin during the last 250000 years (in Japanese, with English summary). Jpn J Palynol 46:81–92

Ogden III JG (1986) An alternative to exotic spore or pollen additions in quantitative microfossil studies. Can J Earth Sci 23:102–106

Stuiver M, Reimer PJ (1993) Extended ^{14}C database and revised CALIB 3.0 ^{14}C age calibration program. Radiocarbon 35:215–230

Stuiver M, Reimer PJ, Bard E, Beck JW, Burr GS, Hughen KA, Kromer B, McCormac G, van der Plicht J, Spurk M (1998) INTCAL98 radiocarbon age calibration, 24000-0 cal BP. Radiocarbon 40:1041–1083

Takahara H, Krivonogov SK, Bezrukova EV, Miyoshi N, Morita Y, Nakamura T, Hase Y, Shinomiya Y, Kawamuro K (2000) Vegetation history of the southeastern and eastern coasts of Lake Baikal from bog sediments since the last interstade. In: Minoura K (ed) Lake Baikal. Elsevier, Amsterdam, p 108–118

Takeuti S, Oda T, Kawai T (1995) Palynostratigraphy of core BDP93-1 from Lake Baikal, southeastern Siberia. Saito HO-on Kai Mus Bull 63:1–7

Sedimentary Fe/Mn Layers in Lake Baikal as Evidence of Past and Present Limnological Conditions

L.Z. Granina, V.D. Mats, O.M. Khlystov, E.L. Goldberg, M.A. Phedorin, S.S. Vorobyeva, and M.Y. Semenov

Summary. As a result of the limnological peculiarities of Lake Baikal, different forms of oxidized Fe and Mn accumulations are widely spread throughout the lake, not only in the uppermost sediments, but also within the sequence of reduced sediments. We present the results of a detailed study of an unusual, oxidized core, up to 2 m long, from the underwater Academician Ridge. This study led us to propose that the Fe/Mn crusts lying deep (several meters) below the bottom surface are not markers of past climatic events, as was suggested previously. They may be the result of a specific diagenetic redistribution of sedimentary Fe and Mn at conditions of slow sedimentation and a sedimentary $C_{org.}$ deficit, which is typical of an oxidation front in a nonsteady-state diagenetic system. Ancient Fe/Mn crusts of subaerial origin, which have been submerged as a result of tectonic events in this area, could serve as the initial material for such a redistribution. A mechanism for the formation of deep oxidized sedimentary sequences in the Academician Ridge region is proposed.

Key words. Lake Baikal, Diagenesis, Fe/Mn crust, Paleoclimate, Redox front

1 Introduction

In Lake Baikal, secondary diagenetic Fe/Mn layers and crusts, usually a few millimeters thick, are wide spread in pelagic sediments. There are buried oxidized layers, which can be recognized by their red–brown color, not only at the water–sediment interface (WSI), but also deeper, within recent reduced sediments. Different types of such relics have been described by Granina et al. (1993). However, in long cores from the underwater Academician Ridge, different, well-preserved Fe/Mn crusts, sometimes up to 1 cm in thickness, were found deep (several meters) below the WSI (Deike et al. 1993, 1997; Khlystov et al. 1999; Mats et al. 2000). Little work has been done on them, although it has

---

Limnological Institute, Siberian Branch of the Russian Academy of Sciences, Ulanbatorskaya str. 3, Irkutsk 664033, Russia

been suggested (Deike et al. 1997) that their formation was connected with climate changes, and thus such crusts have been sometimes dated in specific sedimentary sequences of so-called "yellow" sediments which were oxidized up to 2–4 m below the WSI. At present, the oxidized sedimentary layer in Lake Baikal is usually from a few millimeters to 15 cm thick, and rarely reaches 30 cm in thickness, and therefore the formation of such surprisingly deep oxidized sequences is difficult to explain in the framework of the recent geochemical environment. They have been termed "pre-Baikal clays," and some authors have suggested that they could have been formed in quite different, ancient water bodies—precursors of Lake Baikal—characterized by extremely low sedimentation rates (Galkin and Mizandrontsev 1970). Our study aimed to investigate a long core of this type, and to interpret the results using new data about the structure, development, and peculiarities of the sedimentary sequence on Academician Ridge.

2 Materials, Methods, and Results

A core about 3 m long was sampled at station VER-96-2-6TW (short name st. 6TW) on the underwater hill of Academician Ridge at a water depth of 270 m (Fig. 1). This core characterizes a specific type of sedimentary section which is widespread on the northwest block of Academician Ridge (Mats et al. 2000; Khlystov et al. 1999, 2001). They are found on the underwater elevations, and their upper parts contain deep (up to 2–4 m below the WSI) oxidized sediments. Yellowish-brown oxidized sediments comprise a core up to 170 cm; above a sharp redox interface (RI) these are followed by reduced sediments of a gray color (Fig. 2). At different horizons within the core there are dense Fe/Mn crusts and brown sedimentary layers riched in Fe and Mn oxides (Fig. 2a). We examined 28 subsamples from the st. 6TW core and seven Fe/Mn crusts sampled at horizons from 0.4 to 7.4 m below the WSI at different stations (st.VER-96-1-8GC, 8PC-2, 6 PC, 6 GC, 5X GC, and 1X GC; see Fig. 1) located in the same area. Measurements of the redox potential (Eh), $C_{org.}$ contents, and biogenic silica (BSi), as well as express diatom analyses, were performed by the methods described previously (Grachev et al. 1997; Granina et al. 2000). The concentration of Fe and Mn in the sediments was examined by synchrotron radiation X-ray fluorescent analysis.

The upper 2 m of the core was highly oxidized, and Eh in the dark-brown layers was high (545–635 mV) (Fig. 2b). The sediment $C_{org.}$ content was extremely low, i.e., only 0.5%–1% even within the topmost sedimentary layers (Fig. 3), whereas it is usually 1.3%–3.2% (Granina et al. 2000) in the uppermost sediments of Academician Ridge. Beginning 10 cm below the WSI and deeper, the $C_{org.}$ content ranged 0.07% to 0.80%, and was 0.26% on average. The pattern of vertical Fe and Mn distribution is characterized by maximums corresponding to visually distinguishable dark Fe/Mn layers and crusts (Figs. 2a and 4). There is no Fe and Mn enrichment in the uppermost sediment layer; the highest Fe and Mn contents are found in the lower part of the oxidized zone, 0.5 m above the

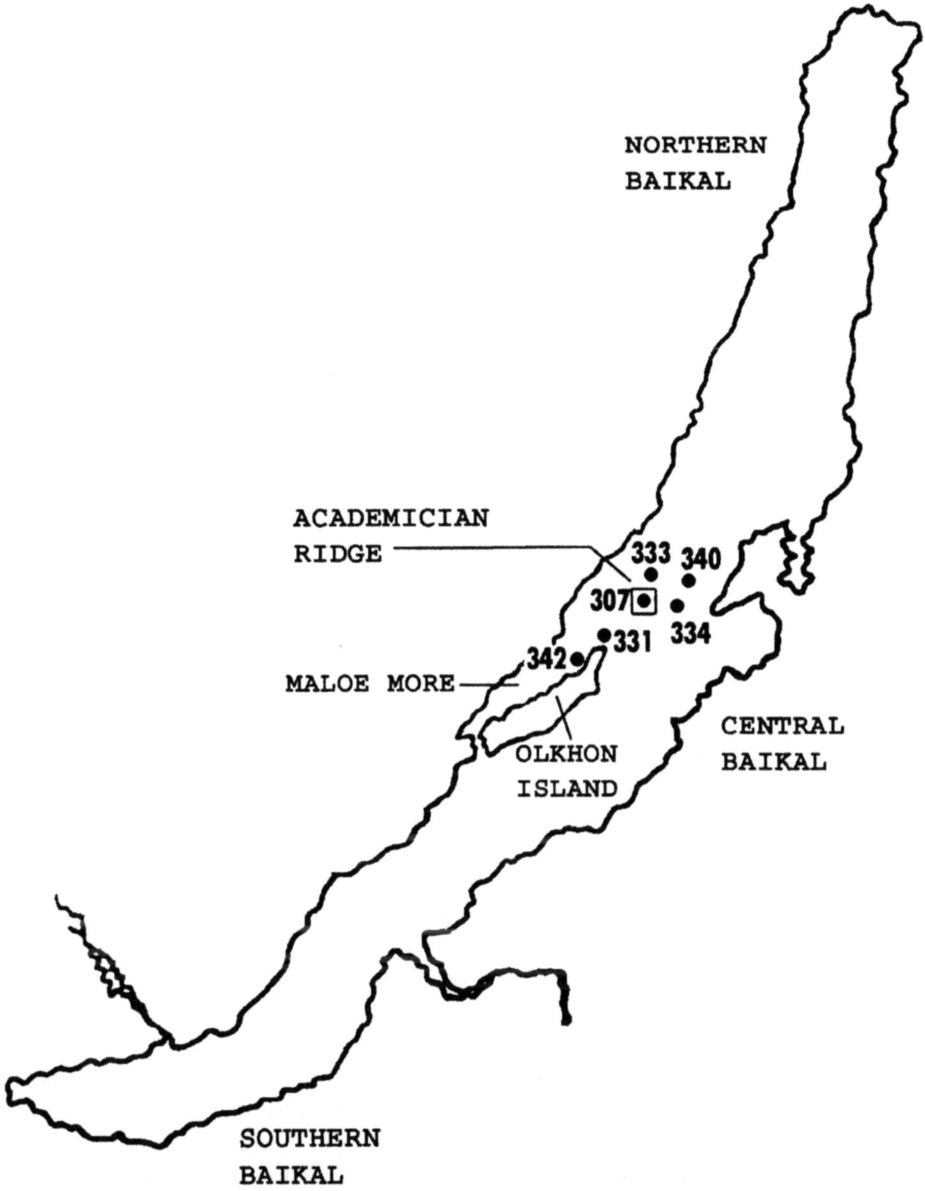

FIG. 1. Sampling sites. The numbers indicate the sites characterized by deeply buried Fe/Mn crusts which were studied by Deike et al. (1997). The *square* shows the location of stations VER-96-6TW, VER-96-1-8GC, 8PC-2, 6 PC, 6 GC, 5X GC, and 1X GC

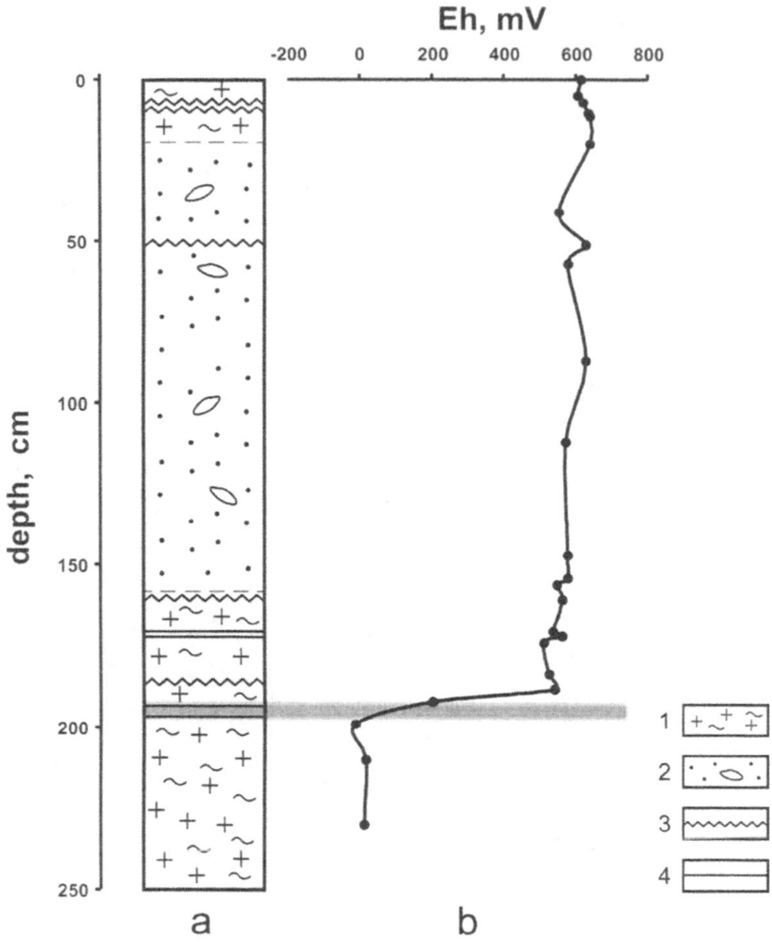

FIG. 2. St. VER-96-2-6TW core: the lithology and pattern of the vertical distribution of (**a**) Fe/Mn layers and crusts, and (**b**) the redox profile. The location of the buried crust is *shaded*. *1*, diatom ooze; *2*, silty clay; *3*, dark-brown layer rich in Fe and Mn; *4*, Fe/Mn crust

RI. Enriched layers and crusts have different Mn/Fe ratios (Table 1), and the interrelation between Fe and Mn in each enrichment also changes along the core (Fig. 4). Crusts at the bottom of the core (numbers 6 and 7 in Table 1) have a low Mn content and are ferriferous. Moving up the core, the Mn content in the crusts increases, reaching a maximum at 171.5 cm depth (number 3 in Table 1). The BSi profile is typical of Baikal sedimentary cores: relatively high (>20%) concentrations in the diatom ooze alternate with low (<10%) BSi contents in silty clays (see Fig. 3). A specific feature of the profile is that the BSi content decreases sharply in layers rich in Fe and Mn (see Fig. 3 and Table 1). This is expected, since Fe/Mn formations are secondary. The same is true of other Fe/Mn

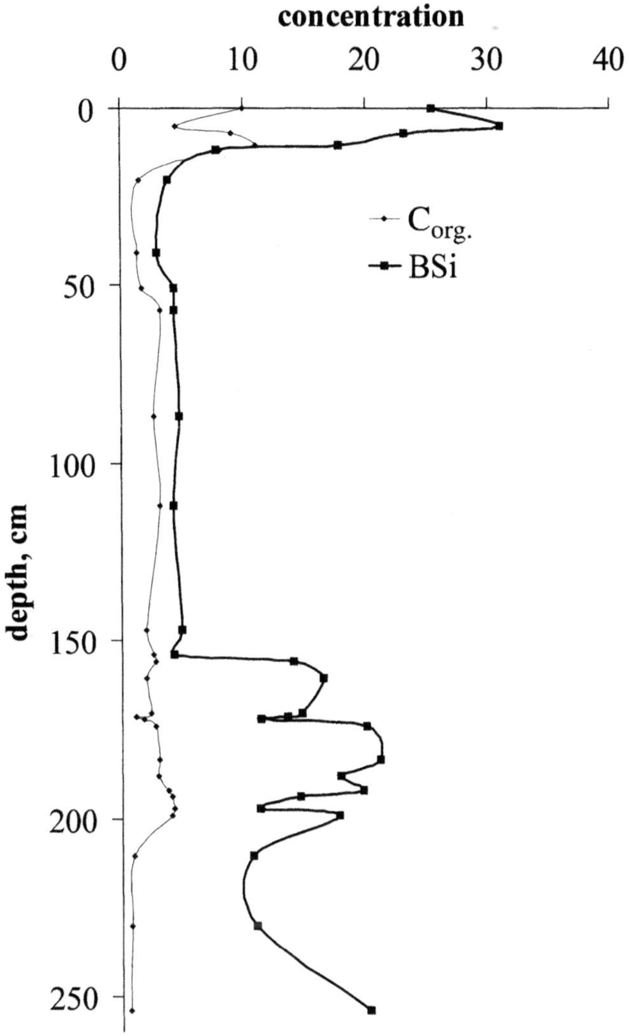

FIG. 3. Profiles of the biogenic silica (*BSi*) (*thick line*) and organic carbon ($C_{org.}$) (*thin line*) contents in the st. VER-96-2-6TW core. The concentrations are in percentages, and are shown as $C_{org} \times 10$ in the figure

crusts found deep below the WSI in this area (Table 1). According to Deike et al. (1997), the crusts are dominated by iron-rich and siliceous amorphous mineral phases, but the silicon is bounded within the lattice. The BSi concentration in Baikal sediments is a reliable indicator of warm and cold climatic periods. The correlation with the oceanic oxygen isotope curve allows us to pick out known climatic isotope stages and substages in the BSi profile (Grachev et al. 1998, and others), and thus conclude/about the age of the sedimentary layer. Thus, the upper peak on the BSi profile (see Fig. 3) corresponds to the Holocene, and the lower interval of increased BSi concentrations (156–200 cm) may be attributed to stage 5a. The relatively low BSi contents within the 210- to 230-cm interval correspond to stage 5b, whereas a new, sharp increase in BSi, starting from 230 cm

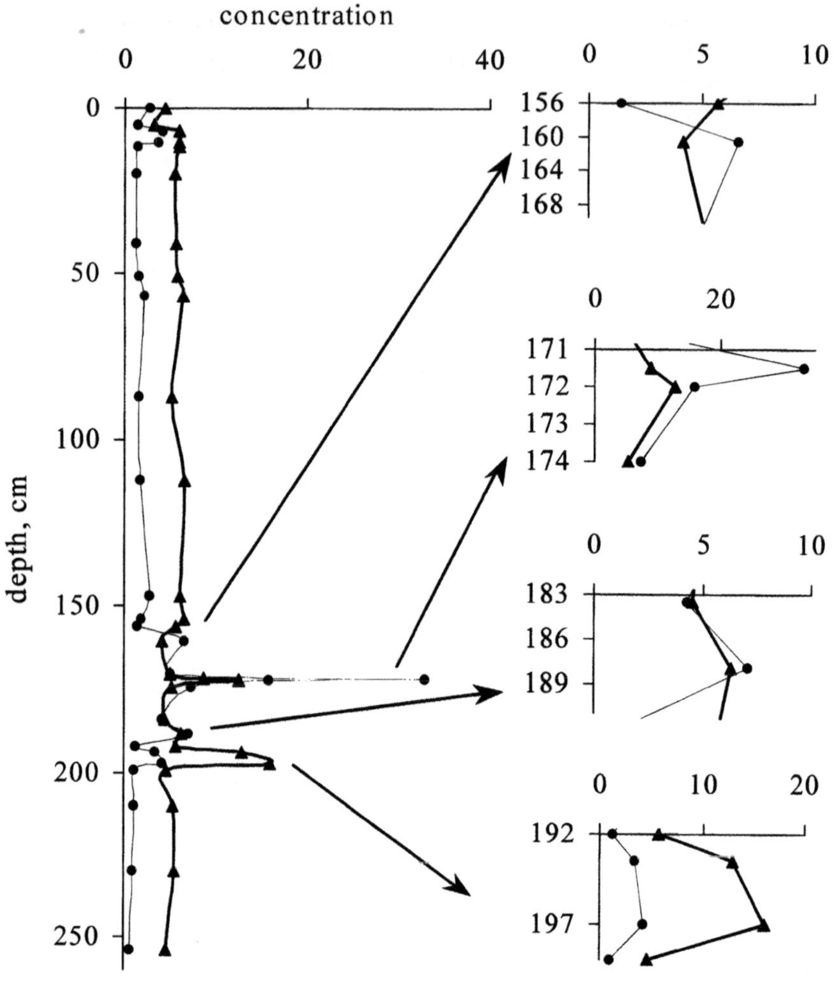

FIG. 4. Profiles of the Fe (*thick line*) and Mn (*thin line*) contents in the st. VER-96-2-6TW core. The concentrations are in percentages, and are shown as Mn × 10 in the figure

depth, may indicate the initial period of stage 5c. Data on diatom analyses confirm this approximate dating. Fossilized diatom species *Stephanodiscus flabellatus*, whose frustules appear in sedimentary intervals which correspond to isotope stage 5e, is the most reliable palaeomarker in Baikal (Grachev et al. 1997, 1998). An absence of such frustules in the samples studied (Table 2) testifies to the younger age of this sedimentary section. Published data (Grachev et al. 1997, 1998), as well as the results of diatom analyses of eight long cores from Lake Baikal (S.S. Vorobyeva, unpublished data, 2002), unambiguously indicate that the set of diatom dominants found in samples I–III (Table 2) corresponds to those typical of stage 5a. In sample V, there are species typical of stage 5c. Thus, following Martinson et al. (1987), we suppose the age of the 156- to 200-cm sedi-

TABLE 1. Geochemical characteristics of st. 6TW core sediments and Fe/Mn crusts

St. 6TW core: intervals bearing Fe/Mn enrichments, contents of Fe and Mn in the enrichment, maximal Mn/Fe ratio, and the horizon of its occurrence

No.	Interval (cm)	Fe (%)	Mn (%)	Mn/Fe	Horizon (cm)	Notes
1	5–10.5	6.0	0.41	0.07	7	EL[a]
2	156–170.5	4.2	0.66	0.16	160.5	EL
3	170–174	8.8	3.31	0.38	171.5	Crust
4		12.5	1.58	0.13	172	Crust
5	183.5–192	6.3	0.71	0.11	188	EL
6	192–199	12.9	0.34	0.03	193.5	Crust
7		16.0	0.42	0.03	197	Crust

Averaged $C_{org.}$ and BSi concentrations and Mn/Fe ratios in deeply buried crusts in comparison with those in the sediments of the st. 6TW core

	$C_{org.}$ (%)	BSi (%)	Mn/Fe
Crusts ($n = 7$)[b]	0.15	2.4	0.19
St. 6TW core ($n = 21$)[c]	0.37	13.1	0.04

[a] EL, enriched layer
[b] Crusts sampled 0.4–7 m below the water–sediment interface at stations 6TW, VER-96-1-8GC, 8PC-2, 6 PC, 6 GC, 5X GC, and 1X GC
[c] Without crusts and enriched layers
n, number of samples studied

TABLE 2. Results of diatom analyses of st. 6 TW core sediments

Number of sample	Horizon (cm)	Composition of dominant diatom complex
I	165	*Stephanodiscus grandis, Aulacoseira baikalensis, Cyclotella minuta, C. baikalensis*
II	180	The same
III	195	*Aulac. baikalensis, Stephanodiscus grandis*
IV	230	*Cyclotella baikalensis, Stephanodiscus grandis*
V	250	*Aulac. baikalensis, Cyclotella minuta, C. baikalensis, Stephanodiscus grandis*

mentary interval to be about 75–85 Kyr, and the age of the sediment at 254 cm deep to be about 100 Kyr. The crusts lying within the 193.5- to 197-cm sedimentary interval were formed about 80–85 Kyr BP.

3 Discussion

The results of recent studies show that in "yellow" oxidized sedimentary sequences there is the same alternation of diatom ooze (characterized by the Holocene–Pleistocene complexes of diatoms) and clayish layers that is typical of the whole of Lake Baikal. Therefore "yellow" oxidized sediments are not as ancient as was once suggested (Galkin and Mizandrontsev 1970, and others). The same alternation is found in the st. 6TW core (see Fig. 3). We suggest that the great thickness of the oxidized zone in this core is because redeposited material from ancient red-colored continental depositions of the upper Miocene/low Pliocene, which spread over Olkhon Island, contributed substantially to the formation of these sediments. The extremely low content of sediment $C_{org.}$ testifies in favor of this supposition. The Miocene–Pliocene cover used to exist everywhere in the region, but it was partly washed out or re-covered by younger deposits. At present it is fixed locally, in particular, on Olkhon Island, in the northern part of Maloe More (Mats et al. 2001), and in some areas of Academician Ridge (Mats et al. 2000). Red soils are colored by waterless or less hydrated iron oxides. During their redeposition in the water column, the pigment changes from its bright red color to brownish or yellow due to hydration of the dye-stuff. The brownish color of particles due to Fe(III) does not change when the particles settle in the oxygenated Baikal water column, and this determines the color of the sediments. Similar changes in the pigmentation of red soils during redeposition are well documented: in central western Pribaikalje, the redeposition of bright red Palaeogene clays has resulted in the formation of yellow Neogene clays (Mats et al. 2001). When particles accumulate in the sediment, the iron is not reduced in conditions of sluggish oxic diagenesis. This is because of the extremely low concentration of $C_{org.}$ (see Fig. 3). Moreover, according to Vykhristyuk (1980), the inert fraction of SOM (sedimentary organic matter) in this type of sediment may reach 100%. The small amount of inert SOM is unable to initiate diagenetic reduction, and therefore the sediment remains oxidized to a great depth (2-4 m), and a relatively high amount of Mn is still preserved in deeply buried crusts (see Table 1). In contrast, the oxidized relics in recent sediments contain almost no Mn owing to their intensive reductive dissolution (Granina et al. 1993).

At st. 6TW, near the RI, there still is some $C_{org.}$ (see Fig. 3) and the lowest Fe/Mn crust is reductively dissolving. The resultant dissolved Fe and Mn are redistributed near the RI forming new enriched layers and crusts (see Fig. 2). This is a typical nonsteady-state diagenetic system in which an oxidation front develops and progresses downward into the sediment, as is the case in the ocean (Wilson et al. 1986). On Academician Ridge, the dissolution/formation of Fe/Mn crusts (layers) within the uppermost sediments takes thousands of years (Granina 2002), and it may take much longer within the oxidized 2-m sedimentary sequence at st. 6TW.

In deeply buried Fe/Mn enrichments, the peaks of Fe and Mn coincide (see Fig. 4), indicating that a slow decomposition of SOM takes place. However, at the 170- to 174-cm interval, these peaks are separated along the vertical line; sediment $C_{org.}$ is completely depleted (see Fig. 3), and the dense Fe/Mn crust is not dissolving because it still contains a high Mn concentration. $C_{org.}$ depletion is also typical of the other crusts studied in this region (see Table 1), since $C_{org.}$ is taken up during the crust's formation. Thus, the Fe/Mn layers and crusts occurring at some horizons within the core (Figs. 2a and 4, and Table 1) reflect different stages of sluggish diagenetic redistribution of Fe and Mn which is typical of an oxidation front in a nonsteady-state diagenetic system.

This interpretation is different from that previously proposed by Deike et al. (1997), who studied similar crusts lying within the sediments at depths from 55 to 313 cm below the WSI. At the six stations studied in this work (see Fig. 1), the crusts were dated at 60–70 Kyr. The authors related the origin of these crusts to long periods of slow sedimentation, which could have taken place during the transition from glacial climates with high terrigenous sedimentation at the end, to interglacial ones, characterized by increased biogenic sedimentation (Deike et al. 1997).

The coordinates of st. 6TW investigated in our study coincide with those of st. 307 studied by Deike et al. (1997). The approximate age of the Fe/Mn crust in the st. 6TW core (80 Kyr) is close to that of crusts reported by Deike et al. (1997) (70 Kyr). However, according to the BSi profile (see Fig. 3) and the results of diatom analyses (see Table 2), the crust at st. 6TW formed during the warm stage 5a, when its burial could not be a result of the climate causing changes in the rate of sedimentation. Its origin may be due to the lack and inertness of SOM, as is the case in the ocean.

However, there is still the question of where the mass of Fe and Mn redistributed within the core came from. An answer can be found in the data on the regional geology of western Pribaikalje. The Paleogene peneplain and a thick weathering crust of a caolinite–laterite type used to exist everywhere in this region. During the formation of the weathering crust, huge masses of Fe and Mn oxides were mobilized. Ore concentrations of Fe and Mn oxides are widespread in the region, and are related to the relics of the peneplain and weathering crust. They form veinlets, crusts, coatings, and infiltration–metasomatic formations within the rocks of the pre-Neozoic basement and lacustrine bog ores, crusts, and nodules in Neogene–Quaternary sediments. The Fe/Mn crusts found in the lower parts of sedimentary cores taken on the northwest block of Academician Ridge (Mats et al. 2000; Khlystov et al. 1999, 2001) are fragments of such formations.

Deike et al. (1993) described the crust at 107.7 cm depth as lying within coarse sandy layers, including rock fragments. At st.VER-96-1-6GC, located 5 km from st. 6TW, at the very bottom of a long core there is a crust about 1 cm thick, which occurs within a subaerial coarse heterogeneous red-colored sandy layer of a spotted structure, with a high amount of iron oxides and carbonates (Mats et al. 2000). We have argued strongly that such crusts could be formed in ancient zones of sedimentation within the united geological structure that used to comprise Olkhon Island and Academician Ridge. Within the borders of this structure, in both the land and underwater parts, there are preserved relics of former lacustrine and subaerial Neozoic formations, which include red-colored sediments and weathering crusts. Until geologically recent times, an apical zone of the underwater Academician Ridge was dry land. However, in the late

Pleistocene, it submerged below the present Baikal level due to tectonic lowering of the Tyiskaya phase (Mats et al. 2000; Khlystov et al. 2001). During the transgression, Neozoic formations have essentially been eroded, and the products of their erosion (including red sediments) have formed part of the upper Pleistocene–Holocene lake sediments. The hypothetical time of these events (the crust age was estimated to be 70–80 Kyr) does not conflict with the Tyiskaya tectonic phase (150–120 Kyr BP), whose youngest pulses became apparent up to the Holocene. Thus, one may suppose that below the reduced sediments in the lower part of the st. 6TW core (deeper 250 m horizon), may be a crust lying in sand which is similar to the crusts described by Mats et al. (2000). This must be gradually dissolving, supplying an upward diffusive flux of Fe and Mn which are redeposited on the geochemical barriers within the upper oxidized part of the sedimentary cross section. If so, the other crusts lying on the tilted surfaces of the northwest block of the ridge near Olkhon Island (see Fig. 1), which were investigated by Deike et al. (1997), could have been formed in a similar way.

Acknowledgments. The authors thank Prof. M.A. Grachev, Director of LIN SD RAS, for his support of this work, and Prof. I.B. Mizandrontsev for useful comments. This study was supported by RFBS grants 99-05-65680 and 03-05-65255.

References

Deike R, Webster D, Callender E (1993) Sedimentary changes and authigenic mineralogy determined from site 307 cores, Academician Ridge, Lake Baikal, Siberia. Russ Geol Geophys 33(10–11):136–147

Deikc RG, Granina L, Callender E, McGee JJ (1997) Formation of iron crusts in Quaternary sediments of Lake Baikal, Russia, and its implications for paleoclimate. Mar Geol 139:21–46

Galkin LM, Mizandrontsev IB (1970) On the genesis of dense clays in bottom sediments of Lake Baikal (in Russian). Trans Zabaikal'skii Branch Geogr Soc SSSR, Cheta 6(4):64–69

Grachev MA, Likhoshway EV, Vorobyeva SS, Khlystov OM, Bezrukova EV, Veinberg EV, Golgberg EL, Granina LZ, Kornakova EG, Lazo FI (1997) Signals of paleoclimates of the upper Pleistocene in sediments of Lake Baikal (in Russian). Russ Geol Geophys 38:957–980

Grachev MA, Vorobyeva GA, Mats VD (1998) Problems of the dating of climatic events of the upper Pleistocene in Siberia and their correlation with Baikalian and global chronicles. In: Luchinskaya IA (ed) Problems of the reconstruction of the natural environment of the Holocene and Pleistocene in Siberia (in Russian). Institute of Archeology and Ethnography, Novosibirsk, p 85–95

Granina L (2002) Origin and dynamics of Fe and Mn crusts in sediments of Lake Baikal. Geochim Cosmochim Acta, Goldschmidt Conference Abstracts, 66(15A):A289

Granina L, Karabanov E, Callender E (1993) Relics of oxidized ferromanganese formations in the bottom sediments of Lake Baikal. In: Horie S (ed) IPPCC Newsl 7:32–39

Granina L, Muller B, Wehrli B, Martin P (2000) Oxygen, iron, and manganese at the sediment–water interface in Lake Baikal. Terra Nostra 9:87–94

Khlystov OM, Mats VD, Granina LZ, Fedorin MA, Goldberg EL (1999) New data on the structure of the N–W block of Academician Ridge in Lake Baikal (in Russian). Geology of Seas and Oceans 1:398–399

Khlystov OM, Mats VD, DeBatist M (2001) Southwest ending of the north Baikalian depression: geological structure and correlation with Neozoic sections of Olkhon Island and BDP-96 and BDP-98 cores (in Russian). Russ Geol Geophys 42:373–383

Martinson DG, Pisias NG, Hayes JD (1987) Age-dating the orbital theory of the ice ages. Development of high-resolution 0 to 300 000-year chronostratigraphy. Q Res 27:1–29

Mats VD, Khlystov OM, De Batist M, Lomonosova TK, Klimanskii AV (2000) Evolution of the Academician Ridge accommodation zone in the central part of the Baikal Rift, from high-resolution reflection seismic profiling and geological field investigations. Int J Earth Sci 89:229–250

Mats VD, Ufimtsev GF, Mandelbaum MM (2001) Cenozoic of the Baikal Rift depression (structure and development) (in Russian). NITS OIGGM, Novosibirsk, 253 p

Vykhristyuk LA (1980) Organic matter of Baikal bottom sediments (in Russian). Nauka, Novosibirsk, p 80

Wilson TRS, Thompson J, Hydes DJ, Colley S, Culkin F, Sorensen J (1986) Oxidation fronts in pelagic sediments: diagenetic formation of metal-rich layers. Science 232: 972–975

Dating Methods Applied to Baikal Sediments

Paleomagnetism and Paleoenvironmental Magnetism Studied on BDP-98 Sedimentary Cores from Lake Baikal

Hideo Sakai[1], Shigehiro Nomura[1], Masae Horii[2], Kenji Kashiwaya[3], Takayoshi Kawai[4], Vadim Kravchinsky[5], and John Peck[6]

Summary. Paleomagnetic and rock-magnetic studies were conducted on two sedimentary cores, BDP-98-1 (200 m in length) and BDP-98-2 (ca. 410 m in length from 191 m to 600 m deep), drilled at the Academician Ridge of Lake Baikal.

The magnetic inclination records assigned to the geomagnetic polarity time scale (GPTS) indicate that BDP-98-1 attained an age of 5 Ma. For BDP-98-2, owing to the long reversed interval between 290 m and 350 m, it became difficult to find a correlation to GPTS. A correlation tested in this study suggests that the basal age is over 11 Ma at 600 m deep. The average sedimentation rate for BDP-98-1 and the upper 50 m of BDP-98-2 is estimated as 4.1 cm/kyr. The sedimentation over the long reversed interval is as high as 29.3 cm/kyr, and below this depth, to 600 m, the rate is about 4.5 cm/kyr.

The magnetic properties of the BDP-98-1 sediment were also studied. The rock-magnetic parameter of anhysteretic remanent magnetization (ARM) susceptibility/susceptibility decreases after about 3 Ma, while the susceptibility increases. Spectral analyses on the susceptibility show that the shift in spectral character from a 100-kyr eccentricity cycle to a 41-kyr obliquity cycle occurred at ca. 3 Ma. These characteristic changes may be correlated with the change in the paleoenvironment in the Baikal region caused by the intensification of northern hemisphere glaciation. After 1.2 Ma, the susceptibility variation increases in amplitude, which may be correlated with the elevation of the Himalayas and Tibetan Plateau, which reached above the snow line at 1.2 Ma.

[1] Department of Earth Sciences, Toyama University, Gofuku, Toyama 930-8555, Japan
[2] Japan Marine Science and Technology Center, Natsushima, Yokosuka 237-0061, Japan
[3] Institute of Nature and Environmental Technology, Kanazawa University, Kakuma, Kanazawa 920-1192, Japan
[4] National Institute for Environmental Studies, 16-2 Onogawa, Tsukuba, 305-8506, Japan (present address: Department of Earth and Environmental Sciences, Graduate School of Environmental Studies, Nagoya University, Furo-cho, Chikusa-ku, Nagoya 464-8602, Japan)
[5] Vinogradov Institute of Geochemistry, Siberian Branch of the Russian Academy of Sciences, Favorsky str. 1A, Irkutsk 664033, Russia
[6] Department of Geology, University of Akron, Akron, OH 44325-4101, USA

Key words. Lake Baikal, Paleomagnetism, Magnetic property, Magnetostratigraphy, Paleoenvironment

1 Introduction

Lake Baikal, in eastern Siberia, is one of the deepest, most voluminous, and oldest freshwater lakes in the world. It is an important and unique site for paleoclimatic studies because of its high latitude, continental-interior setting, and long, continuous stratigraphic record.

In Lake Baikal, several paleomagnetic studies have recently been carried out on cores of lake sediment (e.g., Peck et al. 1996; Sakai et al. 1997, 2000, 2001; Kravchinsky et al. 1998; Horii et al. 2001; Krainov et al. 2001). In this chapter, we show the results of paleomagnetic and rock-magnetic studies on the BDP-98 cores drilled by the International Baikal Drilling Project (BDP) during the winter of 1998. Two of the cores (BDP-98-1 and BDP-98-2) were drilled at Academician Ridge in Lake Baikal (Fig. 1). Near the BDP-98 site, a BDP-96 core 200 m long had been drilled during the winter of 1996.

The drilling of BDP-98-1 (length 200 m) was conducted by piston coring. BDP-98-2 (about 410 m long) was drilled from 191 m to 600 m deep. In the upper portion (depth <277 m), drilling by piston coring was done, and the lower portion (277–600 m) was drilled by rotary coring. The recovery of the core was over 90%. The BDP-98-2 core is the longest that has been drilled at Baikal, so valuable data on the paleoenvironment and the Earth's science were expected. This study had two purposes. One was to examine the magnetostratigraphy to obtain the age-scale of the sedimentary sequence. The other was to study the paleoclimate and paleoenvironment through the magnetic properties of the sediment.

2 Samples for Magnetic Study, Experimental Methods, and Apparatus

The cores were divided into subcores of 2-m length, and each subcore was cut in half lengthwise (split). Samples for paleomagnetic study were taken in plastic cube cases of 10 cm^3 at an interval of 5 or 10 cm. These discrete samples were divided between the Russian, USA, and Japanese groups. In this chapter, we give the data from the Japanese samples (the sampling interval being 15 or 30 cm).

The remanent magnetization was measured using a pass-through-type cryogenic magnetometer (2G Enterprise 760R), and the magnetic cleaning was done by the alternating field (AF) demagnetization method. Magnetic susceptibility was measured with an AGICO KLY-3 meter.

Firstly, the pilot samples from discrete samples were step-wisely AF demagnetized to 90 mT in 5 mT steps. In most of the samples, the secondary magnetization was eliminated by demagnetization to 15 mT, so that the other samples

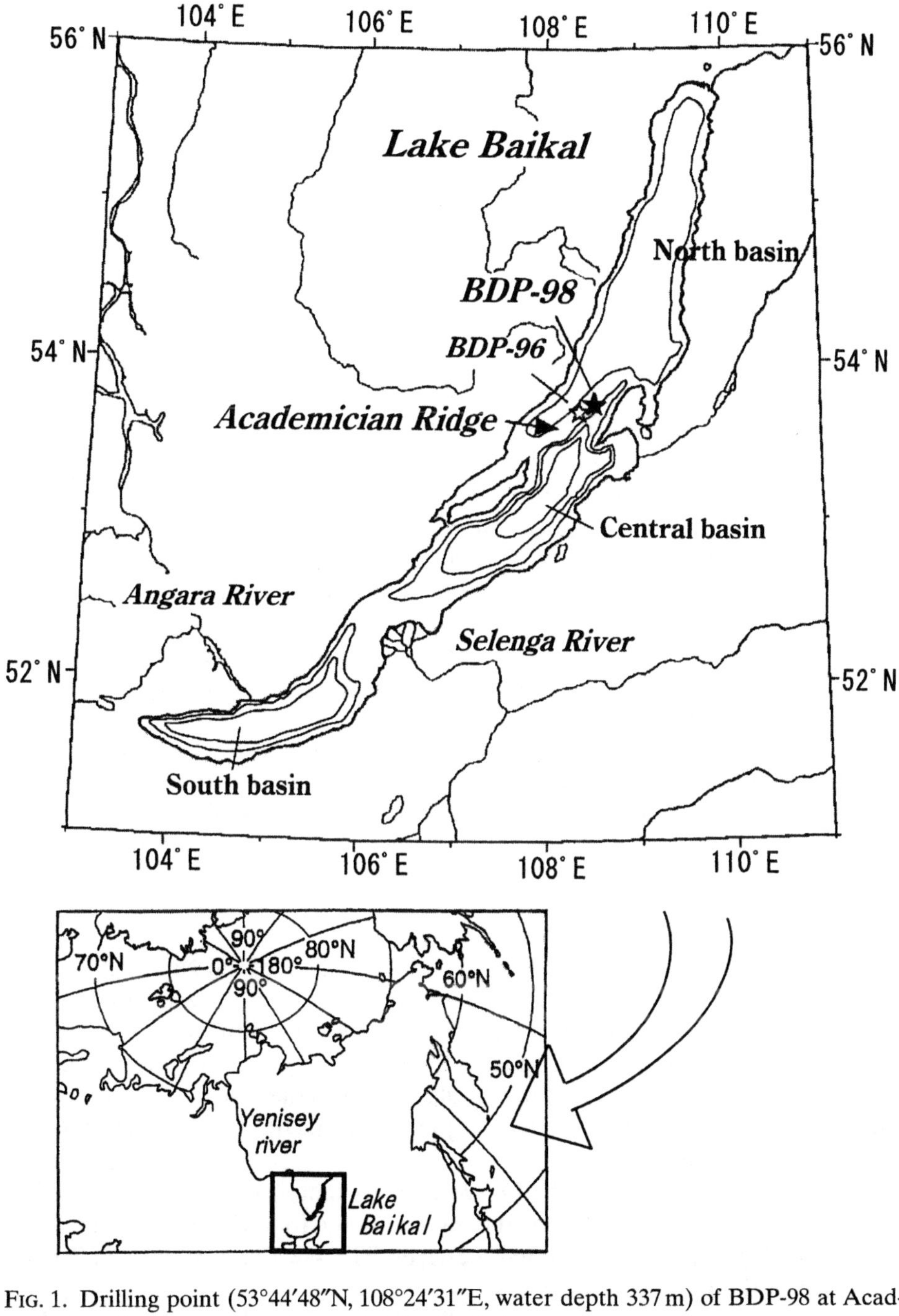

FIG. 1. Drilling point (53°44′48″N, 108°24′31″E, water depth 337 m) of BDP-98 at Academician Ridge in the central part of Lake Baikal

were demagnetized in four steps of 5, 10, 15, and 20 mT. The paleomagnetic data after 15 mT AF demagnetization are mainly used in the following discussion.

3 Paleomagnetic Results and Magnetostratigraphy

Figure 2 shows the direction and intensity of the remanent magnetization of the discrete samples for two cores. We can identify clear polarity reversals both in inclination and declination records.

3.1 BDP-98-1 Core

Figure 3 shows the inclination data of the BDP-98-1 core. The inclination change with depth shows a clear polarity reversal pattern. Comparison with the geomagnetic polarity timescale of Cande and Kent (1995) reveals the geomagnetic polarity epochs of Brunhes normal polarity, Matuyama reversed polarity, Gauss normal polarity, and Gilbert reversed polarity. Most of the geomagnetic events during the above polarity epochs were also identified. This indicates that the BDP-98-1 core covers an age of ca. 5 Ma.

The depth-age scale in Fig. 4 suggests that sedimentation at the rate of 4.1 cm/kyr proceeded fairly constantly. The similar result shown in the lower figure was obtained from the BDP-96 core (see Fig. 1). That is, sedimentation during the past 5 Ma at Academician Ridge has proceeded under a quiet environment with no remarkable tectonic event.

Several short inclination anomalies (reversed and/or narrow inclination zones) found during the Brunhes epoch (Fig. 3) may be correlated to the geomagnetic excursions, as discussed for BDP-96 (Kravchinsky et al. 1998). Further study is necessary to examine whether these are the formerly unknown geomagnetic events.

3.2 BDP-98-2 Core

The core depth is calibrated to the data of A. Tanaka, of the National Institute of Environmental Studies in Japan. Below 277 m depth, owing to the long reversed interval between 290 m and 350 m, it becomes difficult to find a correlation to any reference geomagnetic time-scale. Figure 5 is one reasonable correlation, which suggests that the basal age of the core is over 11 million years. In Fig. 6, the geomagnetic polarity assignment for BDP-98-1 and BDP-98-2 are summarized together.

The magnetostratigraphy of BDP-98-2 needs more information, and will be examined by new and/or additional data, taking the following subjects into consideration: (a) the depth adjustment between BDP-98-1 and BDP-98-2; (b) how to interpret the long reversed region.

Figure 7 shows the variation in sedimentation rate. The rate for the long reversed interval (290–350 m) is high at 29.3 cm/kyr. Then below 350–600 m, the rate becomes about 4.5 cm/kyr, which is comparable with the rate for BDP-98-1 of 4.1 cm/kyr.

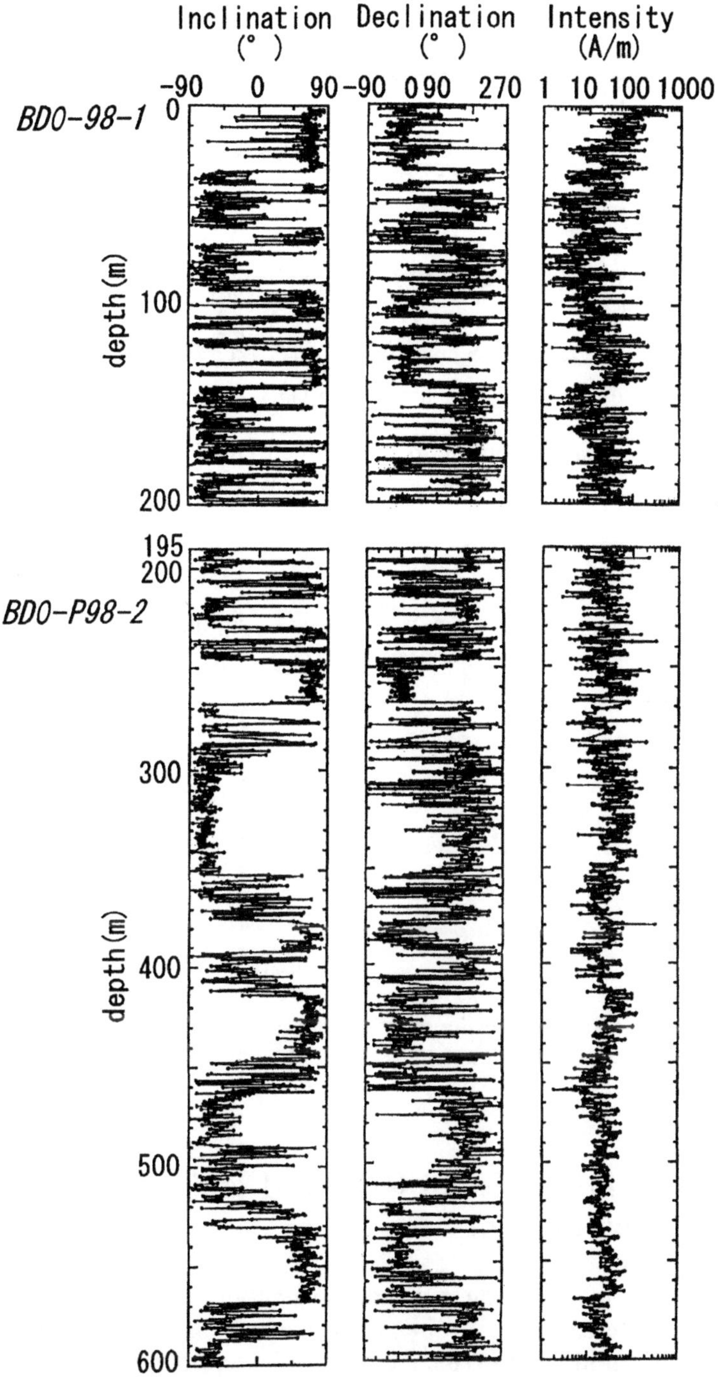

FIG. 2. Variation of inclination, declination, and intensity of remanent magnetization with depth from discrete samples of BDP-98 cores after 15 mT alternating field demagnetization

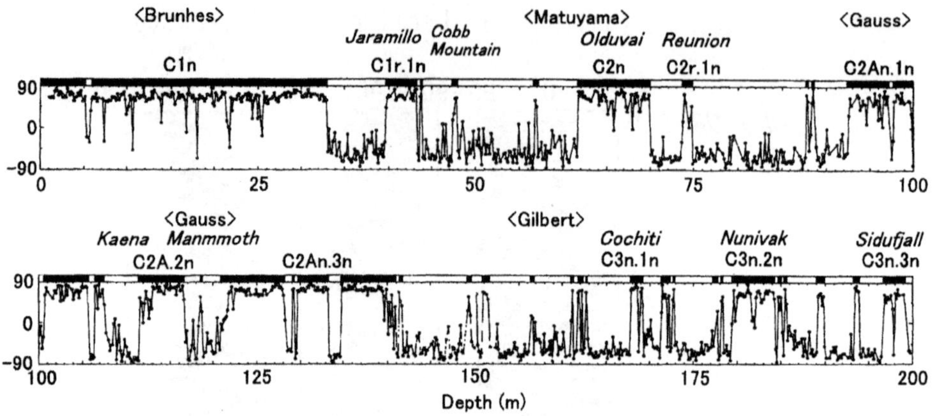

FIG. 3. Variation of inclination with depth for the BDP-98-1 core. In the figure, the geomagnetic polarity time-scale of Cande and Kent (1995) is used

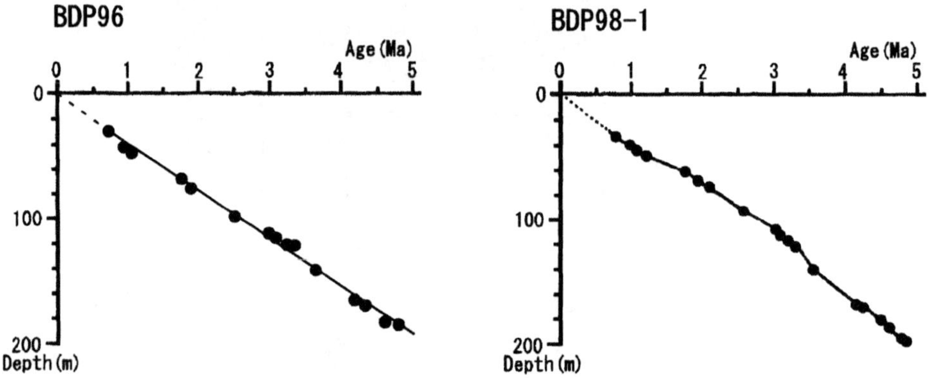

FIG. 4. Depth–age diagram for the polarity boundary of geomagnetic events and epochs for BDP-96 (*left*) and BDP-98-1 (*right*) cores (Sakai et al. 2000)

4 Magnetic Properties of BDP-98-1 and the Paleoenvironment

Figure 8b shows the variation in magnetic susceptibility of BDP-98-1. We can see an increase in susceptibility from ca. 3 Ma, which agrees with the BDP-96 susceptibility shown in Fig. 8a. This increase in susceptibility may be caused by an increase in the magnetic mineral concentration in the lake sediment. Thermal analysis shows that the magnetic mineral responsible for remanent magnetiza-

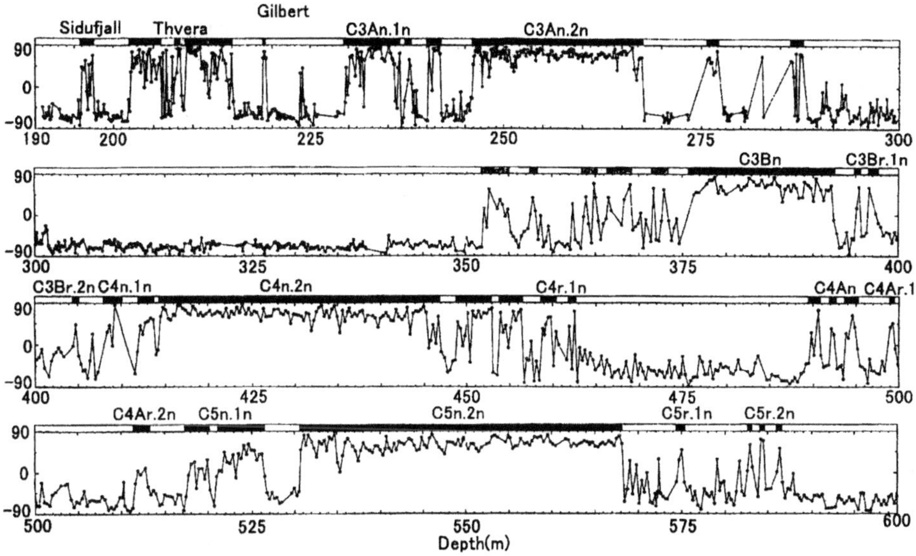

FIG. 5. Variation of inclination with depth for the BDP-98-2 core. In the figure, the geomagnetic polarity time-scale of Cande and Kent (1995) is used

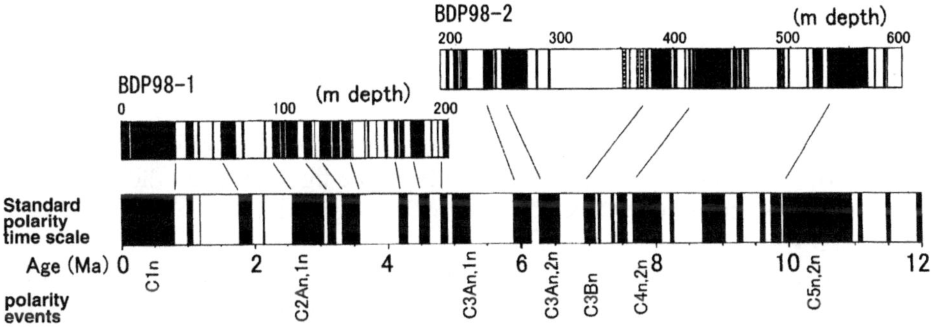

FIG. 6. Depth–age diagram for the polarity boundary of geomagnetic events and epochs for the BDP-98 core. In the lower part, the geomagnetic polarity time-scale of Cande and Kent (1995) is referred.

tion of the sediment around the Academician Ridge is magnetite (Sakai et al. 2000). King et al. (1982) reported that the anhysteretic remanent magnetization (ARM) is sensitive to the grain size of magnetite and that ARM susceptibility/susceptibility has an inverse correlation with the change in the grain size of the magnetite in the sediment. In Fig. 8c, ARM susceptibility/susceptibility shows a decrease from about 3 Ma. We thus conclude that the quantity of magnetite

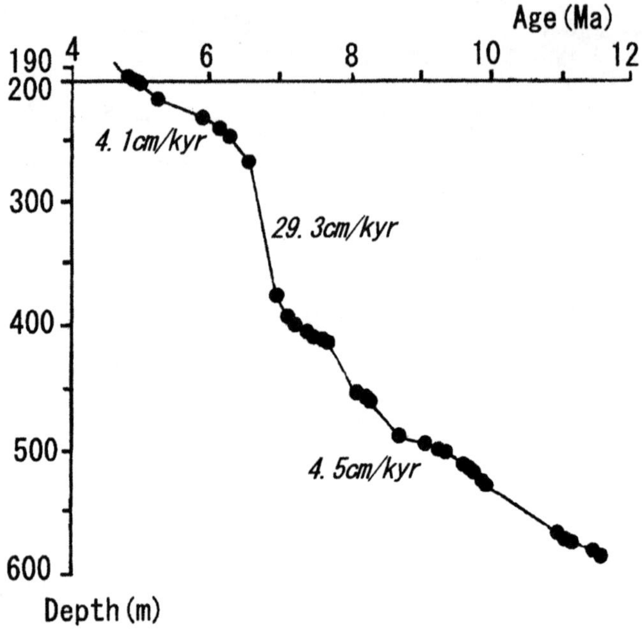

FIG. 7. Assignment of inclination (polarity) records for BDP-98-1 and BDP-98-2 cores with the geomagnetic polarity reversal events

with a fine grain size has increased in the BDP-96 and BDP-98 sedimentary sequences since ca. 3 Ma.

Figure 9 shows the result of a spectral analysis of susceptibility in relation to Milankovich cycles. On the right-hand side of the figure, we see the shift in the periodogram between a 100-kyr eccentricity cycle and a 41-kyr obliquity cycle around 3 Ma.

The above-mentioned changes in the susceptibility and rockmagnetic parameters of lake sediment may have been caused by a paleoenvironmental change around the Lake Baikal area. The intensification of the northern hemisphere glaciation occurred around 3 Ma due to the uplift of the Himalayan and Rocky Mountains (Masuda 1991) and/or the progressive closing of the Panamanian isthmus (Driscoll and Haug 1998). Accordingly, the climate in the Lake Baikal area became cooler, which may have affected the magnetic properties and/or the concentration of fine-grain magnetite in the sediment of Lake Baikal.

Figure 8b shows that the susceptibility and its amplitude of fluctuation increased after 1.2 Ma. Spectral analysis (Fig. 9) shows a shift in the spectral character from a 41-kyr obliquity cycle to 100-kyr eccentricity cycle over this period. The concentration of magnetic minerals in the sediment may have been affected by the change in the sedimentary environment around 1.2 Ma. The probable reason for this change is that the Himalayas and the Tibetan Plateau were raised above the snow line (Yasunari and Seki 1992).

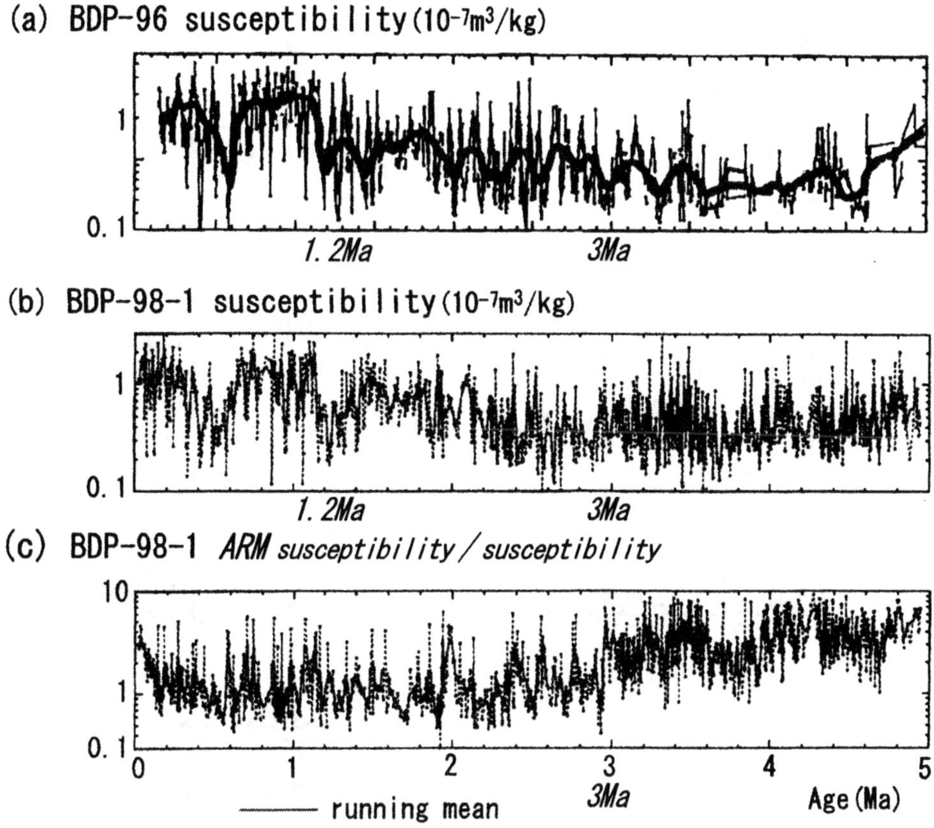

FIG. 8. Variation of susceptibility (BDP-96, BDP-98-1) and ARM susceptibility/susceptibility (BDP-98-1) with age

In summary, a global change in the climate and the environment is clearly recorded in the magnetic properties of Lake Baikal sediment. The mechanism by which these variations in magnetic properties and/or magnetic minerals correlate to the paleoclimate may be not simple. We should consider it as not only a change in the source of the sediment supply, and a change in magnetic minerals, but the effect of bacterial magnetite (Peck and King 1996; Sakai et al. 2000).

5 Concluding Remarks

Paleomagnetic and rock-magnetic (paleoenvironmental) studies were conducted on two BDP-98 sedimentary cores from Lake Baikal.

Magnetostratigraphy indicated that the BDP-98-1 core showed an age of 5 Ma. In the BDP-98-2 core, owing to the long reversed interval between 290 m and

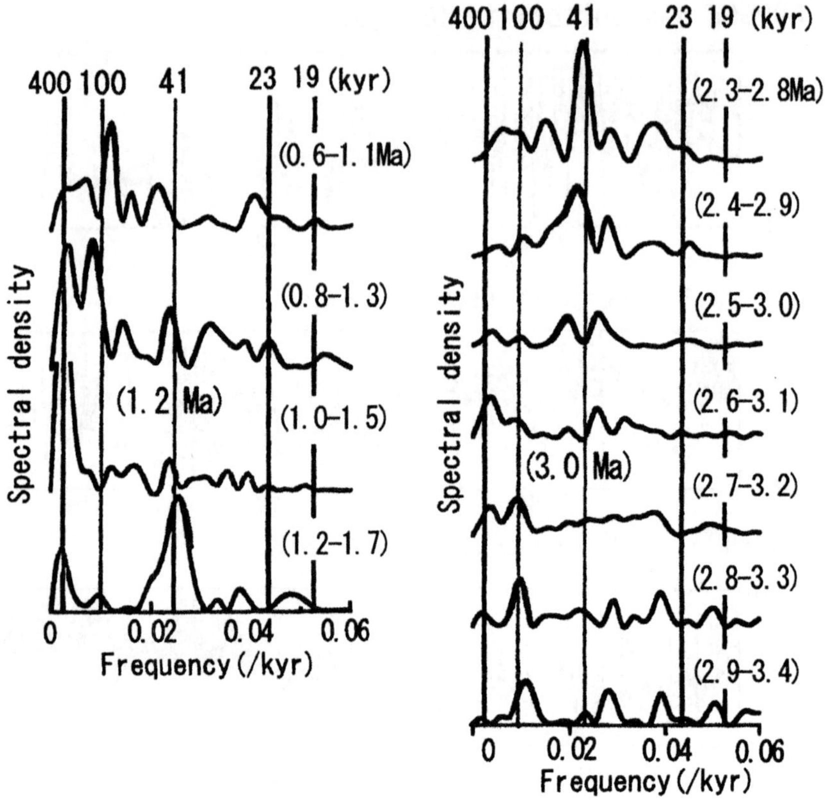

FIG. 9. Spectral analysis of the susceptibility shown in the BDP-98-1 core

350 m depth, it was more difficult to find a correlation to a reference geomagnetic time-scale. A reasonable correlation in this study suggests that the basal age of the core is over 11 million years. The magnetostratigraphy of BDP-98-2 will be examined further by new and/or additional data.

The average sedimentation rate for BDP-98-1 and the upper 50 m of BDP-98-2 is estimated to be 4.1 cm/kyr. Sedimentation over the long reversed interval proceeded at the high rate of 29.3 cm/kyr, and below c.a. 400 m, to 600 m, the rate was about 4.5 cm/kyr.

An analysis of magnetic susceptibility was done on the BDP-98-1 core. The susceptibility increases from about 3 Ma, while the magnetic property of ARM susceptibility/susceptibility decreases. The concentration of fine-grain magnetite in the sediment may have increased since ca. 3 Ma. This may be correlated with a climatic cooling in the Lake Baikal area which correlates with the global event of northern hemisphere glaciation intensification.

The amplitude of the fluctuation in susceptibility increases after 1.2 Ma. The periodogram indicates a shift in spectral character from a 41-kyr obliquity cycle to a 100-kyr eccentricity cycle over this period. The change in the sedimentary environment may have occurred around ca. 1.2 Ma. A probable reason for the change is the elevation of the Himalayas and the Tibetan Plateau above the snow line.

Thus, both the geomagnetic dipole polarity change and the global climatic change are clearly recorded in the magnetization of the Lake Baikal sediment.

Acknowledgments. We acknowledge the BDP (Baikal Drilling Project) members from Russia, the USA, and Japan for their help in this study.

References

Cande SC, Kent DV (1995) Revised calibration of the geomagnetic polarity timescale for the late Cretaceous and Cenozoic. J Geophys Res 100:6093–6095

Driscoll NW, Haug GH (1998) A short circuit in thermohaline circulation: a cause for northern hemisphere glaciation? Science 282:436–438

Horii M, Sakai H, Kashiwaya K, Kawai T (2001) Reconsideration of the age-scale for sediment from Lake Baikal and fluctuations of the magnetic process. Russ Geol Geophys 42:175–185

King J, Banerjee SK, Marvin JM, Ozdemir O (1982) A comparison of different magnetic methods for determing the relative grain size of magnetite in natural materials: some results from lake sediments. Earth Planet Sci Lett 59:404–419

Krainov MA, Kravchinsky VA, Peck JA, Sakai H, King JW, Kuzmin MI (2001) Paleoclimate record of the Lake Baikal sediments with magnetic susceptibility study result. Russ Geol Geophys 42:87–97

Kravchinsky VA, Peck JA, Sakai H, King JW, Nomura S, Tanaka A, Kuzmin MI, Williams D, Kawai T (1998) Magnetostratigraphy scale of the late Cenozoic of central Asia from data obtained from the Baikal Drilling Project (in Russian). In: Dobretsov NL, Kovalenko VI (eds) Geodynamic reorganizations of the lithosphere. Publishing House of the Russian Academy of Science, Siberian Branch, OIGGM, Novosibirsk, p 73–78

Masuda F (1991) Evaluation of the present on the history of paleoclimate change. J Geogr 100:965–987

Peck JA, King JW, Colman SM, Kravchinsky VA (1996) An 84-kyr paleomagnetic study. Earth Planet Sci Lett 140:159–172

Sakai H, Nakamura T, Horii M, Kashiwaya K, Fujii S, Takamatsu T, Kawai T (1997) Paleomagnetic study with ^{14}C dating analysis on three short cores from Lake Baikal. Bull Nagoya Univ Furukawa Mus 13:11–22

Sakai H, Nomura S, Horii M, Kashiwaya K, Tanaka A, Kawai T, Kravchinsky V, Peck J, King J (2000) Paleomagnetic and rock magnetic studies on Lake Baikal sediments—BDP-96 borehole at Academician Ridge. In: Minoura K (ed) Lake Baikal: a mirror in time and space for understanding global change processes. Elsevier, Amsterdam, p 35–52

Sakai H, Nomura S, Horii M, Araki S, Kashiwaya K, Tani Y, Shibata Y, Kawai T, Kravchinsky V, Peck J, King J (2001) Magnetic susceptibility studies on surface sediments of Lake Baikal and Lake Biwa. Russ Geol Geophys 42:339–347

Yasunari T, Seki Y (1992) Role of the Asian monsoon on the interannual variability of the global climate system. J Meteorol Soc Jpn 69:473–487

Exploratory Studies of Dating the Baikal Drilling Project Sediment Core (BDP-96) Using Cosmogenic ^{10}Be: Observations and Implications of Higher ^{10}Be Concentrations in Sediments Older than 2.7Ma

Kazuho Horiuchi[1], Koichi Kobayashi[2], Hideo Sakai[3], Shigehiro Nomura[3], Hiroyuki Matsuzaki[4], Atsushi Tanaka[5], and Yasuyuki Shibata[5]

Summary. The concentrations of ^{10}Be and ^{9}Be were determined in the Baikal Drilling Project 96 hole 1 (BDP-96/hole 1) core using accelerator mass spectrometry at the University of Tokyo, and inductively coupled plasma atomic emission spectrometry at the National Institute for Environmental Studies. The ^{10}Be concentrations showed an increasing upward trend in the core, while the ^{9}Be showed nearly constant values around 2.2 p.p.m. This means that the profile of the $^{10}Be/^{9}Be$ ratio is nearly the same as that of the ^{10}Be concentrations. ^{10}Be concentrations and the $^{10}Be/^{9}Be$ ratio plotted against magnetostratigraphic ages showed that the principal factor controlling the ^{10}Be profiles is the disintegration of the nuclide (half-life 1.5 Myr), while the secondary factor is the dilution effect of glacial clays, which started simultaneously with the Northern Hemisphere glaciation (2.7 Ma). The normal (undiluted) initial ^{10}Be concentration ($^{10}Be^{0}$) and the $^{10}Be^{0}/^{9}Be$ ratio in the BDP-96/hole 1 core are estimated as $16.0 \pm 1.2 \times 10^{8}$ atoms/g and $10.6 \pm 0.8 \times 10^{-9}$ at./at., respectively. These are useful as the initial values for approximate ^{10}Be dating, which seems to provide us with a constraint for "accurate" magnetostratigraphy on BDP cores.

[1] Earthquake Research Institute, University of Tokyo, 1-1-1 Yayoi, Bunkyo-ku, Tokyo 113-0032, Japan
(present address: Department of Earth and Environmental Sciences, Faculty of Science and Technology, Hirosaki University, Bunkyo-cho, Hirosaki, Aomori 036-8561 Japan
[2] Institute of Accelerator Analysis Ltd., 6-270 Ichiridan, Shirasaka, Shirakawa, Fukushima 961-0835, Japan
[3] Department of Earth Science, Faculty of Science, Toyama University, 3190 Gofuku, Toyama 930-8555, Japan
[4] Research Center for Nuclear Science and Technology, University of Tokyo, 2-11-16 Yayoi, Bunkyo-ku, Tokyo 113-0032, Japan
[5] Environmental Chemistry Division, National Institute for Environmental Studies, 16-2 Onogawa, Tsukuba 305-0053, Japan

Key words. Beryllium 10 dating, Magnetostratigraphy, Lake Baikal, Northern Hemisphere glaciation, Central Asia

1 Introduction

Cosmogenic radionuclide ^{10}Be (half-life 1.5 Myr) in sediments, originally produced in the atmosphere of the earth, has been considered as a candidate for radiometric dating of sediments up to 10 Ma (e.g., Amin et al., 1975; Tanaka and Inoue 1979; Bourles et al. 1989). However, only a few attempts have succeeded in this type of dating because environmental factors frequently veil the chronological information preserved in sedimentary ^{10}Be records (e.g., Somayajulu 1977; Bourles et al. 1989). Our preliminary study (Horiuchi et al. 1998) has already shown a down-core decrease in ^{10}Be concentration in a long Baikal Drilling Project (BDP) core. Nevertheless, subsequent studies on the late Quaternary sediments in Lake Baikal show significant climatic control over the fluctuations of ^{10}Be concentrations, although the investigations were restricted to within short (a few hundred thousand year) periods (Horiuchi et al. 1999, 2000, 2001). Here, we examine the 5-Myr record of ^{10}Be and ^{9}Be from a 200-m sediment core from Lake Baikal (BDP-96/hole 1 core) from geochronological viewpoints. The ^{10}Be concentrations and the ^{10}Be/^{9}Be ratio in this core should more or less reflect the decay of the nuclide in three half-lives.

The goal of this work was to explore the possibility of dating BDP cores using both cosmogenic ^{10}Be and magnetostratigraphy. Although magnetostratigraphy has been successful for dating BDP cores (Baikal Drilling Project Group 1998; Sakai et al. 2000), another independent way is needed. This is because obtaining *accurate* ages solely by the reversible directions of detrital remnant magnetization is difficult when drastic changes in sedimentation rates and/or hiatus exist in the sequence. Despite some problems in its strict application to clastic sediments, the disintegration of ^{10}Be is one of the most promising ways to independently confirm the accuracy of the stratigraphic ages. Our investigations of the 5-Myr record of ^{10}Be illuminate problems and solutions for future multiple dating of Lake Baikal sediments.

2 Sediment Core

The BDP-96/hole 1 core was recovered from the deep-water topographic high Academician Ridge of Lake Baikal (53°41′48″N, 108°21′06″E, water depth 321 m) (Fig. 1). The core is 200 m long, composed of hemipelagic sediments (clay, silt, and diatoms), and dated magnetostratigraphically to 5 Ma (Sakai et al. 2000). A lot of palaeoenvironmental work on this core has been published (e.g., Baikal Drilling Project Group 1998; Müller et al. 2001). The sedimentary record of the BDP-96/hole 1 core is one of the present standards for investigating the continental climate of central Asia since the early Pliocene.

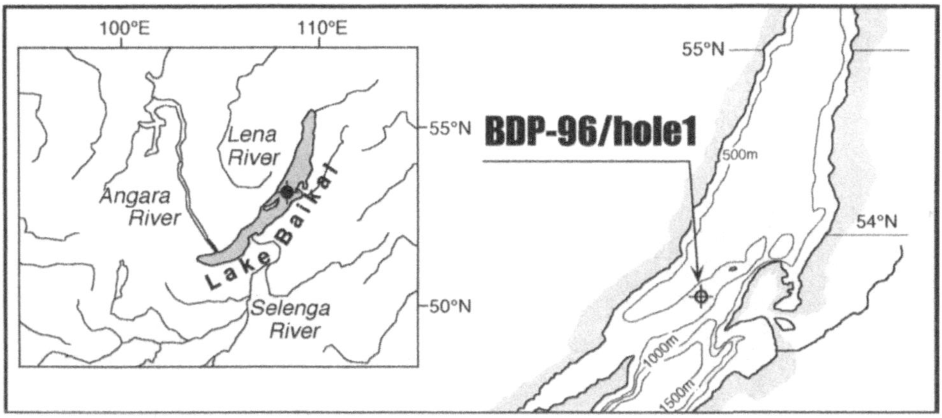

FIG. 1. Map of Lake Baikal showing the location of the core used in this study (BDP-96/hole 1)

3 Analytical Methods

Forty-two samples of depth-averaged (~2 m) sediments, taken at 2- to 8-m intervals, were submitted for ^{10}Be analysis. About 1 g of sediment (a few samples were about 0.3 g) was dissolved in a concentrated $HF/HNO_3/HClO_4$ solution and evaporated to dryness on a hotplate. The residue was dissolved in 0.5 M HCl and brought to a volume of 100 ml. An aliquot of the solution (5 ml) was taken for subsequent ^{9}Be analysis. From 0.4 to 0.8 mg of accurately weighed ^{9}Be carrier was then added to the solution. The solution was evaporated to dryness on a hotplate. Anion and cation exchange separations were then performed to isolate the beryllium fraction. $Be(OH)_2$ was precipitated by adding NH_4(aq.) to the purified solution, which was then converted to BeO by heating in a furnace at 850°C for 15 min, followed by heating at 360°C for 1 h. The BeO was mixed with Ag powder (1 : 1 v/v) for accelerator mass spectrometry (AMS) analysis of the ^{10}Be/^{9}Be ratio using the MALT–AMS system at the University of Tokyo (Kobayashi et al. 2000; Matsuzaki et al. 2000). The measured ratios were calibrated using an internal standard BT97A, which is a diluted ICN standard from K. Nishiizumi, and has a ^{10}Be/^{9}Be ratio of 3.432×10^{-11}. The fully propagated error of the AMS measurement (1σ) was around 5%, and blanks made from a commercially available Be solution showed a ^{10}Be/^{9}Be ratio of about 2×10^{-14} or less. Replicated (triplicate and quadruplicate) ^{10}Be analysis on samples from the BDP-96/hole 1 core, performed at the University of Tokyo between 1997 and 1998, are within the simple experimental errors of each AMS analysis (Fig. 2).

The stable ^{9}Be concentration was determined by using inductively coupled plasma atomic emission spectrometry (ICP–AES) at the National Institute for Environmental Studies. The analytical errors in the ^{9}Be concentration were principally between 5% and 10%.

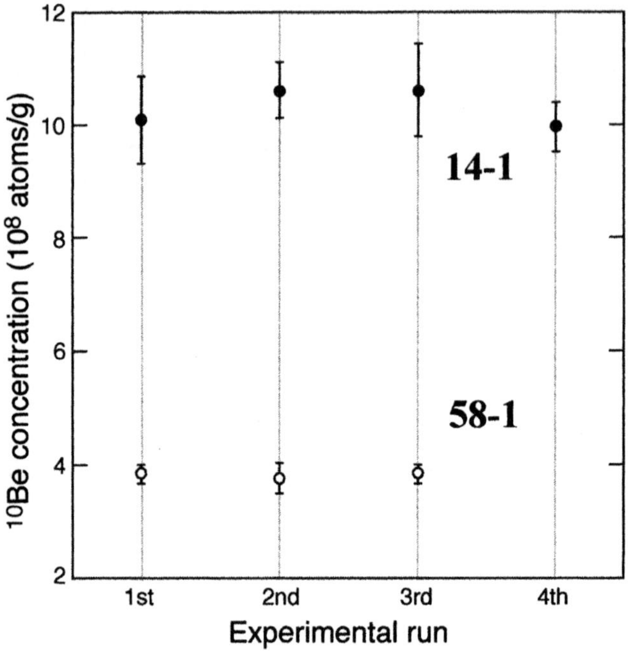

FIG. 2. Replicated ^{10}Be analysis on the BDP-96/hole 1 core at the University of Tokyo in 1997–1998. All the replicated results (of each sample) are the same within the simple experimental errors of AMS analysis (~5%). Each run was restarted from the acid disintegration of sediment samples. The depths of sections 14-1 and 58-1 are 33.05–34.85 m and 112.81–114.535 m, respectively

4 Results

The ^{10}Be concentrations varied between 1.5×10^8 and 11.9×10^8 atoms/g, and tended to increase toward the top of the core (Fig. 3a). On the other hand, the ^{9}Be concentrations are fairly constant at around 2.2 p.p.m. (Fig. 3b). This means that the $^{10}Be/^{9}Be$ ratio corresponds well with the ^{10}Be concentrations (Fig. 3a). Figure 4 shows the ^{10}Be data plotted against the magnetostratigraphic ages. The upward increasing trend of ^{10}Be agrees well with the expected decay of the nuclide between 2.7 and 5.0 Ma (110–200 m), although the correspondence is lower in the profiles dated as being younger than 2.7 Ma (above 110 m). Significant parts of the ^{10}Be profiles, especially 2.7–5.0 Ma, can be regarded as the decay profile of the nuclide, and the temporal information on the BDP-96/hole 1 core is retained in the ^{10}Be record presented here.

5 Discussion and Conclusions

Evidently decay profiles in sedimentary ^{10}Be records require a "constant" initial concentration of ^{10}Be and/or a reservoir making the initial $^{10}Be/^{9}Be$ ratio fairly stable. Successful ^{10}Be dating has been done for ferromanganese crusts on the

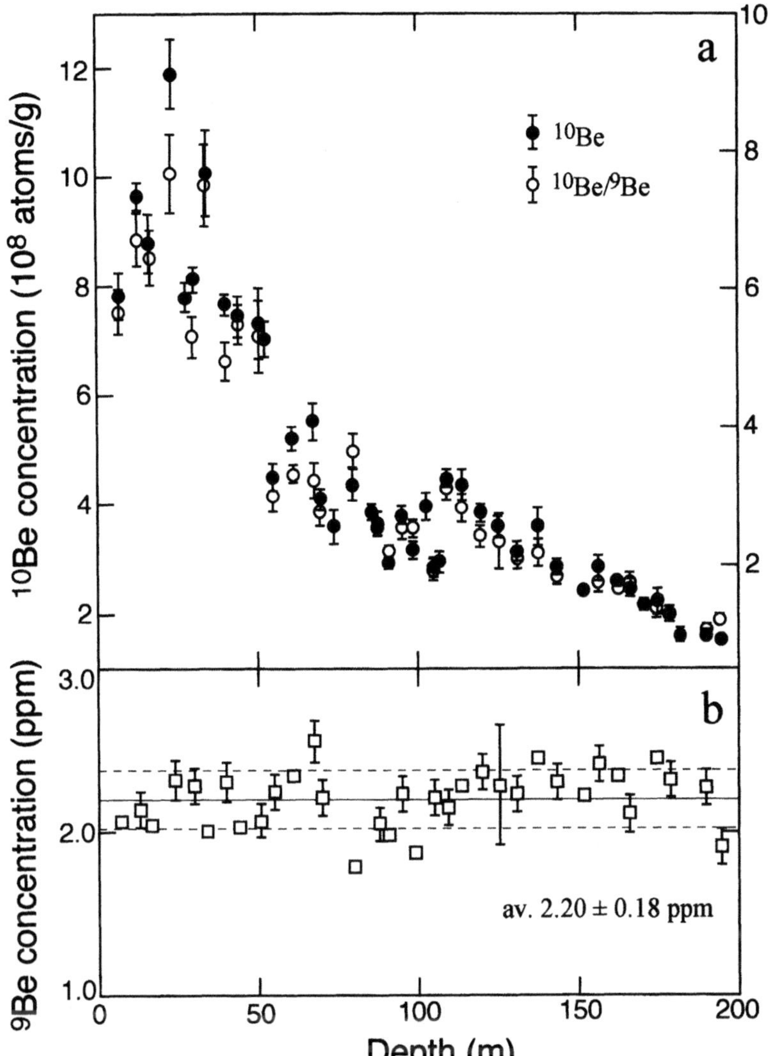

FIG. 3. Cosmogenic ^{10}Be and terrigenous ^{9}Be data plotted against depth in Lake Baikal sediments (BDP-96/hole 1 core). **a** ^{10}Be concentration and ^{10}Be/^{9}Be ratio tending to increase toward the top of the core. **b** ^{9}Be concentration fairly constant around 2.2 p.p.m.

ocean floor (Ku et al. 1982; Segl et al. 1984; Frank and O'Nions 1998). These authigenic materials meet the above requirements for a *stable* ^{10}Be concentration and ^{10}Be/^{9}Be ratio in the bottom water from which the isotopes were precipitated (Ku et al. 1982; Segl et al. 1984). However, cases of clastic sediments are not so simple. Bourles et al. (1989) found a case in central Pacific sediments where the "authigenic" ^{10}Be/^{9}Be ratio showed a worse decay profile than the bulk ^{10}Be/^{9}Be ratio. No clear reason for this discrepancy has yet been presented. However, it would mean that the constancy of the authigenic ^{10}Be/^{9}Be ratio in

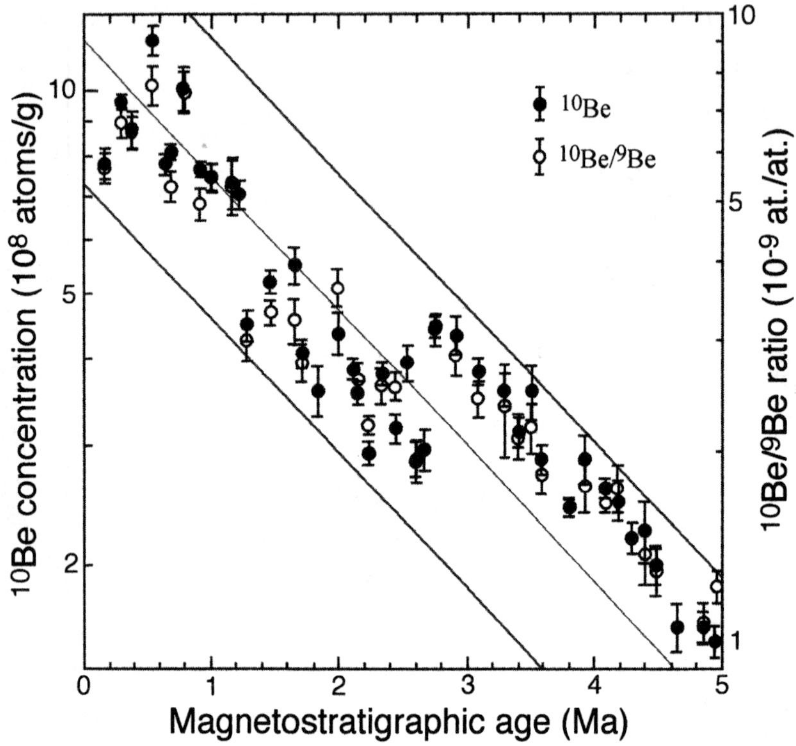

FIG. 4. ^{10}Be and ^{10}Be/^{9}Be ratio plotted against magnetostratigraphic ages. The *gray line* shows the decay trend of ^{10}Be (half-life 1.5 Myr)

clastics was disturbed by changes in the chemistry of the bottom water and/or affected by other ^{10}Be contributions. ^{10}Be contributions to underwater sediments are divided into three factors: direct fallout from the atmosphere, and lateral inputs from water currents and terrestrial clastics (i.e., redistributions of the atmospheric ^{10}Be). The extent of each factor depends on the geographical setting where clastics are deposited.

At Lake Baikal, one expects the contribution of ^{10}Be fallout to come from three sources: atmospheric precipitation, deposition of continental aerosols, and lateral input from fluvially deposited terrigenous materials from the catchment. Based on the significant fluctuations observed in the ^{10}Be/^{9}Be ratios (see Figs. 3a and 4), it is apparent that these sources do not mix well in "time," which can be expected. Consequently, the depositional nature of the ^{10}Be should be considered when interpreting the records of the nuclide in Lake Baikal sediments. In this section, we present the initial ^{10}Be concentration (^{10}Be^{0}) and initial ^{10}Be/^{9}Be ratio (^{10}Be^{0}/^{9}Be) to elucidate the nature of the ^{10}Be deposition that has affected the chronological information since 2.7 Ma. Assuming a time-scale based on magnetostratigraphy, we have calculated the initial ratio for each of the ^{10}Be measurements; the resulting concentrations and ratios are labeled "initial concentrations or ratios." After an examination of the nature of the deposition, we propose certain procedures for ^{10}Be dating which will constrain other stratigraphic methods.

5.1 Possible Causes of the Abrupt Change in ^{10}Be Deposition at About 2.7 Ma

The $^{10}Be^0$ and $^{10}Be^0/^9Be$ maintained fairly constant values (ca. 16×10^8 atoms/g and 11×10^{-9} at./at., respectively) until their abrupt decrease at 2.7 Ma (Fig. 5). Since that time, the lowered $^{10}Be^0$ and $^{10}Be^0/^9Be$ have fluctuated vigorously. Abrupt decreases and vigorous fluctuations of $^{10}Be^0$ have already been found in the late Quaternary sediments of Lake Baikal (Horiuchi et al. 1999, 2000, 2001): ^{10}Be concentrations were high (>10×10^8 atoms/g) throughout the warm (interglacial and interstadial) periods, but decreased steeply to about 6×10^8 atoms/g at the peak of the glaciation. The extent of the $^{10}Be^0$ fluctuations is nearly the

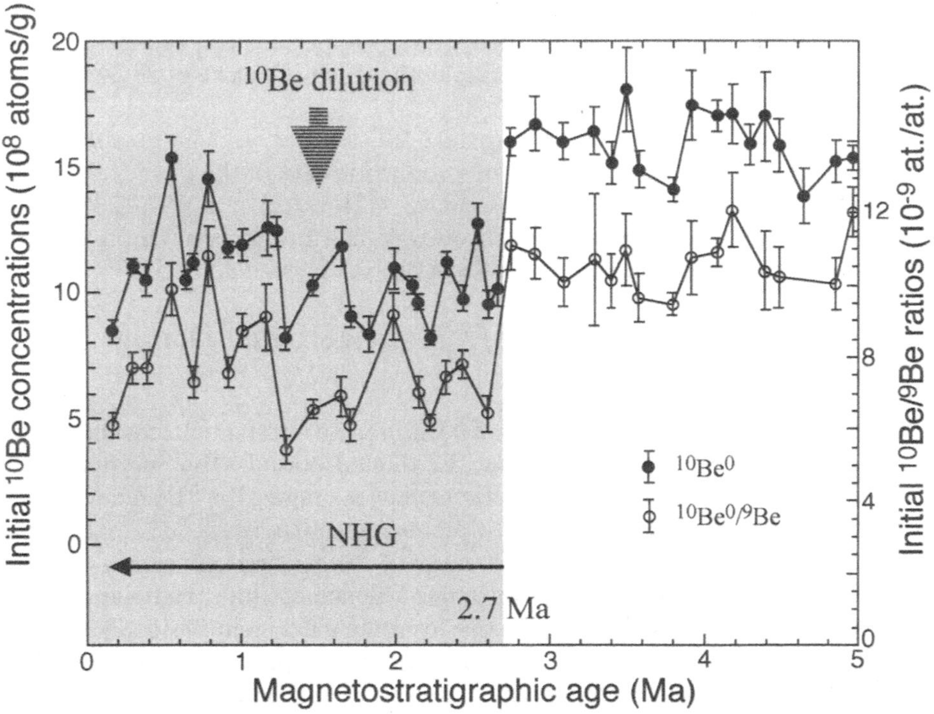

FIG. 5. Initial ^{10}Be concentrations ($^{10}Be^0$) and initial $^{10}Be/^9Be$ ratios ($^{10}Be^0/^9Be$) in Lake Baikal sediments of the last 5 Myr. Assuming a time-scale based on magnetostratigraphy, we have calculated the initial ratio for each of the ^{10}Be measurements. Extensive dilution of the concentrations and ratios started at 2.7 Ma (*shaded area*), which was the same time as the initiation of the Northern Hemisphere glaciation (*NHG*)

same as that of the BDP-96 core (8–15 × 10^8 atoms/g). We believe that these fluctuations in $^{10}Be^0$ have continued during the last 2.7 Myr for the same (or similar) reasons, and have disturbed the decay profile of ^{10}Be in the BDP core. Recently, sedimentary analyses on the late Quaternary ^{10}Be record, with an investigation of trace elements, have revealed that the abrupt decreases were caused by the dilution of atmospherically derived ^{10}Be atoms by terrigenous clays from the bases of the mountain glaciers (Horiuchi et al. 2001; K. Horiuchi et al. 2002, in preparation). Intensive production of clastic materials by the growth of the glaciers is well known at the glacial peaks around the lake (Bezrukova et al. 1991; Karabanov et al. 1998; Back et al. 1999; Goldberg et al. 2001; Horiuchi et al. 2001). Unlike ^{10}Be, which basically originates from the atmosphere and is diluted by terrestrial materials from rocks, ^9Be originates from (i.e., is undiluted by) terrigenous clastics, causing the fluctuations in $^{10}Be^0$ and $^{10}Be^0/^9Be$ to be fairly similar. It is most likely that the abrupt decrease in the $^{10}Be^0$ and $^{10}Be^0/^9Be$ at 2.7 Ma reflects the start of mountain glaciation around Lake Baikal; the rapidity and timing of this start are the same as those of the Northern Hemisphere glaciation (Shackleton et al. 1984, 1995; Prueher and Rea 1998; Haug and Tiedemann 1998; Maslin et al. 1998). The initiation of the glacial condition in the Baikal region is also implied by clay–mineralogical, geochemical, and palynological analyses of the same core at that time (Demske et al. 2000; Müller et al. 2001).

Another possible factor disturbing the profiles might be changes in the atmospheric input of ^{10}Be connected to the production rate of the nuclide. However, this effect is minor in Lake Baikal sediments (Horiuchi et al. 1999, 2000, 2001), and is likely to be counterbalanced by averaging a 2-m section (ca. 50-kyr interval). Thus, the change in the production rate is neglected here.

5.2 *Possible Contribution of ^{10}Be to Accurate Dating of BDP Cores*

Constant $^{10}Be^0$ and $^{10}Be^0/^9Be$ at 2.7–5.0 Ma imply a substantial constancy in the initial values of ^{10}Be except for the ^{10}Be-diluted periods. The "normal" initial values of ^{10}Be and $^{10}Be/^9Be$ can be estimated by averaging the $^{10}Be^0$ and $^{10}Be^0/^9Be$ of 2.7–5.0 Ma: these are $16.0 \pm 1.2 \times 10^8$ atoms/g and $10.6 \pm 0.8 \times 10^{-9}$ at./at., respectively (Fig. 6), Although the mechanism of "substantial constancy" remains unclear, it may be related to the fortunate balance of atmospheric and terrigenous ^{10}Be inputs, the origins of clastics from the catchment basin, and/or other factors, and the "normal" values would be useful for approximate dating of BDP cores. Figure 7 shows the "apparent" ^{10}Be ages based on the disintegration rate of the nuclide and the "normal" values of $^{10}Be^0$ and $^{10}Be^0/^9Be$, indicating their fair agreement with the magnetostratigraphic ages. If we select the youngest ^{10}Be age in a certain period (which means we extract a less diluted value from a group of ^{10}Be-diluted values), both ages are equivalent within experimental error (Fig. 7). The "apparent" ^{10}Be ages seems to be a useful constraint for the magnetostratigraphy of the BDP cores.

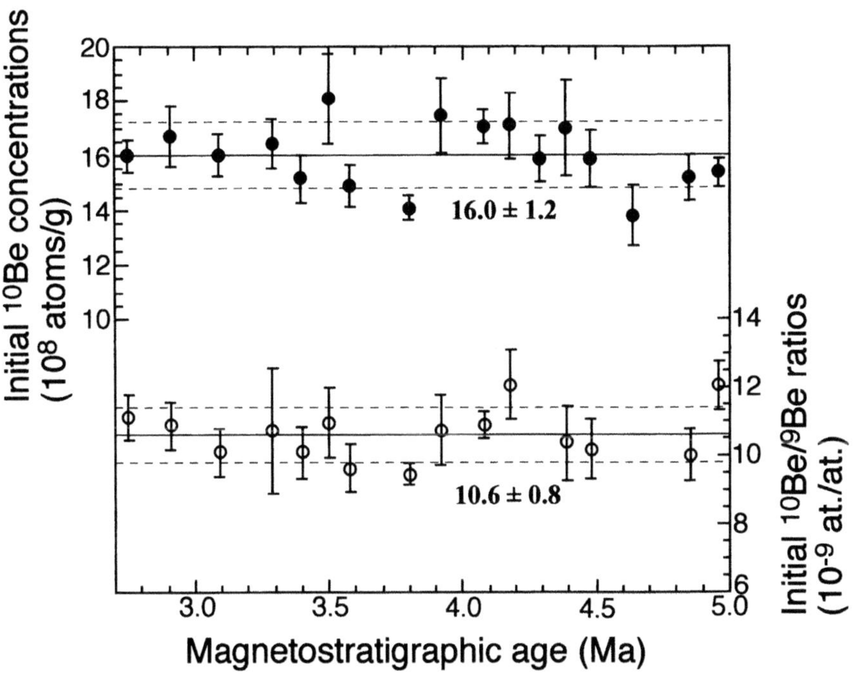

FIG. 6. Initial ^{10}Be concentrations and initial ^{10}Be/^{9}Be ratios in Lake Baikal sediments between 2.7 Ma and 5.0 Ma. Assuming a time-scale based on magnetostratigraphy, we have calculated the initial ratio for each of the ^{10}Be measurements. *Solid circles*, ^{10}Be; *open circles*, ^{10}Be/^{9}Be. In this figure, both the concentrations and ratios show almost constant values: the averages are $16.0 \pm 1.2 \times 10^8$ atoms/g and $10.6 \pm 0.8 \times 10^{-9}$ at./at., respectively. We have used the averages as the "normal" initial values for our ^{10}Be dating (see also Fig. 7)

However, a further improvement in the procedure is essential in order to establish ^{10}Be dating. Normalizing the dilution of ^{10}Be by independent indexes is clearly better than the above assumption of "constant" initial values. Grain-size composition and trace elements in the sediments may be useful for this purpose, as suggested by our previous work (Horiuchi et al. 1999, 2001). Such an attempt and refinement would make ^{10}Be analysis a convenient method for accurately dating BDP cores.

Acknowledgments. We thank T. Kawai and K. Minoura for their interest in this work. We also thank D. Lal and an anonymous reviewer, whose comments have significantly improved this chapter. K. Horiuchi, K. Kobayashi, and H. Matsuzaki thank S. Hatori, C. Nakano, H. Yamashita, and H. Makita for technical support on the MALT–AMS system. This work is cosupported by a grant from the Science and Technology Agency of Japan and by the Ministry of the Education,

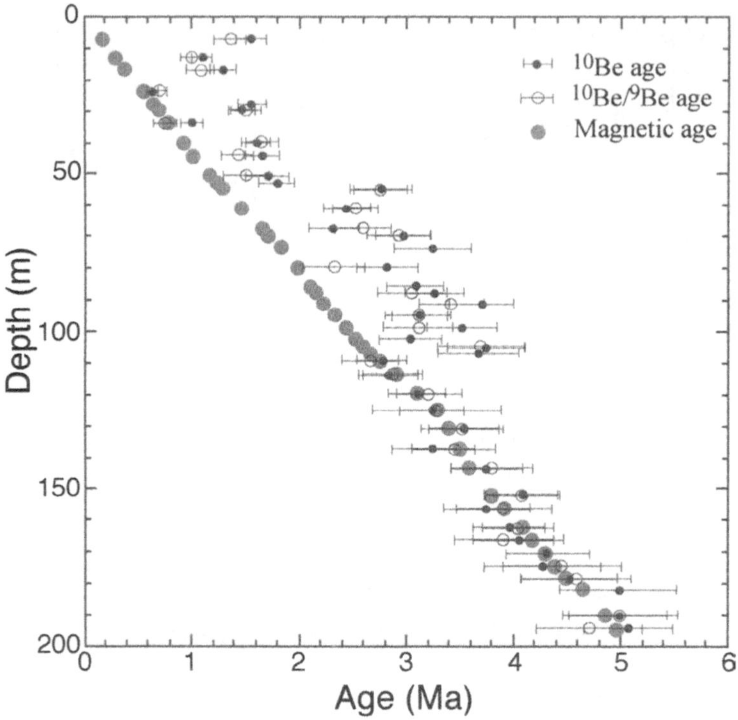

FIG. 7. "Apparent" ^{10}Be (and ^{10}Be/^{9}Be) ages and magnetostratigraphic ages of the BDP-96/hole 1 core. The radiometric ^{10}Be ages are based on the assumption that the averages of the initial ^{10}Be concentrations ($16.0 \pm 1.2 \times 10^8$ atoms/g) and initial ^{10}Be/^{9}Be ratios ($10.6 \pm 0.8 \times 10^{-9}$ at./at.) between 2.7 Ma and 5.0 Ma are the "normal" initial values of the nuclide in the core (see also Fig. 6)

Science, Sports and Culture, Grant-in-Aid for the Encouragement of Young Scientists, No. 08577, 2000.

References

Amin BS, Lal D, Somayajulu LK (1975) Chronology of marine sediments using the ^{10}Be method: intercomparison with other methods. Geochim Cosmochim Acta 39:1187–1192

Back S, De Batist M, Strecker MR, Vanhauwaert P (1999) Quaternary depositional systems in northern Lake Baikal, Siberia. J Geol 107:1–12

Baikal Drilling Project Group (1998) A continuous record of climate changes for the last five million years from the bottom sediments of Lake Baikal, Russ Geol Geophys 39:135–154

Bezrukova YeV, Bogdanov YuA, Williams DF, Granina LZ, Grachev MA, Ignatova NV, Karabanov YeB, Kuptsov VM, Kurylev AV, Letunova PP, Likhoshvay YeV, Chernyayeva GP, Shimarayeva MK, Yakushin AO (1991) Deep-seated change in the ecosystem of northern Lake Baikal during the Holocene. Dokl Akad Nauk SSSR 321:1032–1037

Bourles D, Raisbeck GM, Yiou F (1989) ^{10}Be and ^{9}Be in marine sediments and their potential for dating. Geochim Cosmochim Acta 53:443–452

Demske D, Mohr B, Oberhänsli H (2000) Palaeoclimatic changes from 3.6 to 2.2 Ma BP derived from palynological studies on Lake Baikal sediments. In: Minoura K (ed) Lake Baikal: a mirror in time and space for understanding global change processes. Elsevier, Amsterdam, p 85–89

Frank M, O'Nions RK (1998) Sources of Pb for Indian Ocean ferromanganese crust: a record of Himalayan erosion? Earth Planet Sci Lett 158:121–130

Goldberg EL, Phedorin MA, Grachev MA, Zolotarev KV, Khlystov OM (2001) Geochemical indicators of paleoclimate changes in sediments of Lake Baikal (in Russian). Russ Geol Geophys 42:76–86

Haug GH, Tiedemann R (1998) Effect of the formation of the isthmus of Panama on Atlantic Ocean thermohaline circulation. Nature 393:673–676

Horiuchi K, Kobayashi K, Sakai H, Shibata Y (1998) Beryllium 10 concentration in Lake Baikal sediments during the last 5 million years. EOS Trans AGU 1998 Fall Meeting 79:F477

Horiuchi K, Minoura K, Kobayashi K, Nakamura T, Hatori S, Matsuzaki H, Kawai T (1999) Last-glacial to post-glacial ^{10}Be fluctuations in a sediment core from the Academician Ridge, Lake Baikal. Geophys Res Lett 26:1047–1050

Horiuchi K, Kobayashi K, Oda T, Nakamura T, Fujimura C, Matsuzaki H, Shibata Y (2000) Climate-induced fluctuations of ^{10}Be concentration in Lake Baikal sediments. Nucl Instrum Methods B 172:565–570

Horiuchi K, Goldberg EL, Kobayashi K, Oda T, Nakamura T, Kawai T (2001) Climate-induced variations of cosmogenic beryllium-10 in the sediments of Lake Baikal of the last 150 ky from AMS, SRXFA and NAA data. Nucl Instrum Methods A 470:396–404

Karabanov EB, Prokopenko AA, Williams DF, Colman SM (1998) Evidence from Lake Baikal for Siberian glaciation during oxygen–isotope substage 5d. Q Res 50:46–55

Kobayashi K, Matsuzaki H, Hatori S, Nakano C, Yamashita H, Makita H, Sunohara Y (2000) Multi-nuclide AMS with a middle-sized tandem accelerator of MALT. Nucl Instrum Methods B 172: 75–81

Ku TL, Kusakabe M, Nelson DE, Southon JR, Korteling RG, Vogel J, Nowikow I (1982) Constancy of oceanic deposition of Be as recorded in manganese crusts. Nature 299:240–242

Maslin MA, Li XS, Loutre M-F, Berger A (1998) The contribution of orbital forcing to the progressive intensification of Northern Hemisphere glaciation. Q Sci Rev 17:411–426

Matsuzaki H, Tanikawa M, Kobayashi K, Hatori S (2000) Development of a gas counter for AMS measurement of ^{10}Be and ^{26}Al of cosmic spherules. Nucl Instrum Methods B 172:218–223

Müller J, Oberhänsli H, Melles M, Schwab M, Rachold V, Hubberten H-W (2001) Late Pliocene sedimentation in Lake Baikal: implications for climatic and tectonic changes in SE Siberia. Palaeogeogr Palaeoclimatol Palaeoecol 174:305–326

Prueher LM, Rea DK (1998) Rapid onset of glacial conditions in the subarctic North Pacific region at 2.67 Ma: clues to causality. Geology 26:1027–1030

Sakai H, Nomura S, Horii M, Kashiwaya K, Tanaka A, Kawai T, Kravchinsky V, Peck J, King J (2000) Paleomagnetic and rock-magnetic studies on Lake Baikal sediments: BDP96 borehole at Academician Ridge. In: Minoura K (ed) Lake Baikal: a mirror in time and space for understanding global change processes. Elsevier, Amsterdam, p 35–52

Segl MA, Mangini A, Bonani G, Hofmann HJ, Nessi M, Suter M, Wölfli W, Friedrich G, Plüger WL, Wiechowski A, Beer J (1984) ^{10}Be-dating of a manganese crust from central North Pacific and implications for ocean palaeocirculation. Nature 309:540–543

Shackleton NJ, Backman J, Zimmerman H, Kent DV, Hall MA, Roberts DG, Schnitker D, Baldauf JG, Desprairies A, Homrighausen R, Huddlestun P, Keene JB, Kaltenback AJ, Krumsiek KAO, Morton AC, Murray JW, Westberg-Smith J (1984) Oxygen isotope calibration of the onset of ice-rafting and history of glaciation in the North Atlantic region. Nature 307:620–623

Shackleton NJ, Hall MA, Pate D (1995) Pliocene stable isotope stratigraphy of site 846. Proc ODP Sci Res 138:337–353

Somayajulu BLK (1977) Analysis of cause for the beryllium-10 variations in deep sea sediments. Geochim Cosmochim Acta 41:909–913

Tanaka S, Inoue T (1979) ^{10}Be dating North Pacific sediment cores up to 2.5 million years BP. Earth Planet Sci Lett 45:181–187

Evolution and Biodiversity

Origin and Diversity of the Diatom Genus *Eunotia* in Lake Baikal: Some Preliminary Considerations

David M. Williams and Geraldine Reid

Summary. The benthic diatom flora of Lake Baikal is poorly known and its origins obscure. In this chapter, we outline the distribution of the diatom genus *Eunotia*, in particular taxa related to the unusual species *E. clevei*, and discuss the implications. We believe there are two distinct series of relationships, expressed both geographically and phylogenetically. One, derived from *E. clevei sensu stricto*, extends through the Boreal (with extinctions in Japan and the USA); the other, related to *E. lacusbaikali*, is in China, extending north towards Southeast Asia. These two "routes" have implications for the Lake Baikal flora and the diversity of the *Eunotia* flora in the Far East.

Key words. Lake Baikal, Biogeography, Diatoms (*Eunotia*)

1 Introduction

The diatom flora of Lake Baikal has been studied by a number of people over the last 75 years. The plankton flora is now reasonably well understood, but the benthic flora is less well known. The most comprehensive study of benthic diversity in Lake Baikal was undertaken by Skvortzov and Meyer some 70 years ago (Skvortzov and Meyer (1928) and Skvortzov (1937a); nomenclatural revisions in Meyer (1930) and Skvortzov (1946); for a commentary on Skvortzov's work see Williams and Reid (2001)). Skvortzov presented a preliminary study in 1928 (Skvortzov and Meyer 1928), and 10 years later published a more detailed report on a sample taken from Olhon Gate at a depth of 33 m (Skvortzov 1937a), where he recorded some 304 taxa, 148 of which he described as new to science.

In Britain, joint research links with Lake Baikal were initiated in the late 1980s as a result of the Baikal International Centre for Ecological Research (BICER) agreement organized and supported by the Royal Society. The Natural History Museum, London (BM), University College, London (UCL), and the Limno-

Botany Department, Natural History Museum, Cromwell Road, London SW7 5BD, UK

logical Institute, Irkutsk (LI), have been cooperating on a Darwin Initiative for the Survival of Species (DI) project to study the benthic diatom flora of Lake Baikal, Siberia (Flower and Williams 1999). The DI allowed the first methodical field collection of benthic diatoms to be made. The entire perimeter of the lake (ca. 2000 km) was surveyed at roughly 30-km intervals, with both shallow (ca. 1 m) and deep (ca. 20 m) water samples taken; the deep-water samples proved to have a much more extensive and diverse flora than the shallow-water samples.

Our interest in studying the Lake Baikal benthic diatom flora is to establish its biogeographic relationship with Europe, Northeast Russia, China, and beyond.

2 Benthic Diatom Flora of Lake Baikal

The study on Lake Baikal, initiated by the Darwin Initiative, is far from complete, and revised numbers for taxa present in the flora are as yet unavailable. For the purposes of this chapter, we have used the data from Skvortzov's studies (Skvortzov 1937a). Although his numbers are undoubtedly inaccurate, the data give some idea of the number of taxa involved and allow some perspective on the task of understanding the origins and subsequent evolution of the diatom flora. In the past, doubts have been expressed concerning the usefulness of Skvortzov's studies. However, in the case of Lake Baikal, his numbers of new taxa may well be an underestimate rather than the reverse (see, for example, Mann 1999).

As a general guide, the diatom flora can be divided into taxa considered "cosmopolitan" or "endemic." According to Skvortzov (1937a), there are 304 cosmopolitan taxa and 148 endemics taxa. That is, roughly one-third of the diatom flora is considered unique. This view is strongly dependent on the definitions of cosmopolitan and endemic, on which there seems to be no general agreement.

Skvortzov (1937a) presented an alternative scheme based on an understanding of the origin of the flora (ideas developed by Vereshagin, 1940, after Berg, 1910). He divided the total flora into five "regions," to which he assigned taxa (Table 1). Skvortzov (1937a) did not assign all diatom taxa to a region. The figures presented in Table 1 account for 67% of the total flora. Nevertheless, the "Siberian and subalpine elements" account for nearly 50% of the total diatom flora.

3 Why *Eunotia* C.G. Ehrenberg?

We chose to study the diatom genus *Eunotia* for four reasons:

1. It is a large and diverse genus, with possibly up to 1200 species, most of which occur in acidic environments.
2. It has a global distribution.

TABLE 1. Numbers of diatom taxa assigned to the five "regions of origin" by Skvortzov (1937a). The total number of taxa described by Skvortzov was 304. The percentage of the total is the proportion assigned to a region

Region of origin	Numbers (% of total)	Numbers of taxa in the genus *Eunotia*[a]
(A) Siberian and subalpine elements	150 (49)	1
(B) Tertiary freshwater remains and tropical origin	31 (10)	2
(C) Marine elements or marine relicts	7 (2)	0
(D) Brackish-water species	3 (1)	0
(E) Elements of indistinct origin	15 (5)	0
Not in any region	98 (32)	7

[a] Numbers given for the genus *Eunotia* only

3. There are many endemic taxa. For instance, in a recent study on tropical diatoms of South America, Metzeltin and Lange-Bertalot (1998) found 49 new taxa, including a new subgenus. The implication seems to be that *Eunotia* is prone to diversification.
4. The genus is also of significance for diatom evolution, appearing as a possible precursor to all the raphid diatoms. We will not deal further with this topic here, but see Kociolek (2000) and Reid and Williams (2001).

The diversity of *Eunotia* in Lake Baikal can be compared with the total number of diatoms (Table 2). Correcting for current nomenclature, there are 30 cosmopolitan taxa, of which 12 are known as fossils, and 11 endemic taxa, of which 4 are known as fossils. That is, roughly one-third of the taxa are endemic, which is similar to the entire flora of Lake Baikal. In addition, half of the known taxa occur as fossils.

Viewed from Skvortzov's (1937a) alternative grouping, one species of *Eunotia* can be placed in the "Siberian and Subalpine elements" group, two species in the "Tertiary freshwater remains and tropical origin" group, and no species in any of the other categories.

Within the genus *Eunotia* there is the large and distinctive species *E. clevei* Grun. ex Cleve. Our studies have focused on this species, as it is an atypical *Eunotia* and has at least three closely related taxa present in Lake Baikal (Table 3). Our revision of the Lake Baikal species will be published elsewhere (Reid and Williams 2001; Williams and Reid, in preparation), but here we present a summary of their distributions along with a few light micrographs (Figs. 1–3).

4 Distribution of *Eunotia clevei* and its Relatives

4.1 Boreal Taxa

4.1.1 *E. clevei* Cleve ex Grunow (Figs. 1a, 2e)

E. clevei has a reasonably wide distribution globally, but it would be inaccurate to describe it as cosmopolitan. It is distributed sparingly throughout the Boreal

TABLE 2. Partial checklist of *Eunotia* species from Lake Baikal. Taxon names in bold occur as fossils. Data from Skvortzov and Meyer (1928), Meyer (1930), Skvortzov (1937a), and Tschermissinova (1971, 1973). Names have been updated where possible using Krammer and Lange-Bertalot (1991). Approximately 30 taxa are thought to be cosmopolitan and 11 endemic, suggesting roughly 33% endemism

Cosmopolitan (30 taxa)	Endemic (11 taxa)
E. arcus	*E. baicalensis*
E. bigibba	***E. clevei* var. *baikalensis***
E. bilunaris	*E. clevei* var. *hispida*
E. bilunaris var. *mucophila*	***E. lacusbaikali***
E. clevei[a]	*E. "lunaris"* var. *baikalensis*[b]
E. delicatula	*E. monodon* var. *baicalensis*
E. dinodon	***E. monodon* var. *latestriata***
E. dinodon* var. *diminuta	***E. polygliphoides* var. *tuncaensis***
E. exigua	*E. praerupta* var. *curta f. elongata*
E. faba	*E. sublitoralis*
E. formica	*E. sudetica* var. *baicalensis*
E. lavrovii	
E. monodon	
E. papilo	
E. pectinalis	
E. pectinalis var. *ventralis*	
E. pectinalis var. *minor f. impressa*	
E. pectinalis var. *minor*	
E. pectinalis var. *undulata*	
E. polygliphoides	
E. praerupta	
E. praerupta* var. *bidens	
E. praerupta* var. *inflata[c]	
E. revolute	
E. serra	
E. submonodon[a]	
E. sudetica	
E. triodon	
E. veneris	
E. veneris var. *obtusiuscula*	

[a] Indicates taxa placed in Skvortzov's category "Tertiary remains and tropical origin"

[b] This taxon was described by Jasniskij (1936). The one-line drawing Jasniskij provides suggests that this taxon may well be a curved species of *Synedra*, probably *S. cyclopeum*

[c] Indicates the single taxon placed in Skvortzov's category "Siberian and subalpine elements"

TABLE 3. Taxa thought to be closely related to *Eunotia clevei*

Cosmopolitan	Endemic
E. clevei	*E. clevei* var. *baikalensis*
	E. hispida
	E. lacusbaikali

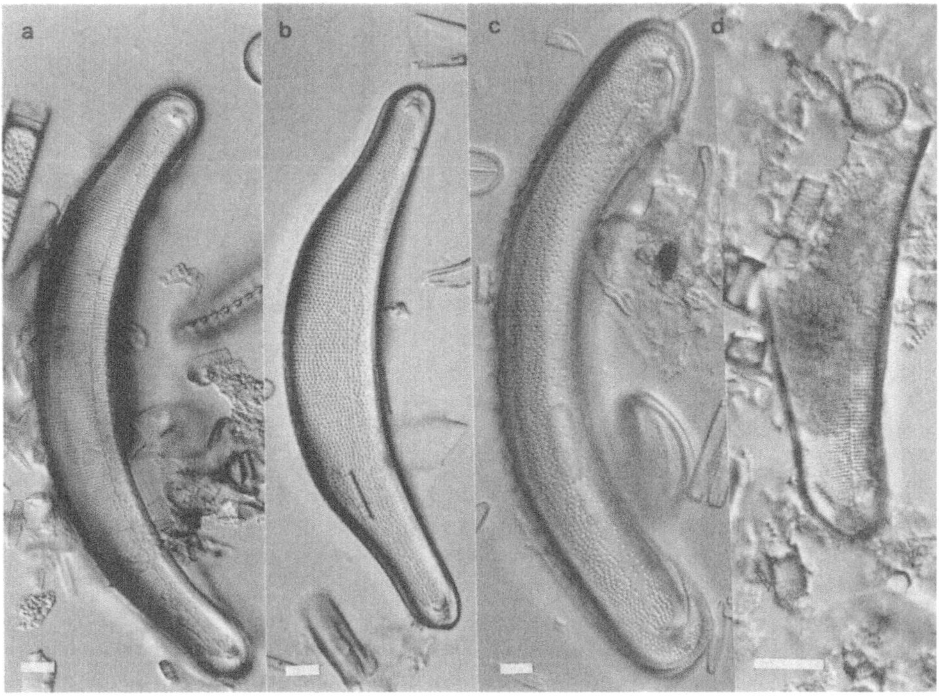

FIG. 1. **a** *Eunotia clevei* Cleve ex Grunow, Lake Baikal, Elokhin (54.46323N, 108.55405E), small granite rocks, water depth 1 m, collected June 29, 1997 (BM 100353). **b** *Eunotia clevei* var. *baikalensis* Skvortzov, Lake Baikal, Olhon Gate (Ol'khonskie Vorota) (53.01538 N, 106.92979 E), silt sample, water depth 33 m, collected June 27, 1997 (BM 100334). **c** *Eunotia lacusbaikali* Skvortzov, Lake Baikal, Buguldeika (52.58418N, 106.07620E), stones, 20 m, collected June 26, 1997 (BM 100322, 100324–5). **d** *Eunotia* cf. *clevei* Cleve ex Grunow, from Shang-wang, Shantung Province, China (BM 88875). *Bars* 10 µm

in Scandinavia, Northern Europe, and Russia, usually in large lakes. More precise details are given in Reid and Williams (2001), but of note is its presence in Lake Ladoga (Hustedt in Schmidt 1913 and Saarnisto and Grönland 1996), Lake Onega (Wislouch and Kolbe 1927), Lake Baikal (Reid and Williams 2001; Skvortzov 1937a), and Lake Husvogol (Edlund et al. 2000). Of significance is its fossil record. Fossil specimens have been noted from Lake Baikal, in the Tunkin

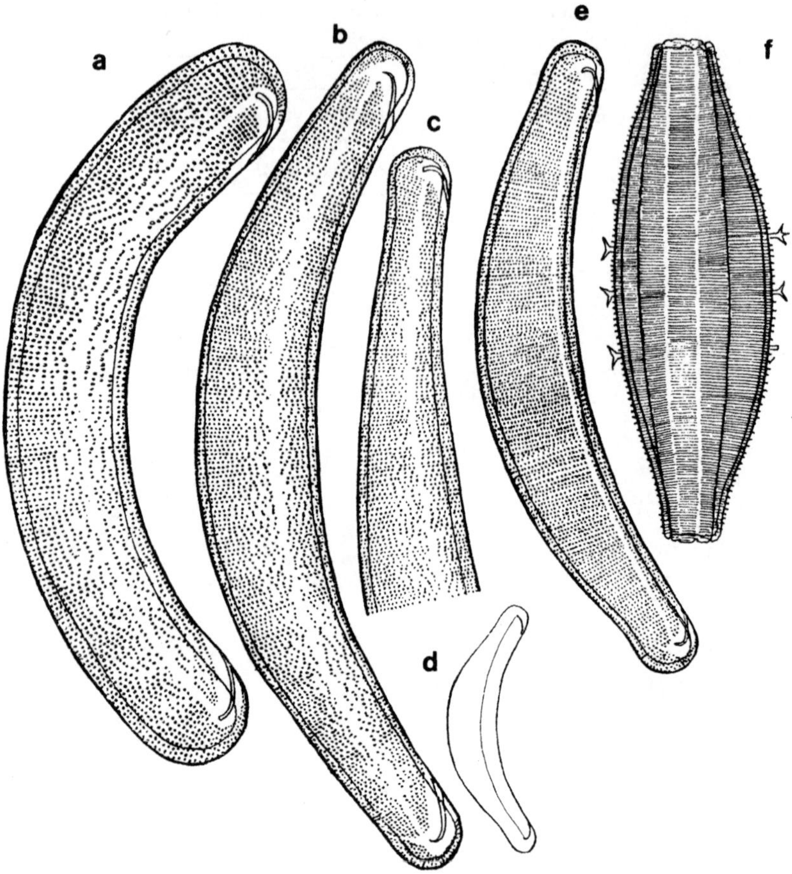

FIG. 2. Reproduction from Skvortzow (1937a), Plate 4 Figs. 2 and 4–9, showing illustrations of some species of *Eunotia* from Lake Baikal. **a** *Eunotia lacusbaikali* Skvortzov. **b–d** *Eunotia clevei* var. *baikalensis* Skvortzov. **e** *Eunotia clevei* Cleve ex Grunow. **f** *Eunotia hispida* (Skvortzov) Reid and Williams

Hollow deposit (Tscheremissinova 1971, 1973), from Japan (Hustedt in Schmidt 1913; specimens from Setanai, Japan), and from the USA (New Jersey, Washington, and Oregon, personal observation; Patrick and Reimer 1966). Fossil specimens have been recorded from China (Skvortzov 1937b, Voigt 1937), but we suspect these belong to another, different species (see below).

4.1.2 *Eunotia clevei* var. *baikalensis* Skvortzov (Figs. 1b, 2b–d)

This taxon (which should be considered a species rather than a variety; Williams and Reid, in preparation) is recorded only from Lake Baikal, fossil and Recent (Loginova and Khursevich 1990).

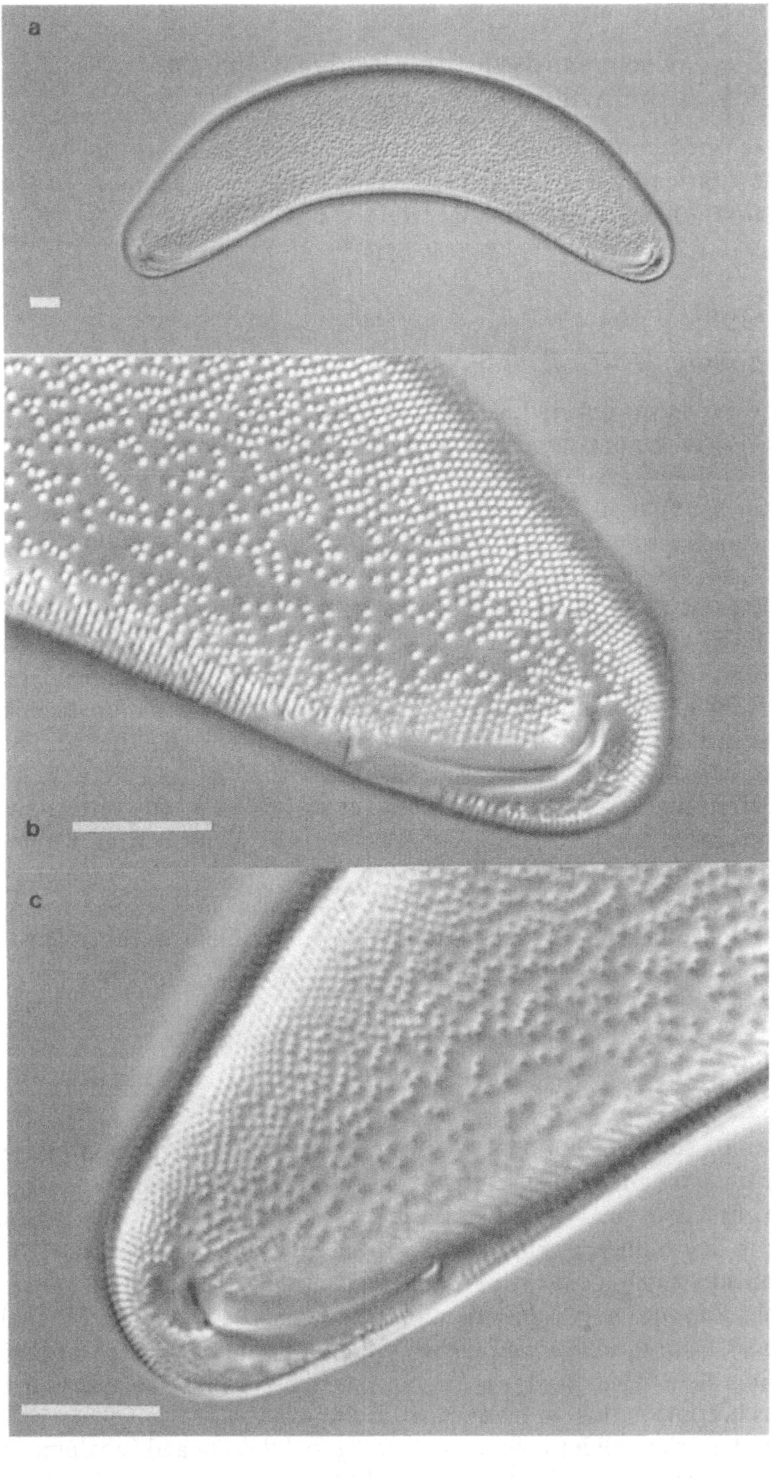

FIG. 3. *Eunotia reimerii* sp. nov., West River, China (Holotype, slide Schulze 1596, ANSP (Philadelphia Academy of Sciences)). **a** Whole valve. **b** Polar view with raphe curved into the margin. **c** Polar view with raphe curved into helictoglossa. *Bars* 10 μm

4.1.3 *Eunotia hispida* (Skvortzov) Reid and Williams (Fig. 2f; Reid and Williams 2001, Figs. 3, 6–9)

We have presented a detailed study of this species elsewhere, but to summarise, while it is primarily restricted to Lake Baikal, we have examined subfossil specimens from Lake Ladoga, Finland (Reid and Williams 2001).

4.2 Asian Taxa

4.2.1 *Eunotia lacusbaikali* Skvortzov (Figs. 1c, 2a)

This species occurs only in Lake Baikal. While we believe that it is closely related to *E. clevei*, its general appearance suggests that it has more features in common with several new fossil species of *Eunotia* described from China: *E. clevei* var. *obliquestrata* Qi et al. [=: *E. clevei* var. *sinica* Qi et al.] from Qi-au Island in the Lingding waterway (Qi et al. 1986), and "Borehole 47 in the Jianghan Plain, Hubei" (Shi 1991); *E. pseudoclevei* Wang and *E. botuliformis* Wang (both from the "Delta of Pearl River, near Jiangmen City on the west bank of Xijangh," Wang 1998, p 37).

In addition, we have had the opportunity to examine a specimen from "West River, China," that differs from all species of *Eunotia* so far described, but belongs within this Chinese group (Fig. 3). As we feel sure that China harbors an unknown diversity, we have decided to describe this taxon as new and give a brief description below, in spite of the fact that there is only one known specimen. We are actively searching for specimens of *Eunotia* from Chinese fossil deposits for electron microscope study.

We have also examined specimens from material first studied by Skvortzov (1937a) and Voigt (1937) and, although these specimens were named *E. clevei*, we believe they may belong to an undescribed species (Fig. 1d).

5 Conclusions

Partitioning species by geography is problematic. The distinction between cosmopolitan and endemic is usually too crude to be of any real value. Skvortzov's (1937a) five divisions, while helpful in establishing possible relationships, also appear to be insufficient. For example, species that fall into Skvortzov's category Siberian and subalpine elements could be considered as cosmopolitan as well as endemic; *Eunotia clevei* is found in Sweden, Finland, Russia, and Mongolia (and as a fossil, as in the USA and Japan) (Reid and Williams 2001). *Eunotia clevei* has a Boreal distribution at present, but was clearly more widespread in the past and has become extinct in many parts of the world.

Other taxa that fall into Skvortzov's category Siberian and subalpine elements can be understood as truly endemic: *E. clevei* var. *baicalensis* and *E. lacusbaikali* are (probably, but see Reid and Williams 2001; Williams and Reid, in preparation) found only in Lake Baikal.

Other species are relevant to the general problem of explaining the distribution of *Eunotia* in the Far East. Although there are only a few relevant species, they may be sufficient to suggest fruitful areas of further investigation. If *E. clevei* and the taxa above are indeed a monophyletic group, as we suspect, then their relationships "spread" in two directions: one, derived from *E. clevei sensu stricto*, occurring in the Boreal but extinct in Japan and USA; the other derived from *E. lacusbaikali* extending from Lake Baikal into China and north towards Southeast Asia. These two routes have implications for the Lake Baikal flora and the diversity of the *Eunotia* flora in the Far East.

While our interest is in establishing the origins of the *Eunotia* flora, the most pertinent question is how these taxa are related phylogenetically. A second question relates to explaining their current, as well as their past, distribution. Answers to these questions will come from obtaining more specimens in relevant geographical areas (Northeast China, Mongolia, and Northeast Russia), and developing a credible working theory of relationships.

Appendix

Description of a New Species of Eunotia

Eunotia reimerii Williams and Reid, sp. nov. (Figs. 1–3).
Holotype: ANSP Schulze 1596 (slide marked "*Epithemia*").
Type locality: "West River, China."
Valve lunate, gently tapering toward each pole (Fig. 3a). Dorsal margin convex, ventral margin concave. Poles rounded. Short raphe present at each pole, reflexed toward the margin from the valve surface (Fig. 3b); large helictoglossa present on valve face, no evidence of rimoportulae (Fig. 3b–c). Striae irregular and scattered at sternum, more regular toward the valve margins (at margin 12–15 in 10 μm, toward sternum 6–8 in 10 μm); striae composed of large puncta. Sternum ill defined but present, oriented toward, and more or less parallel with, the ventral margin. Length 150–175 μm, breadth 40–50 μm.

Valvae lunatus, ad apices leniter contractae, 150–175 μm longae, 40–50 μm latae. Dorsalis marginem convexus, ventralis marginem convexus. Apices rotundatus praebens curtus raphe, grandis helictoglossa. Striae punctate insolitus ad sternum, 6–8 per 10 μm, regulares ad marginem 12–15 per 10 μm.

References

Berg L (1910) Fauna of Lake Baikal and its origin. Biol Zh 1:10–45

Edlund MB, Soninkhishig N, Williams RM, Stoermer EF (2000) Morphology and taxonomy of *Eunotia clevei*. Diatom Res 15:209–219

Flower RJ, Williams DM (1999) The Natural History Museum 1999 Meeting on the Biodiversity of Lake Baikal and a Workshop on the Taxonomy of Lake Baikal Diatoms. Combined Reports. ECRC Report 62, 27 p

Jasniskij V (1936) Novye I interesnye vidy diatomovykh vodoroslej iz oz Bajkala. Bot Zh 21:689–703

Kociolek JP (2000) Valve ultrastructure of some Eunotiaceae (Bacillariophyceae), with comments on the evolution of the raphe system. Proc Calif Acad Sci 52:11–21

Krammer K, Lange-Bertalot H (1991) Bacillariophyceae: Centrales, Fragilariaceae, Eunotiaceae. In: Ettel H, Gerloff J, Heynig H, Mollenhauer D (eds) Süßwasserflora von Mitteleuropa 2. Fischer, Stuttgart, Jena

Loginova LP, Khursevich GK (1990) New representatives of the fossil fauna and flora of Belorussia and other regions of the USSR: collected scientific papers. Navuka i Tekhnika, Minsk, p 146–176

Mann DG (1999) The species concept in diatoms. Phycologia 38:437–495

Metzeltin D, Lange-Bertalot H (1998) Tropical diatoms of South America. I. Icon Diat 5

Meyer K (1930) Einfuhrung in die Algenflora des Baicalsees. Byull Mosk Ova Ispyt Prir Odt Biol 39:179–396

Patrick R, Reimer CW (1966) The diatoms of the United States, exclusive of Alaska and Hawaii. Academy of Natural Sciences of Philadelphia, Monograph 13, 688 p

Qi Y-Z, Lin L-Y, Zhang Z-A, He R-R (1986) Studies on fossil diatoms in the sediments of Lingding Gulf, South China Sea. Proceedings of the 8th International Diatom Symposium, Koeltz Scientific Books, Koenigstein, p 579–591

Reid G, Williams DM (2001) On *Eunotia clevei* var. *hispida* Skvortzov from Lake Baikal. Proceedings of the 16th International Diatom Symposium, Amvrosiou Press, Athens, p 89–106

Saarnisto M, Grönland T (1996) Shoreline displacement of Lake Ladoga—new data from Kilpolansaari. Hydrobiologia 322:205–215

Schmidt A (1913) Atlas der Diatomaceen-Kunde. O.R. Reisland, Leipzig

Shi Z-X (1991) New taxa of fossil diatoms from borehole 47 in the Jianghan plain, Hubei. Acta Micropaleontal Sin 8:449–459

Skvortzov BW (1937a) Bottom diatoms from Olhon Gate of Baikal Lake, Siberia. Phil J Sci 62:293–377

Skvortzov BW (1937b) Neogene diatoms from eastern Shantung. Bull Geol Soc China 17:193–204

Skvortzov BW (1946) Novye I malo izvestnye vidy Algae, Flagellatae, Phycomicetae iz Azii, Ameriki, Afriki, a takzhe s ostrovov Aponii I Cejlona, olisannye v 1931–45 g.g. s. 18 tablicami risunkov. [Species novae et minus cognitae Algarum, Flagellatarum et Phycomicetarum Asiae, Africae, Americae et Japoniae nec non Ceylon anno 1931–1945 descripto et illustrato per tab. 1–18]. Proc Harbin Soc Nat Hist Ethnogr Bot 2:1–34

Skvortzov BW, Meyer K (1928) A contribution to the diatoms of Baikal Lake. Proc Sungaree River Biol Stat 1(5):1–55

Tscheremissinova EA (1971) Species novae et curiosae Bacillariophytorum e neogeno regionis baicalensis. Nov Sist 8:52–66

Tscheremissinova EA (1973) Diatom flora of Neogene deposits in the Baikal region (Tunkin trough). Izdatelstvo Nauka Sibirskoe Otdelenie, Novosibirsk, 68 p

Vereshagin GY (1940) Proiskhozhdenie I istoriya Baikal, ego fauny I flory [Origine et Histoire du Baïkal, de sa Faune et de sa Flore]. Tr Baïk Limnol Stan 10:73–239

Voigt M (1937) Fossil diatoms of China. China J 27:310–315
Wang G-R (1998) Holocene diatoms from the delta of Pearl River, South China. Acta Palaeontal Sin 37:322–326
Williams DM, Reid G (2001) A bibliography of the botanical work of Boris V. Skvortzov (1896–1980), with commentary on the publications concerning diatoms (Bacillariophyta). Bull Nat Hist Mus (Bot) 31(2):89–106
Wislouch SM, Kolbe RW (1927) Materialy po diatomovym Onezhskogo I Lososinskogo ozer. Tr Olenetskoj Nauch Ehksped 5:3–76

Diatom Succession in Upper Miocene Sediments of Lake Baikal from the BDP-98 Drill Core

GALINA K. KHURSEVICH[1,2], EUGENE B. KARABANOV[2,3], MIKHAIL I. KUZMIN[3], DOUGLAS F. WILLIAMS[2], ALEXANDER A. PROKOPENKO[2,4], and SVETLANA A. FEDENYA[1]

Summary. On the basis of the diatom analysis of lacustrine sediments recovered by the BDP-98 drill core in the depth interval 600–221 m, eight successive periods in the development of the Lake Baikal diatom flora during the late Miocene have been distinguished. They differ by the specific generic and species composition of the diatoms, and clearly reflect the sedimentation conditions in the ancient paleobasin (the gradual facial changes from distal deltaic sequences to fine hemipelagic deposits). The late Miocene diatom flora of Lake Baikal is characterized by the presence in its composition of the biochronologically important extinct genus *Mesodictyon* Theriot & Bradbury, and by the evolutionary appearance, blooming, and extinction of the new genus *Mesodictyopsis* Khursevich, Iwashita, Kociolek & Fedenya, gen. nov., as well as by the occurrence of new freshwater extinct taxa of *Alveolophora*, *Concentrodiscus*, and *Actinocyclus*. Moreover, the extinct species are typical not only within the centric diatom genera mentioned above, but also within certain pennate ones.

Key words. Diatoms, Succession, Late Miocene, Lake Baikal

1 Introduction

The new 600-m drill core BDP-98 from the Academician Ridge of Lake Baikal recovered a continuous sedimentary record of the past 10 Ma (Baikal Drilling Project Members 2000, 2001). The entire 10-Ma lacustrine section contains more

[1] Institute of Geological Sciences, National Academy of Sciences of Belarus, Minsk 220141, Republic of Belarus
[2] Baikal Drilling Project, Department of Geological Sciences, University of South Carolina, Columbia, SC 29208, USA
[3] Vinogradov Institute of Geochemistry, Siberian Branch of the Russian Academy of Sciences, Favorsky str. 1A, Irkutsk 664033, Russia
[4] United Institute of Geology, Geophysics and Mineralogy, Siberian Branch of the Russian Academy of Sciences, Novosibirsk 630090, Russia

or less abundant diatoms, and thus allows the Plio-Pleistocene diatom succession described previously (Khursevich et al. 2000, 2001a, b, c) to be extended into the Miocene. Eleven biostratigraphic diatom zones have been distinguished in the Lake Baikal deposits from the BDP-96-1 and BDP-96-2 cores over the Pliocene–Pleistocene interval from 5.0 to 0.8 Ma. The zone boundaries are defined mainly by first appearance datum (FAD) and last appearance datum (LAD) of the new extinct species in Lake Baikal sediments. A total of 31 local diatom assemblage zones (LDAZ) were identified in the deposits of the Brunhes chron. They reflect the dramatic glacial/interglacial fluctuations of a paleoclimate. The orbitally tuned age model allows the zonal diatom assemblages to be compared with individual marine isotopic stages (MIS), and even substages, during the Brunhes chron. New results on diatom succession revealed in the upper Miocene sediments of Lake Baikal from the BDP-98 drill core allow us to trace the major trends in the evolution of nonmarine centric diatoms in this ancient, continuously existing, freshwater basin of the world for the last 10 Ma.

2 Materials and Methods

The new Baikal Drilling Project (BDP) borehole BDP-98 was drilled at the Academician Ridge of Lake Baikal at 53°44′48″N and 108°24′34″E in a water depth of 333 m (Fig. 1). The borehole penetrated to a sediment depth of 670 m. Continuous core sampling was performed to a depth of 600 m. Core recovery was 95%. The sediments of the BDP-98 core consist of biogenic diatomaceous ooze and terrigenous clay. Clay layers alternate with diatom-enriched layers throughout the section. The age model for the BDP-98 section is based on a correlation of event/reversal boundaries with the reference polarity timescale (Cande and Kent 1995). Paleomagnetic data from this section permit the sediments lying at a depth of 600 m to be dated at 10.3 Ma (Baikal Drilling Project Members 2000). The upper Miocene deposits were laid down in the BDP-98 core in the depth interval between approximately 600 and 221 m according to the preliminary age model for this core. Permanent diatom slides were prepared with identical volumes of material according to the method described by Grachev et al. (1997). Diatoms were enumerated along vertical and horizontal transects of permanent slides in 50 or 200 fields of view, depending on diatom concentration. Chrysophyte cysts and sponge spicules were also counted in slides (Fig. 2). The distributions of dominant individual planktonic diatom genera were quantitatively determined for the interval of core depth between 600 and 221 m. Elsewhere, ranked abundances in million (mln) valves per gram of dry sediment were used. Specimens were examined using oil immersion light microscopy (Ergaval brightfield, NA = 1.25, Zeiss, Germany) and scanning electron microscopy (JEOL JSM-35C, JEOL, Japan).

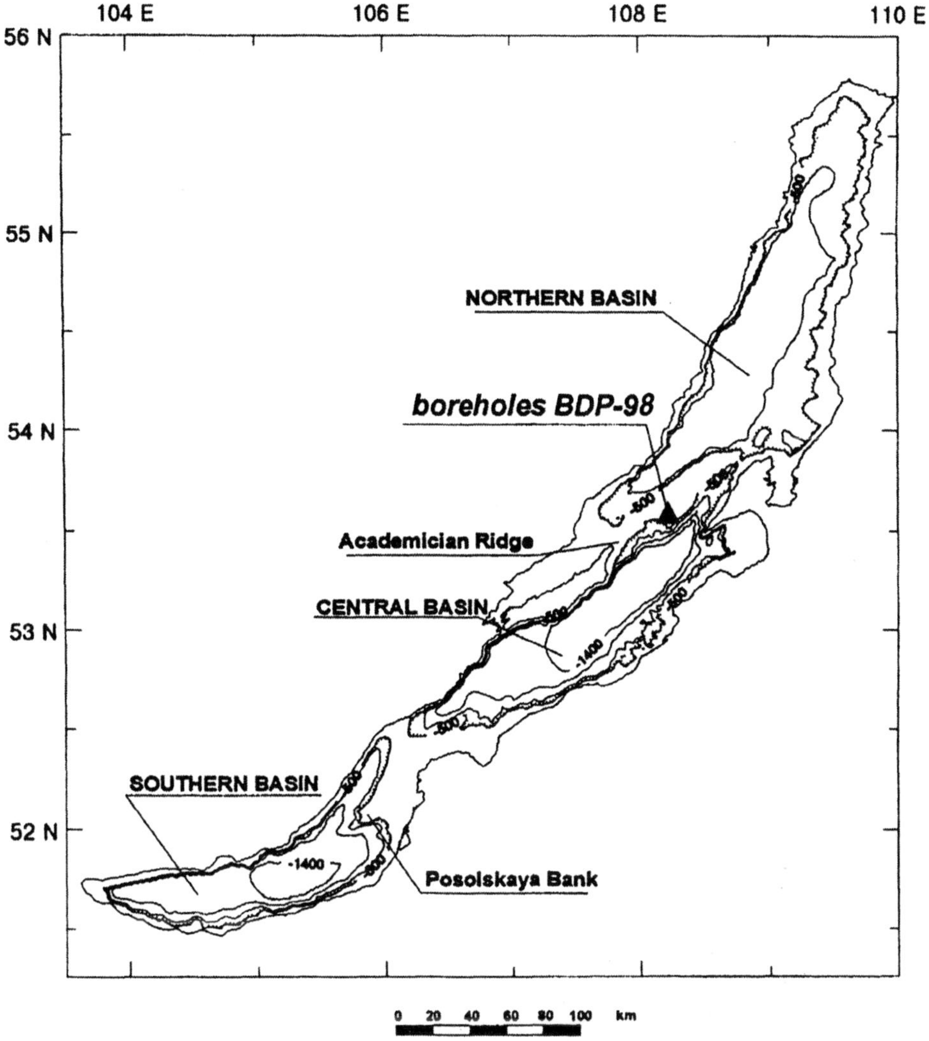

FIG. 1. Bathymetric map of Lake Baikal showing the location of the BDP-98 drill core at the underwater Academician Ridge

3 Diatom Record in Upper Miocene Sediments of Lake Baikal

The profile of diatom abundance and generic composition change recorded in sediments of the BDP-98 drill core allow us to distinguish eight successive periods in the development of the Lake Baikal diatom flora during the late Miocene. These are identified by the Roman numerals I–VIII, from the base of core upward (Fig. 3).

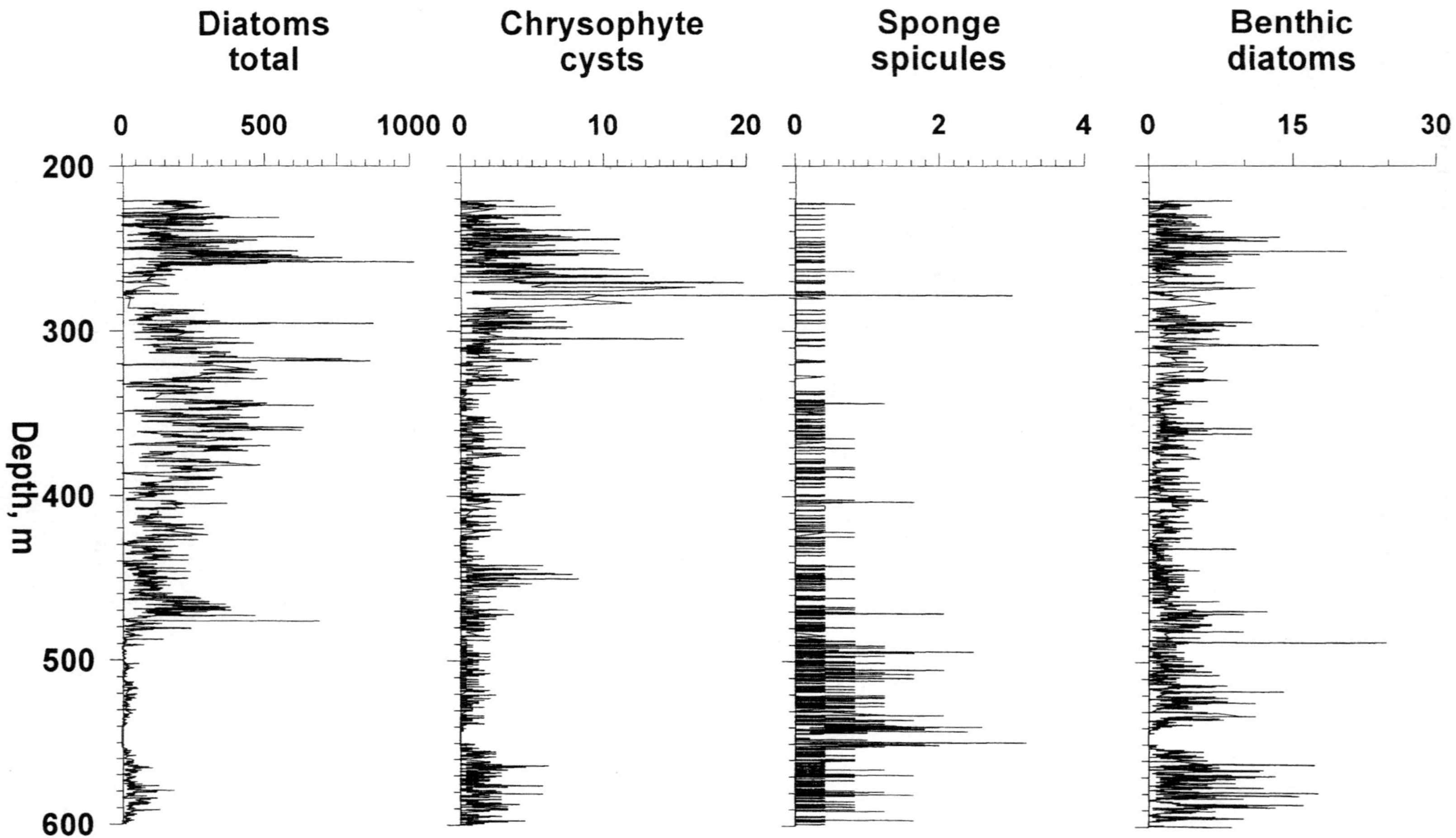

FIG. 2. Distribution of total diatoms, chrysophyte cysts, sponge spicules, and benthic diatoms in the sediments from the BDP-98 drill core in the depth-interval 600–221 m. Abundances are given in mln valves, cysts, and spicules per gram of dry sediment

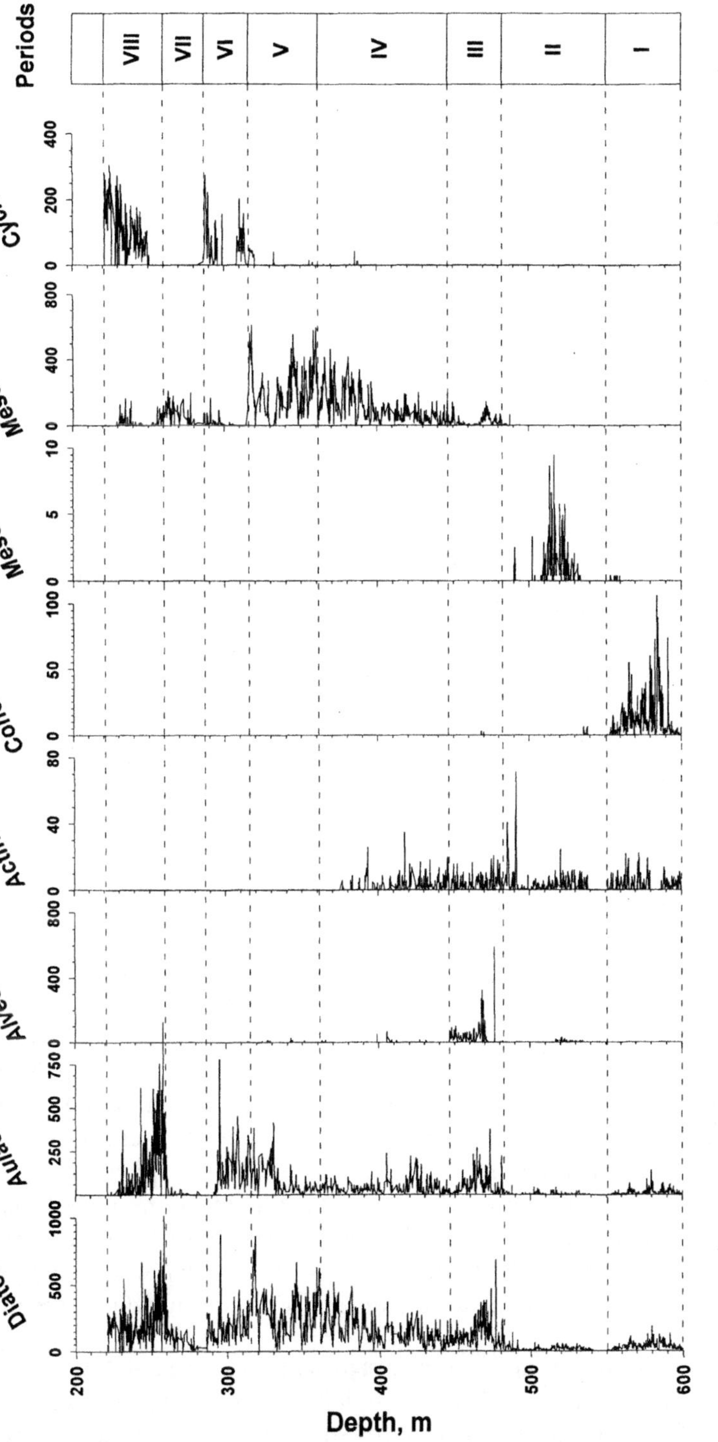

FIG. 3. Distribution of total diatoms and dominant planktonic diatom genera in the sediments from the BDP-98 drill core in the depth-interval 600–221 m. Abundance is given in mln valves per gram of dry sediment. *Periods*, the main periods in the development of the Lake Baikal diatom flora during the late Miocene

The sediment interval lying in the lower part of the section studied (from 600 to 550m core depth) is characterized by a comparatively poor concentration of diatom valves, most of which are corroded and fragmented. Among the planktonic diatoms, the extinct species of the genus *Concentrodiscus* Khursevich, Moisseeva & Sukhova showed the highest amount (up to 106.6mln valves per gram) at a depth of 584.45m, and these are important biostratigraphic markers of this interval. They are represented by *Concentrodiscus kuzminii* Khursevich & Fedenya and *C. indigenus* Khursevich & Fedenya (Khursevich et al. 2002).

Moreover, a rather high content of shallow-water benthic (mainly epiphytic) diatoms (from 30% to 72% of the total spectra) is fixed in many samples within this interval. These are represented by taxa of *Achnanthidium*, *Martyana*, *Fragilariforma*, *Staurosira*, *Staurosirella*, *Tabellaria*, *Tetracyclus*, *Eunotia*, *Navicula sensu lato*, *Amphora*, *Cymbella*, and *Gomphonema*, many of which belong to extinct species. In general, the planktonic (*Concentrodiscus, Actinocyclus, Aulacoseira)*-benthic/epiphytic diatom community (period I in the development of the diatom flora) characteristic of the lower part of the BDP-98 core, as well as the lithological evidence (high contents of silt, sand, gravel, plant debris, and wood fragments), indicate that the sediments of the 600- to 550-m interval were deposited in a near-shore, shallower environment with high inputs of terrigenous clastic material from the Paleo-Barguzin river delta.

The next interval (550–481 m core depth) is distinguished by a lower amount of diatoms (from 0.2 to 191.7mln valves per gram) and by the highest number of sponge fragments (up to 3.2mln spicules per gram at a depth 549.08m) (see Fig. 2). Among these, the abundance of the *Concentrodiscus* species is reduced to a minimum. The taxa of *Aulacoseira* were not numerous. The maximum occurrence of the extinct genus of *Mesodictyon* Theriot & Bradbury is typical for this depth-interval. The most important biostratigraphic value is from *Mesodictyon nativus* Khursevich & Fedenya (Khursevich et al. 2002) (as much as 9.5mln valves per gram).

Moreover, the planktonic species of *Actinocyclus* and benthic diatoms here achieve their highest quantity: as much as 71.48mln valves per gram of dry sediment for *Actinocyclus* at a depth 491.01 m, and as much as 24.8mln valves per gram for benthic taxa at a depth 487.71 m. The genus *Actinocyclus* Ehrenberg is represented by two new extinct species in the section studied, and benthic diatoms are mostly represented by the same genera as mentioned above. Thus, the planktonic (*Actinocyclus*, *Mesodictyon*)-benthic/epiphytic diatom community is characteristic of period II in the development of the Lake Baikal diatom flora during the late Miocene. The lithological composition of the 550- to 481-m interval in BDP-98 is similar to the lithology of the previous interval (with a high content of coarse-grained material), and suggests continued riverine input to the drill site.

The next core depth interval (481–446m) is characterized by a considerable diminution in the number of benthic/epiphytic taxa, and an essential increase in the frequency of planktonic species (to 96%–99% in most cases) in the general composition of diatoms. Planktonic genera of *Aulacoseira* Thwaites and *Alve-*

olophora Moisseeva & Nevretdinova belong to dominants within this interval. *Alveolophora* sp. aff. *A. jouseana* (Moisseeva) Moisseeva has a maximum value in this part of the section (up to 590.4 mln valves per gram of dry sediment at a depth of 476.53 m), whereas only single specimens of *A. antiqua* (Moisseeva) Moisseeva are found. *Aulacoseira* species are represented mainly by *A. praegranulata* (Jousé) Simonsen (the range of forms). Moreover, this depth-interval is marked by the appearance of the new genus *Mesodictyopsis* Khursevich, Iwashita, Kociolek & Fedenya, and by the constant presence of the *Actinocyclus* species mentioned above. The changes noted in the composition of the diatom community (period III in the development of the diatom flora) (see Fig. 3) and in the lithology of the depth-interval considered (in particular, a decreasing trend in the coarse fraction content) testify to the deepening of the basin and a gradual increase in the distance between the BDP-98 drill core site and the terrestrial sediment source.

In the core depth interval between 446–361 m, the average diatom content rises. This is caused by a clear increase in amount of the genus *Mesodictyopsis*, represented by four new species. Among these, *Mesodictyopsis academicus* Khursevich, Iwashita, Kociolek & Fedenya and *M. singularis* Khursevich, Iwashita & Fedenya occur not only within this interval, but also in the next one. The other members of this genus, *Mesodictyopsis medius* Khursevich & Iwashita (424–403 m) and *M. peculiaris* Khursevich, Kociolek & Fedenya (366–361 m) have a limited distribution. As far as the species of *Actinocyclus* are concerned, they disappear completely in the section studied in deposits corresponding to the end of the characterized interval. The taxa of *Aulacoseira* are common in this part of core. In general, period IV in the development of Lake Baikal diatom flora during the late Miocene is distinguished by the planktonic (*Mesodictyopsis, Aulacoseira, Alveolophora*) community, with small amounts of benthic species. This diatom community and the lithological characteristics (the progressive decline in the content of sand laminae and plant debris) over the 446- to 361-m core depth interval are indicative of the further extention and deepening of the ancient basin.

The next interval (361–315 m core depth), corresponding to period V in the development of diatoms, is dominated by two species of *Mesodictyopsis* and to a lesser degree by the *Aulacoseira* species. *Mesodictyopsis academicus* is most abundant (as much as 597.43 mln valves per gram) at a depth 360.24 m, and *M. singularis* (as much as 613.5 mln valves per gram) at a depth 318.18 m. The typical prevailing planktonic (*Mesodictyopsis*, *Aulacoseira*) community in this interval is evidence for the continuous deepening of the basin and a more or less deep water environment at the BDP-98 drill site during this period.

The next core depth interval (315–286 m, period VI in the development of the diatom flora) is clearly recognized by the following peculiarities: (a) by the mass occurrence of representatives of the genus *Aulacoseira*, with *Aulacoseira* sp. reaching a maximum quantity of up to 782 mln valves per gram of dry sediment at a depth of 295.44 m; (b) two essential peaks of members of *Cyclotella* belonging to the group "*C. iris* Brun & Héribaud"; (c) the appearance of new species within the genus *Mesodictyopsis* (*M. baicalensis* Khursevich, Iwashita, Kociolek & Fedenya, with a content as high as 166.92 mln valves per gram at a depth 291.04 m). This dominant planktonic (*Aulacoseira*, *Cyclotella*, *Mesodictyopsis*) diatom community testifies to the formation of deep water environments at the BDP-98 core site.

The core depth interval between 286 and 260 m is distinguished by an abrupt fall in abundance of diatoms (up to 6.2 mln valves per gram of dry sediment at a depth of 276.67 m), and by a clear rise in amount of chrysophytes algae (up to 38.4 mln cysts per gram at a depth of 278.33 m) (see Fig. 2). The monodominant planktonic diatom community represented by the genus *Mesodictyopsis* is characteristic of this depth range (period VII in the development of diatoms) (see Fig. 3). *Mesodictyopsis baicalensis* shows the highest peak in the lower part of the section studied (197 mln valves per gram at a depth of 277.93 m), while the new taxon *M. similis* Khursevich & Fedenya has a maximum abundance in the upper part (210 mln valves per gram at a depth of 263.16 m). Changes noted in the diatom composition reflect less favorable paleoecological conditions in the ancient basin as compared with the previous interval. This probably coincides with the increased aridity in southeast Siberia at the beginning of the Messinian period, which appears to be related to the rise of the Tibetan Plateau during 7–5 Ma (Ruddiman and Kutzbach 1989; Raymo and Ruddiman 1992).

Lastly, the core depth interval 260–221 m, corresponding to period VIII in the development of the Lake Baikal diatom flora during the late Miocene, is marked by repeated shifts in the species composition of diatoms. The lower part of this interval is dominated by *Aulacoseira* (mainly by the range of *A. praegranulata* forms, which achieved as much as 997 mln valves per gram of dry sediment at a depth of 258.34 m), the middle part by *Aulacoseira* in association with taxa of *Cyclotella* and *Mesodictyopsis*, and the uppermost part by *Cyclotella* species from the group "*C. iris*," which had a maximum of 307.8 mln valves per gram at a depth of 224.95 m. It is known that the local diatom assemblage zone of *Cyclotella iris* var. *insueta* Khursevich was found in the lowermost part of the Pliocene sedimentary sequence in the BDP-96-1 core (Khursevich et al. 2000, 2001a). The planktonic diatom communities of the interval considered, found in diatom-rich fine sediments, indicate a stable hemipelagic environment during this period.

4 Biostratigraphic Significance of Extinct Members of Freshwater Centric Diatoms from the BDP-98 Core (Comparative Aspect)

The late Miocene diatom flora of Lake Baikal is represented by abundant species of *Aulacoseira*, *Alveolophora*, *Actinocyclus*, *Concentrodiscus*, *Mesodictyon*, *Mesodictyopsis*, and *Cyclotella*.

Aulacoseira species have a long age-range (from the middle Eocene to recent time) and a wide geographic distribution, and hence are not very good stratigraphic markers. The genus *Alveolophora* is extinct, and its age range is restricted to the late Eocene (?)–Miocene. *A. antiqua* occurred in the BDP-98 drill core, and was first described from the lower-middle Miocene sediments of the Transbaikal area (Moisseeva and Nevretdinova 1990). Hence, the geological range of this species may be extended from the early to late Miocene. As far as *Alveolophora* sp. aff. *A. jouseana* is concerned, this taxon is similar to that found in Miocene lacustrine deposits from the western USA (Bradbury 1984). *A. jouseana* is a constant component in the early–middle Miocene diatom communities of the Far East, the Transbaikal area, and Belarus (Moisseeva and Nevretdinova 1990; Moisseeva 1995; Rylova et al. 1999), as well as in the late Miocene diatom assemblage of Kamchatka (Moisseeva 1992).

Freshwater extinct representatives of *Actinocyclus* are known from the late Oligocene to the Pliocene in different regions of the world (Krebs et al. 1987; Khursevich 1994, 1995; Krebs 1994; Temniscova-Topalova et al. 1994; Krebs and Bradbury 1995). However, their maximum diversity and abundance have been reported from middle Miocene sections of the western USA (Krebs et al. 1987; Krebs and Bradbury 1995) and from middle and upper Miocene profiles in Russia (Khursevich 1994). Two new taxa of *Actinocyclus* found in the upper Miocene sediments of Lake Baikal extend the geographic range of fossil nonmarine species of this genus.

The genus *Concentrodiscus* and its type species *C. abnormis* Khursevich, Moisseeva & Sukhova were first described from the middle–upper Miocene diatomaceous rocks of the upper Sulbanian depression within the Transbaikal area (Khursevich et al. 1989). Later a new member of this genus, *Concentrodiscus variabilis* Khursevich & Chernyaeva, was found from the Miocene diatomites within the Amalat plateau of the Transbaikal area (Khursevich and Chernyaeva 1994). Two new representatives of *Concentrodiscus* recorded in the BDP-98 core differ from the species mentioned above in certain distinct morphological features (Khursevich et al. 2002).

The genus *Mesodictyon* is restricted to the late Miocene and is widespread and diverse in lacustrine deposits of that time (Krebs et al. 1987; Krebs 1994). In the western regions of the USA, the Chalk Hills formation containing the various abundant species of *Mesodictyon* has been dated radiometrically or by the fission track as between 10 and 7 Ma (Theriot and Bradbury 1987; Theriot 1990). The Ch'orora formation in Ethiópia in which several taxa of *Mesodictyon* were identified has a radiometric age range of 10.7–10.5 Ma (Fourtanier 1987; Fourtanier and Gasse 1988). The Namora formation (northern Peru), with a common occurrence of *Mesodictyon* is also late Miocene in age based on radiometric analyses (7.2 ± 0.6 Ma) (Fourtanier et al. 1993). *Mesodictyon* has also been reported by Khursevich (Rylova et al. 1999) in the upper Miocene section at Lozy in Belarus. The presence of the new fossil species *Mesodictyon nativus* in the deposits of the BDP-98 drill core is the first find of this genus within Asia. Thus, the genus *Meso-*

dictyon is one of the important biochronological markers and can be used for cross-continental correlation of lacustrine sedimentary sections.

The genus *Mesodictyopsis* is known at present only from the upper Miocene deposits of Lake Baikal. It is represented by six extinct species, which have a useful biostratigrafic value. This genus is similar to the genus *Mesodictyon* in the position of cribrum inside the loculus, but differs in the presence of marginal fultoportulae with three satellite pores, in the availability of the valve face fultoportulae, and in the various locations of rimoportula(e) on the valve surface. The appearance, blooming, and extinction of the new genus *Mesodictyopsis* is a characteristic feature of the late Miocene stage in the development of the Baikalian diatom flora.

The genus *Cyclotella* Kützing ex Brébisson is represented in the studied section of the BDP-98 core mainly by taxa from the group "*C. iris*." A large quantity and infraspecific variations of *Cyclotella iris* have been recorded in the upper Miocene sediments within the Massif Central in France (Serieyssol 1981, 1984; Serieyssol and Gasse 1991) and the Sofia basin in Bulgaria (Ognjanova-Rumenova 1996). It is intersting to note that the monodominant *Cyclotella* community, consisting of *C. iris* with varieties, was found in the lower part of the BDP-96-1 core (200–193 m), which corresponds to 5.0–4.8 Ma (Khursevich et al. 2001a). Above this core depth interval, the *C. iris* group disappears completely in the sedimentary sequence from the BDP-96 core.

5 Conclusions

1. We have documented for the first time eight successive periods in the development of the Lake Baikal diatom flora during the late Miocene, reflecting the definite sedimentation conditions in the paleobasin (gradual change in deposits from distal deltaic facies to fine undisturbed hemipelagic sediments in the depth interval 600–221 m, in the BDP-98 core).
2. The late Miocene age of the diatom flora studied is based mainly on the occurrence of the biochronologically important extinct genus *Mesodictyon* (various species of this genus are known only from the late Miocene localities around the world), and is also confirmed by the paleomagnetic age model for the BDP-98 section.
3. The evolutionary appearance, blooming, and extinction of the new genus *Mesodictyopsis* is a distinctive peculiarity of the late Miocene stage in the development of the Lake Baikal diatom flora.

Acknowledgments. This work was supported by the National Scientific Foundation of the USA (NSF) grants EAR-93-1720401 and EAR-96-14770, and the Siberian Branch of the Russian Academy of Sciences. Cores for this study were made possible through the Baikal Drilling Project (BDP) supported by NSF and the International Continental Scientific Drilling Program (ICDP), the Russian

Academy of Sciences, the Russian Ministry of Geology, the Science and Technology Agency (STA) of Japan. The authors wish to express their gratitude to B.N. Khakhaev, L.A. Pevzner, and V.V. Kochukov, of the Nedra Drilling Enterprise, for organizing and providing the deep drilling of Lake Baikal sediments. We also thank A.N. Gvozdkov and N. Soshina, of the Institute of Geochemistry, for technical assistance with the BDP-98 core and smear slide preparation.

References

Baikal Drilling Project Members (2000) Paleoclimatic record in the late Cenozoic sediments of Lake Baikal (600m deep-drilling data). Russ Geol Geophys 41:3–32

Baikal Drilling Project Members (2001) The new BDP-98 600-m drill core from Lake Baikal: a key late Cenozoic sedimentary section in continental Asia. Quat Int 80–81:19–36

Bradbury JP (1984) Fossil *Actinocyclus* species from freshwater Miocene deposits in China and the United States. In: Mann DG (ed) Proceedings of the 7th International Diatom Symposium. Koeltz, Koenigstein, p 157–172

Cande SC, Kent DF (1995) Revised calibration of the geomagnetic polarity time scale for the late Cretaceous and Cenozoic. J Geophys Res 100:6093–6095

Fourtanier E (1987) Diatomées Néogènes d'Afrique: Approche Biostratigraphique en Milieux Marine (Sud-ouest Africain) et Continental. Thèse de Doctorat de L'Université Paris VI, Paris, p 365

Fourtanier E, Gasse F (1988) Premiers jalons d'une biostratigraphie et évolution des diatomées lacustres d'Afrique depuis 11 Ma. C R Acad Sci Paris 306:1401–1408

Fourtanier E, Gasse F, Bellier O, Bonhomme MG, Robles I (1993) Miocene non-marine diatoms from the western Cordillera basins of northern Peru. Diatom Res 8:13–30

Grachev MA, Likhoshwai EV, Vorobyova SS, Khlystov OM, Bezrukova EV, Veinberg EV, Goldberg EL, Granina LZ, Kornakova EG, Lazo FI, Levina OV, Letunova PP, Otinov PV, Pirog VV, Fedotov AP, Iaskevich SA, Bobrov VA, Sukhorukov FV, Rezchikov VI, Fedorin MA, Zolotarev KV, Kravchinsky VA (1997) Signals of the Paleoclimates of the upper Pleistocene in the sediments of Lake Baikal. Russ Geol Geophys 38:957–980

Khursevich GK (1994) Evolution and phylogeny of some diatom genera of the class *Centrophyceae*. Mem Calif Acad Sci 17:257–267

Khursevich GK (1995) Evolution of freshwater centric diatoms within the Eurasian continent. In: Marino D, Montresor M (eds) Proceedings of the 13th International Diatom Symposium. Biopress, Bristol, p 507–520

Khursevich GK, Chernyaeva GP (1994) A new species of the genus *Concentrodiscus* (Bacillariophyta) from the Miocene sediments of the Transbaikal area. Bot Zh 79:107–109

Khursevich GK, Moisseeva AI, Sukhova GA (1989) *Concentrodiscus*--a new genus of the family *Stephanodiscaceae* (Bacillariophyta). Bot Zh 74:1660–1661

Khursevich GK, Karabanov EB, Williams DF, Kuzmin MI, Prokopenko AA (2000) Evolution of freshwater centric diatoms within the Baikal rift zone during the late Cenozoic. In: Minoura K (ed) Lake Baikal: a mirror in time and space for understanding global change processes. Elsevier, Amsterdam, p 146–154

Khursevich GK, Fedenya SA, Karabanov EB, Prokopenko AA, Williams DF, Kuzmin MI (2001a) Late Cenozoic diatom record from Lake Baikal sediments. In: Economou-

Amilli A (ed) Proceedings of the 16th International Diatom Symposium. University of Athens, Athens, p 451–460

Khursevich GK, Karabanov EB, Prokopenko AA, Williams DF, Kuzmin MI, Fedenya SA, Gvozdkov AN, Kerber EV (2001b) Detailed diatom biostratigraphy of Lake Baikal sediments during the Brunhes chron and climatic factors of species formation. Russ Geol Geophys 42:108–129

Khursevich GK, Karabanov EB, Prokopenko AA, Williams DF, Kuzmin MI, Fedenya SA, Gvozdkov AN (2001c) Insolation regime in Siberia as a major factor controlling diatom production in Lake Baikal during the past 800000 years. Quat Int 80–81:47–58

Khursevich GK, Fedenya SA, Kuzmin MI, Karabanov EB, Williams DF, Prokopenko AA (2002) Morphology of new species of *Concentrodiscus* and *Mesodictyon* (Bacillariophyta) from the upper Miocene deposits of Lake Baikal. Algologia 12:361–370

Krebs WN (1994) The biochronology of freshwater planktonic diatom communities in western North America. Mem Calif Acad Sci 17:485–499

Krebs WN, Bradbury JP (1995) Geologic ranges of lacustrine *Actinocyclus* species, western United States. US Geol Surv, Prof Pap 1543-B:53–73

Krebs WN, Bradbury JP, Theriot E (1987) Neogene and Quaternary lacustrine diatom biochronology, western USA. Palaios 2:505–513

Moisseeva AI (1992) The genus *Alveolophora* Moisseeva & Nevretdinova. In: Gleser ZI, Makarova IV, Moisseeva AI, Nikolaev VA (eds) The diatoms of the USSR, fossil and recent. Nauka, St. Petersburg, p 85–87

Moisseeva AI (1995) Subdivision of the Neogene continental sediments of the Far East based on diatoms. Stratigr Geol Korrelatsiya 3:92–103

Moisseeva AI, Nevretdinova TL (1990) New family and genus of freshwater diatom algae (Bacillariophyta). Bot Zh 75:539–544

Ognjanova-Rumenova NG (1996) *Cyclotella iris* Brun & Héribaud--a group from the upper Miocene sediments of the Sofia basin, Bulgaria. Geol Carpathica 47:301–310

Raymo ME, Ruddiman WF (1992) Tectonic forcing of late Cenozoic climate. Nature 359:117–122

Ruddiman WF, Kutzbach JE (1989) Forcing of late Cenozoic north hemisphere climate by plateau uplift in southern Asia and west American. J Geophys Res 94:18409–18427

Rylova T, Yakubovskaya T, Khursevich G (1999) Palaeobotanical evidence for correlating the stratigraphy of the Neogene deposits of Belarus. Acta Palaeobot Suppl 2:359–363

Serieyssol K (1981) *Cyclotella* species of late Miocene age from St. Bauzile, France. In: Ross R (ed) Proceedings of the 6th Symposium on Recent and Fossil Diatoms. Koeltz, Koenigstein, p 27–42

Serieyssol K (1984) *Cyclotella iris* Brun & Héribaud. In: Mann DG (ed) Proceedings of the 7th International Diatom Symposium. Koeltz, Koenigstein, p 197–212

Serieyssol K, Gasse F (1991) Diatomées néogènes du Massif Central Français: quelques faits biostratigraphiques. C R Acad Sci Paris 312:957–964

Temniskova-Topalova D, Valeva M, Ognjanova-Rumenova N (1994) Nonmarine biostratigraphy of some genera of the class Centrophyceae from southern Bulgaria. Mem Calif Acad Sci 17:301–310

Theriot E (1990) New species of *Mesodictyon* (Bacillariophyta: Thalassiosiraceae) in late Miocene lacustrine deposits of the Snake River basin, Idaho. Proc Acad Nat Sci Philadelphia 142:1–19

Theriot E, Bradbury JP (1987) *Mesodictyon*, a new fossil genus of the centric diatom family Thalassiosiraceae from the Miocene Chalk Hills formation, western Snake River plain, Idaho. Micropaleontology 33:356–367

Late Pliocene Spongial Fauna in Lake Baikal (from Material from the Deep Drilling Core BDP-96-1)

E. WEINBERG[1], I. WEINBERG[1], S. EFREMOVA[2], A. TANICHEV[1], and Y. MASUDA[3]

Summary. Spicule analyses of part of the core from the BDP-96-1 bore hole corresponding to the late Pliocene were performed. The samples were dated by paleomagnetic analysis. An analysis of species composition allowed us to describe 40 sponge species, 14 of which belong to present-day Baikal sponges, while the other 26 species do not currently exist in Lake Baikal. According to the quantitative parameters of sponge spicules and species content, sponge fauna evolution in the late Pliocene was divided into three stages: (1) 3.1–2.9 Ma, corresponding to warm-water conditions; (2) 2.9–2.5 Ma, when the warm-water fossil species disappear, suggesting a sharp cooling down; (3) 2.5–2.1 Ma, when the sponge fauna was restored, but did not reach the full diversity of the first period. The dates obtained for sponges correspond well to palynological and diatom analyses of this core.

Key words. Sponge fauna, Spicule analysis, Late Pliocene, Lake Baikal

1 Introduction

The Lake Baikal spongial fauna is represented by two families: cosmopolitic Spongillidae, and endemic Lubomirskiidae. From the family Spongillidae, five sponges species which inhabit shallow water have been reported (Masuda et al. 1999; Efremova 2001a). From the family Lubomirskiidae, 13 sponge species and one subspecies, which are all distributed from 3 to 900 m, have been reported (Rezvoy 1936; Efremova 2001b). Spicules, which are skeletal needles of the sponges, occur abundantly in bottom sediments of Lake Baikal, together with diatoms. The fossil sponge spicules in the bottom sediments of Lake Baikal have

[1] Limnological Institute, Siberian Branch of the Russian Academy of Sciences, Ulanbatorskaya str. 3, Irkutsk 664033, Russia
[2] Biological Institute of St. Petersburg University, Oranienbaumskoye sch.2, Stary Peterhof, St. Petersburg 198904, Russia
[3] Department of Biology, Kawasaki Medical School, Matsushima, Kurashiki, Okayama 701-0192, Japan

not been extensively studied except for several repoerts by Martinson (1936, 1938, 1948), who studied sponge spicules in ancient pre-Baikalian deposits and in the sediments of the Baikalian depression itself in the 1930s and 1940s. The development of modern drilling and sampling techniques has given us new ways to study lake history. Some papers about Baikalian sponges spicules in the sediments of the underwater Academician Ridge have been published recently (Weinberg 2001; Weinberg et al. 1997, 1999a, b). These showed that sponge spicules are a promising subject for paleoclimatic reconstructions and a study of the evolution of spongial fauna in the past. The aim of this research was to study species composition and quantitative characteristics of spongial fauna in the Late Pliocene, as well as its response to the climatic changes which occurred in this period. For this, we selected part of a core from the hole BDP-96-1, from a depth of 90–114 m, which corresponded to 2.1–3.1 Ma. Within this interval, according to diatom and pollen studies on the sediments in a 200-m core, there was a period of considerable cooling which did not appear in known Siberian climatic scales (Kuzmin et al. 1997; Bezrukova et al. 1999; Karabanov 1999).

2 Materials and Methods

Core BDP-96-1, which was sampled from the top of the underwater Academecian Ridge in Lake Baikal from the station with coordinates 53°41′48″N and 108°2′106″E. The samples for quantitative spicule content and species composition studies were collected every 30–40 cm, which corresponds to a temporal resolution of ca. 10000 years. The samples were prepared by a method described previously (Weinberg et al. 1999a), and analyzed by light and electronic microscopy. A total of 60 samples were analyzed. The systematic identification of the sponges was done after taking spicule morphology into account (Simpson 1983), since this is very diverse and often subject to intraspecific variability. For reliable sponge identification, another important feature, the skeletal architecture, was also observed. Because we only have isolated spicules in the sediments, systematic identification was somewhat difficult. We based our studies on the photographs and preparations of spicules of recent species from Baikal sponge collection in the joint Russian-Japanese expeditions of the last 5 years. By comparing spicules samples from the core with more than 1000 photographs of spicules of recent species, we identified the spicules in the case. Some of the spicules were difficult to correlate with any recent sponge species. These are referred to as "fossil species" in this chapter.

3 Results

In our samples, the family Lubomirskiidae is represented by the following species: *Lubomirskia fusifera* (Soukatschoff 1895); *L. abietina* (Swartschzewsky 1902); *Baikalospongia bacillifera* (Dybowski 1880); *B. intermedia* (Dybowski 1880); *B. i. profundalis* (Rezvoy 1936); *Swartscheskia papyracea* (Dybowski

1880); *Lubomirskia incrustans* (Efremova 2001); *Baikalospongia recta* (Efremova 2001); *B. martinsoni* (Efremova 2001); *Rezinkovia echinata* (Efremova 2001); *R. arbuscula* (Efremova 2001); *B. fungiformis* (Makushok 1927). Thus, in the core sediments we found all the sponges species inhabiting Lake Baikal at present day (Efremova 2001a, b) except for *Lubomirskia baicalensis* (Pallas 1771) and *Swartschewskia irregularis* (Swartschewski 1902). We also found samples with a distinctive morphology which was difficult to relate to present species (Fig. 1). We related ten of them to the genus *Lubomirskia*, eight to the genus *Baikalospongia*, and one to the genus *Swartschewskia*.

The family Spongillidae is represented by four genera: *Spongilla* Lamarck, *Ephydatia* Lamouroux, *Eunapius* Gray, *and Trochospongilla* Vejdovsky (Efremova 2001a). Two species (*Spongilla* sp. and *Ephydatia* sp.) from the sediments are related to the present-day Spongillidae, and eight species which we related to this family occur only in the sediments (Fig. 2).

Sediments aged 3.19–3.12 Ma (including six of the samples studied) are characterized by a high spicule concentration, on average 9400 spicules/g. In the different samples of this period, 13–31 species of the family Lubomirskiidae occurred, 2–16 of which were fossil species, and 3–8 belong to the family Spongillidae; 1–6 of these are also fossil (Fig. 2). The dominant forms are *Lubomirskia* sp.5, *Rezinkovia* sp., and *Ephydatia* sp., and *Baikalosponga fungiformis*. These sponge spicules in the sediments are generally large.

In sediments aged 3.10–3.099 Ma (including two of the samples studied), the total spicule abundance was 20300 spicules/g. These are 17 or 18 spicules which relate to the family Lubomirskiidae, and 8–12 of these are fossil species. Spicules of the family Spongillidae numbered 8–11 species. The dominant forms were *Spongilla* sp.2, *Ephydatia* sp., *Ephydatia* sp.1, and *Spongilla* sp.4. These sponges spicules are not as large as in the previous interval. Microscleres of Spongillidae also occur.

The sediments aged 3.090–3.02 Ma (including six of the samples studied) are characterized by a low spicule concentration, on average 1500 spicules/g, and by a decrease in sponges species diversity. There are 4–12 species of the family Lubomirskiidae, and 3–6, of the family Spongillidae in different samples from this period. The number of fossil species decreases to 0–3. The main forms are *Ephydatia* sp., *Ephydatia* sp.2, *Baikalospongia* sp.4, and *Baikalospongia. fungiformis*. The sponge spicules are generally very large.

In sediments aged 3.0–2.90 Ma (including seven of the samples studied), the spicule concentration increases again to 14500 spicules/g. There are 8–20 species of the family Lubomirskiidae, of which 3–8 are fossils, and 4–10 species of Spongillidae, of which 2–8 are fossils. Spicules of *Ephydatia* sp., *Ephydatia* sp.3, *Spongilla* sp., *Lubomirskia incrustans*, and *Baikalospongia fungiformis* dominate. Microscleres of Spongillidae also occur here.

In sediments corresponding to 2.8–2.5 Ma (including 23 of the samples studied), the spicule concentration is 5500 spicules/g. A dramatic change in the species composition occurs. Sponges spicules of the family Lubomirskiidae dominate. There are 5–22 in different samples from this period, of which 1–7 are

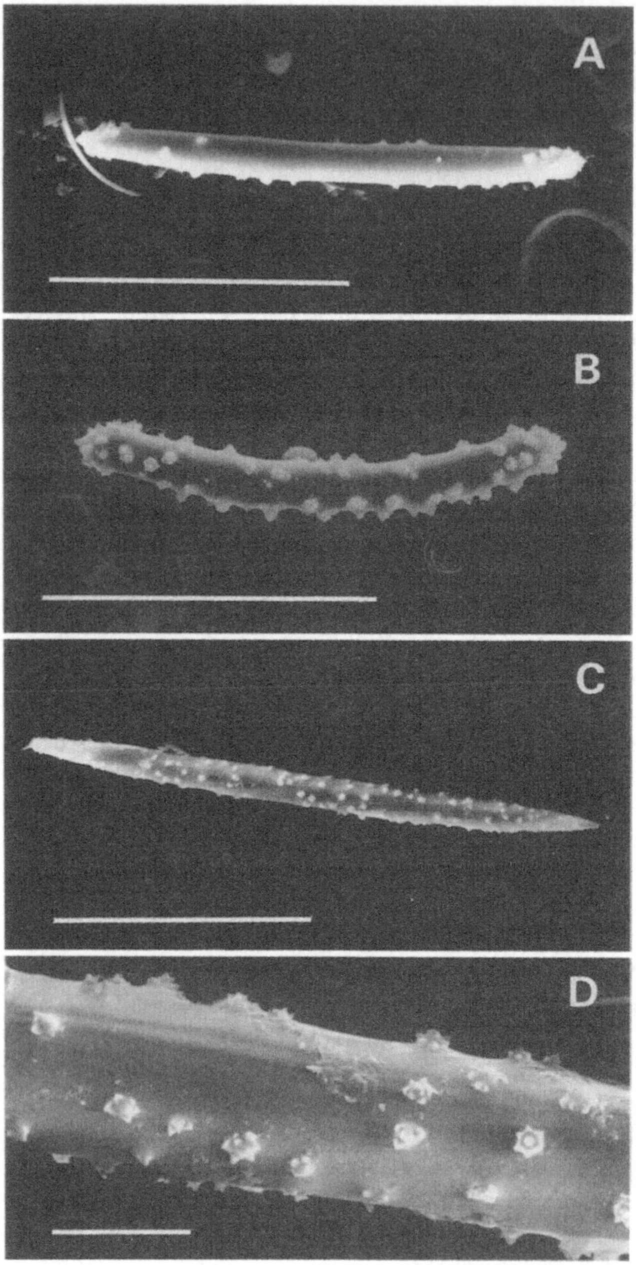

FIG. 1. Sponge spicules of fossil species in the sediments of Academician Ridge, Lake Baikal (BDP-96-1). **A, B** Sponges spicules of the family Lubomirskiidae (scale 0.1 mm). **C** Sponge spicule of the family Spongillidae (scale 0.1 mm). **D** The same at a larger magnification (scale 10 μm).

FIG. 2. Distribution of spicules of different sponge species in Academician Ridge sediments, Lake Baikal (BDP-96-1)

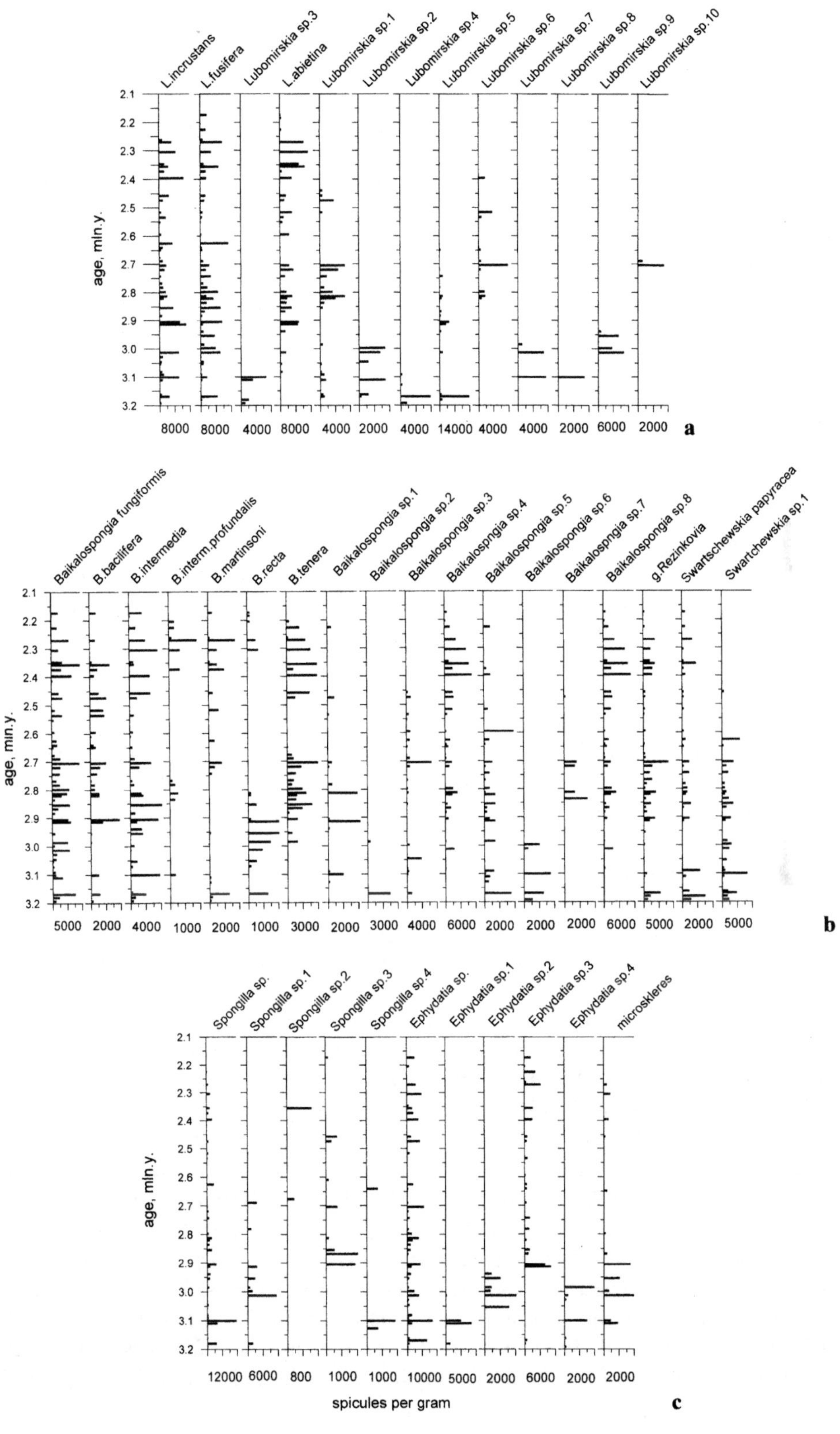

L.incrustans
L.fusifera
Lubomirskia sp.3
L.abietina
Lubomirskia sp.1
Lubomirskia sp.2
Lubomirskia sp.4
Lubomirskia sp.5
Lubomirskia sp.6
Lubomirskia sp.7
Lubomirskia sp.8
Lubomirskia sp.9
Lubomirskia sp.10
age, mln.y.
8000 8000 4000 8000 4000 2000 4000 14000 4000 4000 2000 6000 2000
a
Baikalospongia fungiformis
B.bacilifera
B.intermedia
B.interm.profundalis
B.martinsoni
B.recta
B.tenera
Baikalospongia sp.1
Baikalospongia sp.2
Baikalospongia sp.3
Baikalospngia sp.4
Baikalospongia sp.5
Baikalospongia sp.6
Baikalospngia sp.7
Baikalospongia sp.8
g.Rezinkovia
Swartschewskia papyracea
Swartchewskia sp.1
age, mln.y.
5000 2000 4000 1000 2000 1000 3000 2000 3000 4000 6000 2000 2000 2000 6000 5000 2000 5000
b
Spongilla sp.
Spongilla sp.1
Spongilla sp.2
Spongilla sp.3
Spongilla sp.4
Ephydatia sp.
Ephydatia sp.1
Ephydatia sp.2
Ephydatia sp.3
Ephydatia sp.4
microskleres
age, mln.y.
12000 6000 800 1000 1000 10000 5000 2000 6000 2000 2000
spicules per gram
c

fossils. The dominant forms are *B. fungiformis*, *L. incrustans*, *L. abietina*, and *L. fusifera.* Here only five species of Spongillidae occur, among which *Ephydatia* sp. dominates. The spicules are large and the diatom content is low. In sediments corresponding to 2.56 Ma, sponge spicules do not occur.

Sediments corresponding to 2.5–2.1 Ma (including 16 of the samples studied) contain an average of 4600 spicules/g. The species composition changes, and the content of spicules of the family Spongillidae increases to 1–8 species. The family Lubomirskiidae is represented by 2–17 species, of which 1–5 are fossils. The dominant forms are *Baikalospongia* sp., *Ephydatia* sp., and *Lubomirskia* sp. 6. Short-term oscillations in abundance and in species composition are observed.

4 Discussion and Conclusions

The quantitative characteristics of the spicule composition in the sediments of the period studied showed a gradual decrease in abundance up the section (Fig. 3) and short-term oscillations were observed. The tendency of sponges spicules to decrease up the cross section of the core correlates well with the decrease in diatom frustules, and reflects a known trend of climate cooling in the northern hemisphere in the Late Pliocene (Karabanov et al. 2000).

Short-term oscillations of the quantitative sponge spicule content in Lake Baikal sediments, as has been reported previously (Weinberg et al. 1997, 1999a), are related to the oscillations of other components (diatom algae, pollen, biogenic silica, etc.), and this rhythm due to the climatic oscillations caused by astronomical factors (Grachev et al. 1998). Low quantities of these biogenic components corresponds to the so-called "cold" periods, and high quantities to "warm" ones.

It is interesting that in the sediments with the poorest diatom algae content, the size of the sponge spicules is unusually large, which considerably increases the mean sizes of present-day spicules. In present day spongial fauna, the majority of species have spicules sizes within the range 180–240 μm except for new species such as *Rezinkovia echinata* and *Baikalospongia recta*, with spicules which reach a length of 300 μm or more. At the same time, spicules occurring in sediments with a low diatom content are longer than 300 μm, and often reach a size of 700 μm (Fig. 4). This might be explained by the concurrence of silicon in diatom algae and sponges (Maldonado et al. 1999).

According to Martinson (1938), taking into account the quantitative ratio of silicon spicules of two sponges families in Lake Baikal sediments, it is possible to get an idea of the relationship between Siberian and Baikalian biological complexes, and consequently of the characteristic of the water body in the period of sediment formation. The analysis of the distribution of sponge species in the BDP-96-1 core allowed us to see three periods of Lake Baikal spongial fauna development within the time-interval studied (see Fig. 3). The first is 3.1–2.9 My BP. Here, all 40 spicule types occurred: 14 recent ones, and 26 "fossil" ones. The amount of spicules of the family Spongillidae reached almost 60% of the total.

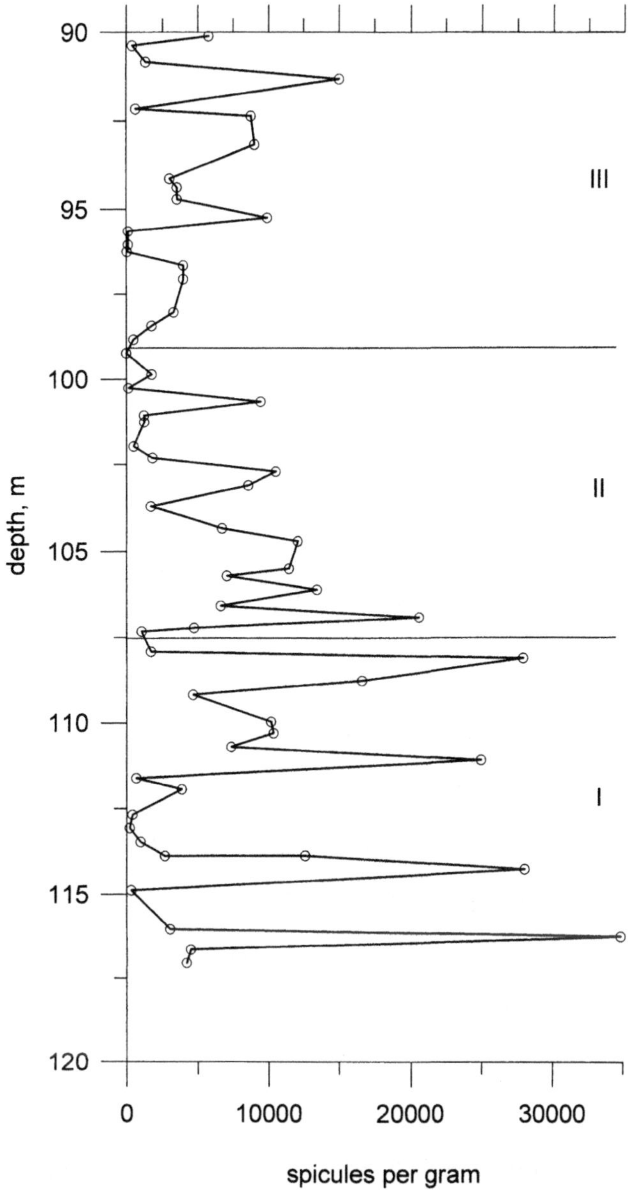

FIG. 3. Quantitative content of sponge spicules in Academician Ridge sediments, Lake Baikal (BDP-96-1)

In general, for the first period, a high spicules concentration, a rich species composition, and an abundance of representatives of the cosmopolitan Spongillidae fauna and of fossil species are characteristic. Since Spongillidae are traditionally considered to be inhabitants of shallow water areas, and are a more thermophilic species than Lubomirskiidae, we can suppose that the conditions were characterized by a warm water body with relatively shallow water compared with the first period. The presence of pollen of moderately thermophilic flora in the core

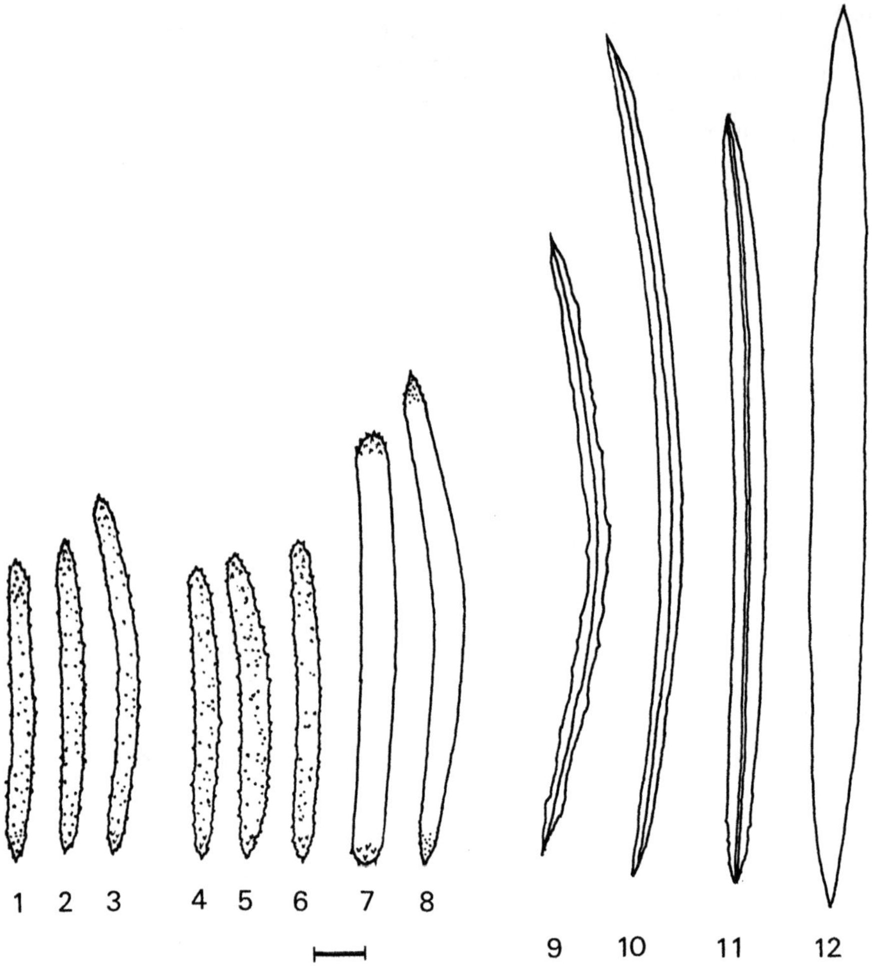

FIG. 4. Ratio of sizes of present day and fossil sponge spicules. Modern spicules: 1–6, *Baikalospongia intermedia*; 7, *Baikalospongia fungiformis*; 8, *Rezinkovai echinata*. 9–12, fossil spicules (scale = 40 μm)

sediments (hemlock, hazel, nut-tree, elm, lime) indicates a warm climate in that period (Bezrukova et al. 1999; Demske et al. 1999).

The second period in the development of spongial fauna (2.9–2.5 Ma) is characterized by a considerable decrease in spicule concentration in the core sediments, and by a dramatic change in the species composition. Here, 26 species occur, 11 of which are fossils. In the sediments of that period, when sponge spicules of the family Lubomirskiidae dominate, they make up 85%–100% of total number of spicules. Spicules of the family Spongillidae are never more than 15%, and in some intervals this decreases to zero. Many fossil spicules disappear. In sediments aged 2.6–2.5 Ma, sponge spicules are either not found, or are in extremely small amounts, and only a few spicules from the family Lubomirskiidae occur. These are *B. fungiformis*, *B. bacillifera*, and *Rezinkovia arbuscula*. Thus, during the second period, the majority of spicules of thermophilic species of the family Spongillidae, and the majority of fossil spicules which probably also belonged to thermophilic species, disappear. The age of the sediments of that period is within the late Pliocene, and coincides with the Gauss–Matuyama magnetic inversion (2.82–2.48 Ma). In this time-interval, a considerable change occurs in the Baikal regional vegetation composition and structure. This was due to a dramatic cooling and climate aridization (Bezrukova et al. 1999). The Lake Baikal diatom flora shows this cooling by the disappearance of heat-loving species of the genus *Aulacoseira*, and later, in 2.6–2.5 Ma, by the extinction of warm-water genera *Stephanolopsis* and *Tertiarius* (Khursevich et al. 1998).

The third period is 2.5–2.1 Ma. It is characterized by short-period oscillations of the abundance and species composition of the spongial fauna. On average, the spicule content in the sediments increases compared with the second period. There are 23 species, 8 of which are fossils. Spicules of thermophilic species of the family Spongillidae increase to 10%–30% of the total amount. However, the percentage of spicules of fossil species decreases. Thus, after the dramatic decrease in numbers of sponges and species diversity found in the second period, the reconstitution of the numbers and species composition of spongial fauna begins in the third period. However, this does not reach the abundance which preceded the cool period. In general, the species composition of the sponge fauna of this period is close to present-day one. Data from palynological analysis (Bezrukova et al. 1999) suggest that in the time-period following the climate cooling, the forest landscapes of the first part of the late Pliocene begin to dominate again. However, the proportion of *Tsuga* sp. (hemlock) and broad-leaf trees decreases considerably. The diatom flora (Khursevich et al. 1998) suggests numerous oscillations in abundance (as for sponges spicules), and consequently the unstable character of the climate at that time.

Thus, in Lake Baikal, analyses of the species composition of spongial fauna and the quantitative characteristics of sponge spicules in the late Pliocene sediments show that the spongial fauna reacted to climatic changes by variations in both number and species. Up to 2.5 Ma, the species composition of spongial fauna was quite different from that of the present day. Spicules of the cosmopolitan family Spongillidae, as well as spicules of extinct species of the family Lubomirskiidae, were abundant. During the significant cooling which occurred 2.5 Ma, the most dramatic changes in both numbers and species composition of spongial fauna took place. Later, the composition of the spongial fauna became closer to that of the present day, but it never reached the abundance that preceded the climate cooling.

Acknowledgments. The authors sincerely thanks their German colleagues Dr. Hedi Oberhansli and Carsten Eckert, who provided the opportunity to prepare samples for spicule analysis in the Alfred Wegener Institute for Polar and Marine Research in Potsdam. This study was supported by RFBR grant No. 01-04-97229.

References

Bezrukova EV, Kulagina NV, Letunova PP, Shestakova O (1999) Way of changes of Baikal region vegetation and climate during last 5 My (by data of palynological studies of Lake Baikal sediments). Russ Geol Geophys 40:739–749

Demske D, Müller J, Eckert C, Nowaczyk N, Mohr B, Oberhänsli H, Hubberten HW, Melles M (1999) A sedimentological and palynological record of Lake Baikal at the Pliocene–Pleistocene boundary. A preliminary report. Int Project Paleolimnol Late Cenozoic Climate 12:95–100

Efremova SM (2001a) Sponge (Porifera). Index of animal species inhabiting Lake Baikal and its catchment area. Lake Baikal, vol 1. Nauka, Novosibirsk, p 179–192

Efremova SM (2001b) New genus and new sponge species of the family Lubomirskiidae Rezvoy, 1936. Index of animal species inhabiting Lake Baikal and its catchment area. Lake Baikal, vol 2. Nauka, Novosibirsk

Grachev MA, Vorobyova SS, Likhoshway YEV, Khlystov OM, Bezrukova EV, Veinberg EV, Goldberg EL, Granina LZ, Kornakova EG, Lazo FI (1998) A high-resolution diatom record of the palaeoclimates of East Siberia for the last 2.5 My from Lake Baikal. Q Sci Rev 17:1101–1106

Karabanov EB (1999) Geological structure of Lake Baikal sedimentary bed and reconstructions of climate changes of Central Asia during the late Cenozoic. PhD thesis, p 72

Karabanov EB, Kuzmin MI, Williams DF, Khursevich GK, Bezrukova EV (2000) Global coolings in Central Asia in the late Cenozoic by the sediments record from Lake Baikal. DAN 370:61–66

Khursevich GK, Karabanov EB, Prokopenko AA, Williams DF, Kuzmin MI, Fedenya SA (1998) Pliocene–Pleistocene geochronology and biostratigraphy of Lake Baikal bottom sediments: new results of deep water drilling. International Symposium on Paleoclimates and the Evolution of Paleogeographic Situations in the Geological History of the Earth. Petrozavodsk, Russia, p 87–88

Kuzmin MA, Grachev MA, Williams D, Kawai T, Horie Sh, Oberhänsli H (1997) Permanent paleoclimate record of the last 4.5 Ma from Lake Baikal (first information). Russ Geol Geophys 38:1021–1023

Maldonado M, Carmona MC, Uriz MJ, Cruzado A (1999) Decline in Mesozoic reef-building sponges explained by silicon limitation. Nature 401:785–788

Martinson GG (1936) Distribution of sponge spicules in the deep drilling hole near Posolsk settlement on Lake Baikal. Dokl AN SSSR 1:261–264

Martinson GG (1938) Fossil spongial fauna of Tertiary pre-Baikal deposits. Dokl AN SSSR 21:212–214

Martinson GG (1948) Residues of animal organisms in Lake Baikal bottom sediments. Dokl AN SSSR 59:229–302

Masuda Y, Itskovich V, Veinberg EV, Efremova SM (1999) Perspective studies of freshwater sponges in Lake Baikal. Berliner Geowiss 30:329–332

Rezvoy PD (1936) Freshwater sponges. USSR fauna. Sponges 2:124

Simpson TL (1983) The cell biology of sponges. Springer, NewYork, Berlin, Heidelberg, Tokyo, p 300

Weinberg E (2001) The sponge fauna of Lake Baikal in the late Pliocene. Russ Geol Geophys 1:130–137

Weinberg EV, Khlystov OM, Vorobyuva SS, Kornakova EG, Levina OV, Efremova SM, Grachev MA (1997) Distribution of sponge spicules in sediments of the underwater Academician Ridge of Lake Baikal. Berliner Geowiss 20:141–145

Weinberg EV, Vorobyova SS, Eckert C (1999a) Spicule analyses: probable perspectives of its use in the studies of Lake Baikal bottom sediments. Russ Geol Geophys 40: 1258–1261

Weinberg E, Eckert C, Mehl D, Mueller J, Masuda Y, Efremova S (1999b) Extant and fossil spongiofauna from the underwater Academician Ridge of Lake Baikal (Se Sibiria). Mem Queensl Mus 44:651–657

Limnology in Lake Baikal and Lake Khubusgul

A Conceptual Model of Sedimentation Processes for a Hydrogeomorphological Study in Lake Baikal

SHINYA OCHIAI[1] and KENJI KASHIWAYA[2]

Summary. Analytical results for sediments from Academician Ridge in Lake Baikal show that the mineral grain size was large during interstadials, while it was small during stadials. The grain size distribution can be classified into four log-normal groups (A, B, C and D in size order). The mean grain size of the sediment is mainly dominated by the group A content. The group A content decreases considerably during stadials, leading to a small mean size.

Numerical calculations of sedimentation based on a two-dimensional advective diffusion model suggest that the grain size distribution of lake sediments is governed by (1) the grain size distribution of input sediment from outside the lake, and/or (2) flow velocity, and/or (3) lake water level.

Key words. Lake sediment, Physical properties, Sedimentation processes

1 Introduction

It has been widely accepted that the causes of glacial–interglacial change have been closely related to fluctuations in insolation resulting from variations in the orbit of the Earth. This is supported by information from marine sediments, ice cores, loess sediments, etc. (e.g., Shackleton et al. 1990; Ding et al. 1995). Information from lake sediments has also contributed to reveal orbit-related fluctuations (Kashiwaya et al. 1991). In Lake Baikal, analyses of diatom abundance (Colman et al. 1995) and the physical properties of long sediment cores (BDP-96, BDP-98) from Academician Ridge (Kashiwaya et al. 1998, 2001) indicate that climatic change in the continental interior is also related to the insolation.

[1] Department of Earth Sciences, Kanazawa University, Kakuma, Kanazawa 920-1192, Japan
[2] Institute of Nature and Environmental Technology, Kanazawa University, Kakuma, Kanazawa 920-1192, Japan

Lake sediments have some advantages for recording earth-surface processes in detail, including hydrological records which are limited in land areas. If it can be assumed that endogenic processes (crustal movement, etc.) are constant and/or ignored and the physical properties of rock materials are constant, then the sediment discharge from catchments is related to the external forces which erode and deform the earth surface (Kashiwaya 1994). The external forces are related to hydrological conditions such as precipitational intensity and the tractive force in rivers. Generally, sediment discharge and the grain size of eroded sediment are controlled by the tractive force. Some eroded sediment may be transported to downstream lakes. Therefore, external forces (precipitational intensity, tractive force, etc.) may be reconstructed from the physical properties of the lake sediments (grain size, sedimentation rate, etc.).

Information on grain size distribution may include not only external forces, but also limnological processes in the lake. Therefore, it is necessary to clarify what grain-size distribution is related to hydrological conditions in order to reconstruct past external forces. Firstly, we discuss in detail the grain size distribution of sediments from Academician Ridge during the past 250kyr. Secondly, we carry out numerical experiments on sedimentation processes with a model which considers Academician Ridge in order to discuss the dominant factors governing grain size distribution.

2 Area Studied and Samples

Sediment core VER98ST6 (10m), discussed here, was obtained from Academician Ridge in 1998. Academician Ridge (about 300m in depth) is shallower than the surrounding area, and is located far from large river mouths, indicating that this area is not directly influenced by turbidity current and riverine flow (Fig. 1). Grain size was measured every 1cm above and every 2cm below the point in the core which was 1.6m deep. The grain size distribution of mineral grains was measured with a laser diffraction particle size analyzer (SALD-2000J, Shimadzu, Kyoto, Japan) after the removal of biogenic silica and organic material (Nakagawa 2000). The mineral grain size is thought to reflect fluctuations in precipitation (Shimada et al. 2002). A tentative age scale for this core was determined assuming that grain size fluctuation is matched to the insolation curve at 65°N in July (Laskar et al. 1993) and the climatic time lag is ignored.

3 Grain Size Distribution

Figure 2 shows the analytical results for mean grain size for VER98ST6. The mean grain size becomes larger during interstadial periods and smaller during stadial periods, particularly in isotope stages 2, 4, and 5d. Measurement error (standard deviation) for mean grain size, which is estimated by measuring the

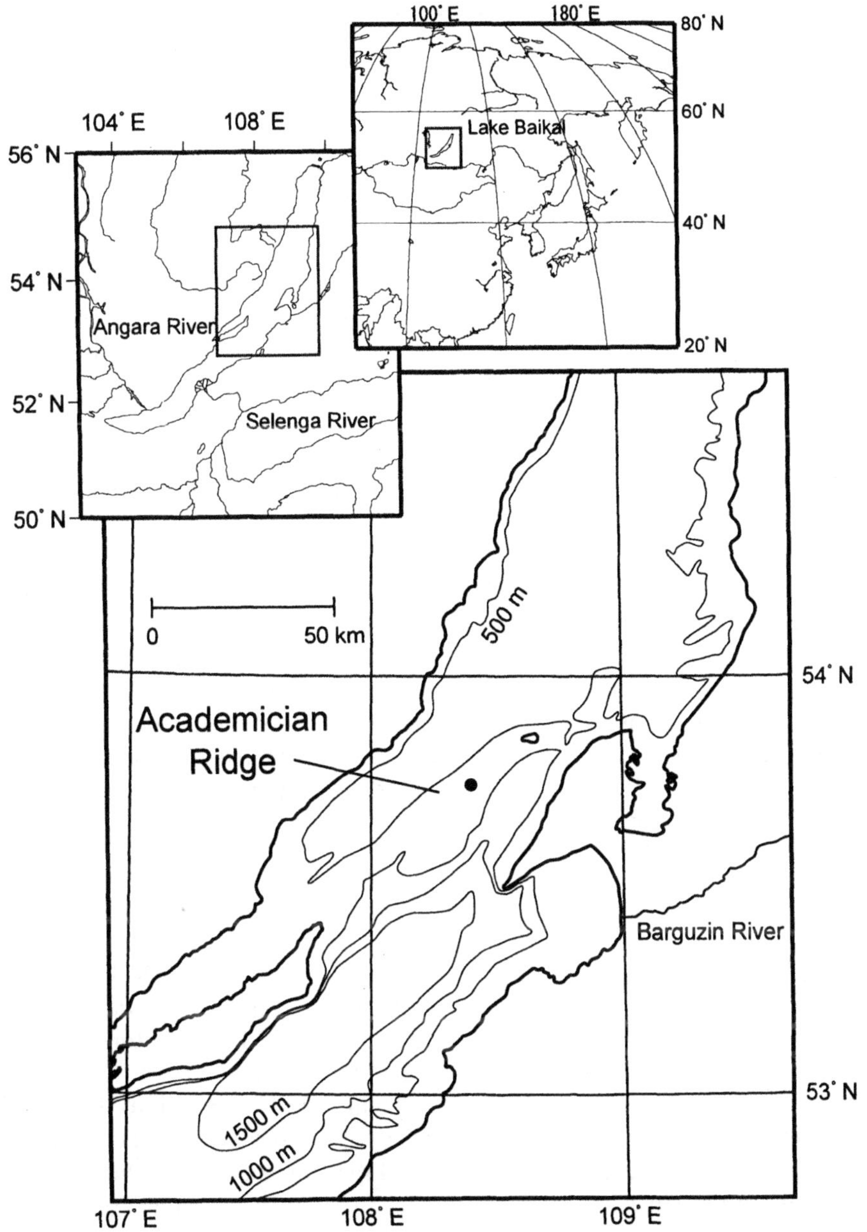

FIG. 1. Locations of Lake Baikal and the sampling point of the VER98ST6 core (*solid circle*)

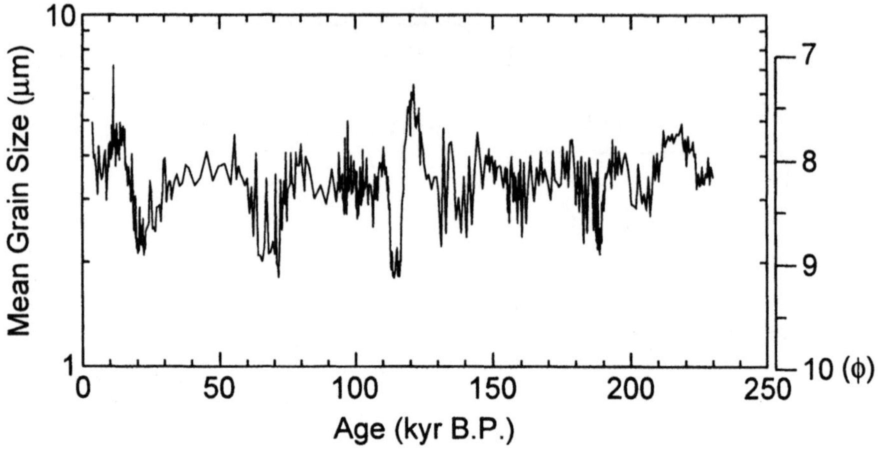

FIG. 2. Fluctuations in mean mineral grain size in VER98ST6 during the past 250 kyr

same sample 16 times, is about 0.05 ϕ. This indicates that a fluctuation with an amplitude greater than 0.1 ϕ is significant.

Here, we introduce another expression for grain size distribution in order to discuss grain size variation in more detail. The grain size distribution in sediment is thought to consist of some groups which relate to the form of transportation (traction, saltation, or suspension) in flowing water (Inokuchi and Mezaki 1974; Chikita 1990). We will classify grain size distribution into log-normal groups expressed in the following equation with three parameters:

$$f(x) = p \cdot \exp\left\{ -\ln 2 \cdot \frac{(x-m)^2}{w^2} \right\} \tag{1}$$

where p is the peak height, m is the mean, and w is the half-width at the half-maximum of the group. The set of most suitable parameters (p, m, and w) was calculated for all samples with the Simplex method, which is an optimization algorithm. The method of grouping is based on Minami et al. (1986). The grain size distribution of mineral grains of VER98ST6 may be classified into three or four log-normal groups (A, B, C, and D, in size order). Typical examples of this classification are shown in Fig. 3. The synthesized distributions of the four groups are almost the same as the original ones, suggesting that the classification is valid. In interstadial periods, grain size distributions consist of these four groups, while in stadial periods, group A disappears and the distributions consist of three groups.

Mean grain size, M_a, is expressed as

$$M_a = m_A c_A + m_B c_B + m_C c_C + m_D c_D \tag{2}$$

where m_A, m_B, m_C, and m_D are the mean size of each group, and c_A, c_B, c_C, and c_D are the content of each group. The mean size of each group is almost constant

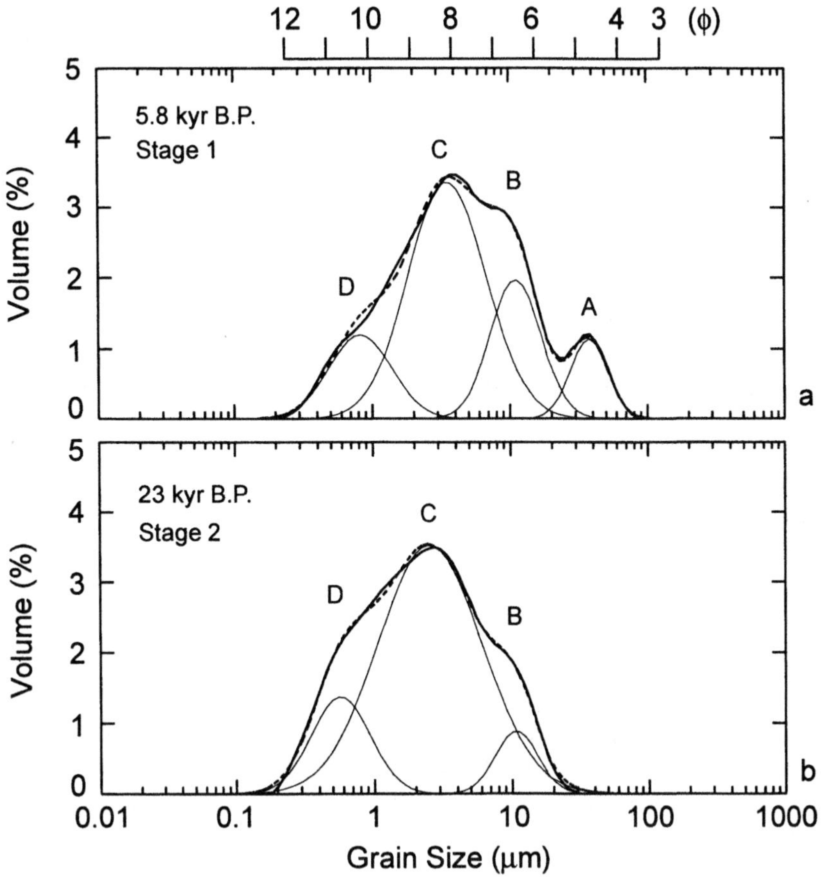

FIG. 3. Typical examples of the classification of grain size distribution in VER98ST6. **a** 5.8 kyr BP; **b** 23 kyr B.P. *Heavy solid lines* indicate the original distribution. *Heavy dashed lines* indicate the synthesized distribution from classified groups. *Light solid lines* indicate classified log-normal groups

over 250 kyr (Fig. 4). Measurement errors for m_A, m_B, m_C, and m_D are 0.16 ϕ, 0.09 ϕ, 0.12 ϕ, and 0.16 ϕ, respectively. This indicates that any change in the mean grain size is mainly controlled by the content of each group. Changes in the content of these groups are shown in Fig. 5. Measurement errors for c_A, c_B, c_C, and c_D are 3.00%, 9.26%, 11.76%, and 5.57%, respectively. According to Eq. 2, mean grain size is largely influenced by the content of group A, because m_A is much larger than the others. As shown in Fig. 2, mean size becomes small during stadials. This is mainly because the content of group A decreases considerably during stadials. This is important when considering the hydrological conditions between stadials and interstadials.

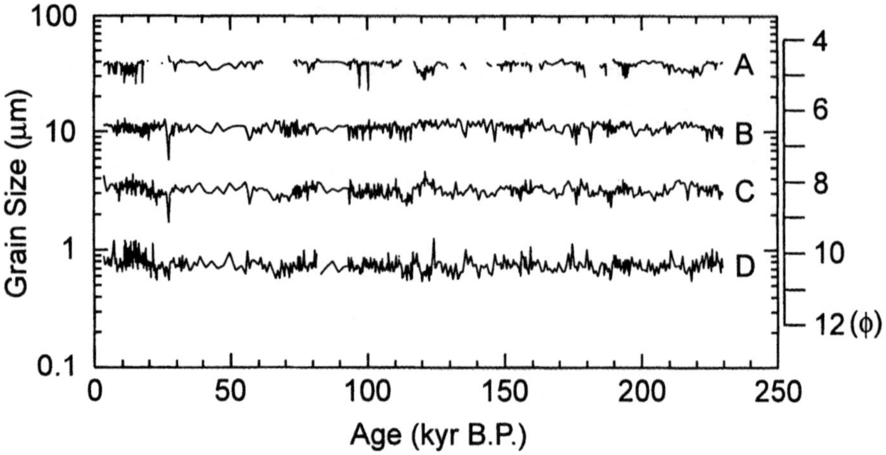

FIG. 4. Fluctuations in the mean size of the four groups (m_A, m_B, m_C, m_D) during the past 250 kyr

4 Sedimentation Model

In this chapter, we perform numerical experiments on sedimentation processes with a simplified conceptual model mainly to discuss the dominant factors governing the grain size distribution of the lake sediment, and in order to find simulation methods to express actual sedimentation processes on Academician Ridge in the next step.

Because Academician Ridge is topographically isolated, sedimentation processes around there are expected to be relatively simple. We therefore introduce an ideal two-dimensional (cuboid) lake as a first approximation model, assuming that (1) there are no convection currents and no vertical flows in the lake, and sediment is transport only by horizontal flow, diffusion, and settling, and (2) the horizontal flow velocity and diffusion coefficient are constant. Thermal conditions (thermal stratification) also may influence sedimentation processes (e.g., Chikita et al. 1995). Lake Baikal is often thermally stratified in the summer (Shimaraev et al. 1994), which may have some effect on the calculation. This problem will be discussed in future work.

Generally, temporal changes in sediment concentration in the two-dimensional lake are expressed in the following equation:

$$\frac{\partial c}{\partial t} = -u\frac{\partial c}{\partial x} - v\frac{\partial c}{\partial y} + k_x\frac{\partial^2 c}{\partial x^2} + k_y\frac{\partial^2 c}{\partial y^2} - w_s\frac{\partial c}{\partial y} \tag{3}$$

where x is the distance from the lake shore, y is the depth from the lake surface, c is the sediment concentration, u and v are the x and y components of flow velocity, respectively, k_x and k_y are the eddy diffusion coefficients with respect to

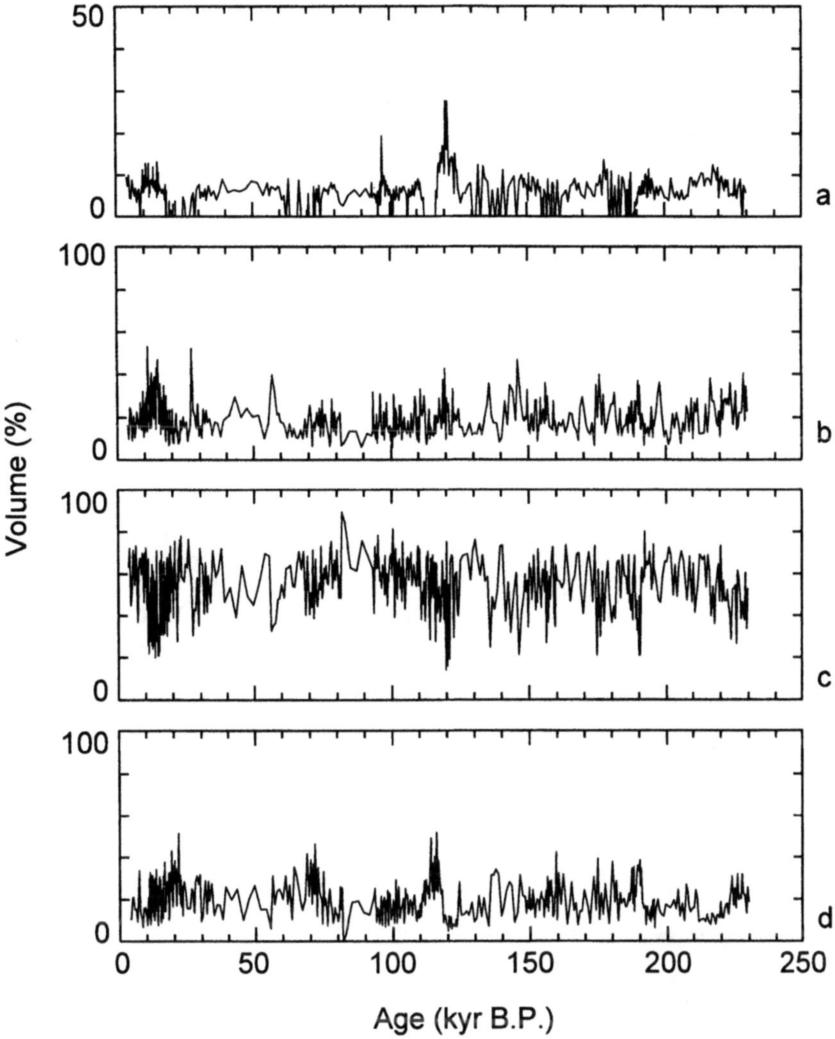

FIG. 5. Fluctuations in the content of the four groups (c_A, c_B, c_C, c_D) during the past 250 kyr. **a** Group A; **b** group B; **c** group C; **d** group D

the x and y directions, respectively, and w_s is the settling velocity of the sediment. Based on the assumptions mentioned above, Eq. 3 is simplified as

$$\frac{\partial c}{\partial t} = -u\frac{\partial c}{\partial x} + k_x\frac{\partial^2 c}{\partial x^2} + k_y\frac{\partial^2 c}{\partial y^2} - w_s\frac{\partial c}{\partial y} \tag{4}$$

Equation 4 was solved numerically by the finite-difference method. It is assumed that the horizontal length of the lake is 100 km (referring to the width of Lake Baikal), and that the depth is 200, 300, or 400 m (considering the depth of

Academician Ridge, i.e., about 300 m). It is also assumed that the grid sizes (x and y) are 1000 m and 10 m, respectively. We assumed circulation along the lake shore. To express this with a two-dimensional model, we assumed that flows circulate through two planes: flows turn in opposite directions at the edges (Fig. 6). We define the x-axis of one plane as x = 0–100 km, and of the other plane as x = 100–200 km. Then, the point x = 200 km is equivalent to x = 0 km. The horizontal and vertical eddy diffusion coefficients are 10–$10^4\,m^2/s$ and 10^{-5}–$10^{-3}\,m^2/s$, respectively in the ocean (Wadachi 1987). In Lake Baikal, Kipfer et al. (2000) estimated a vertical eddy diffusion coefficient of 1.7×10^{-4} to $6.3 \times 10^{-4}\,m^2/s$ on the basis of inverse modeling and tracer data. In this study, we assume a horizontal eddy dif-

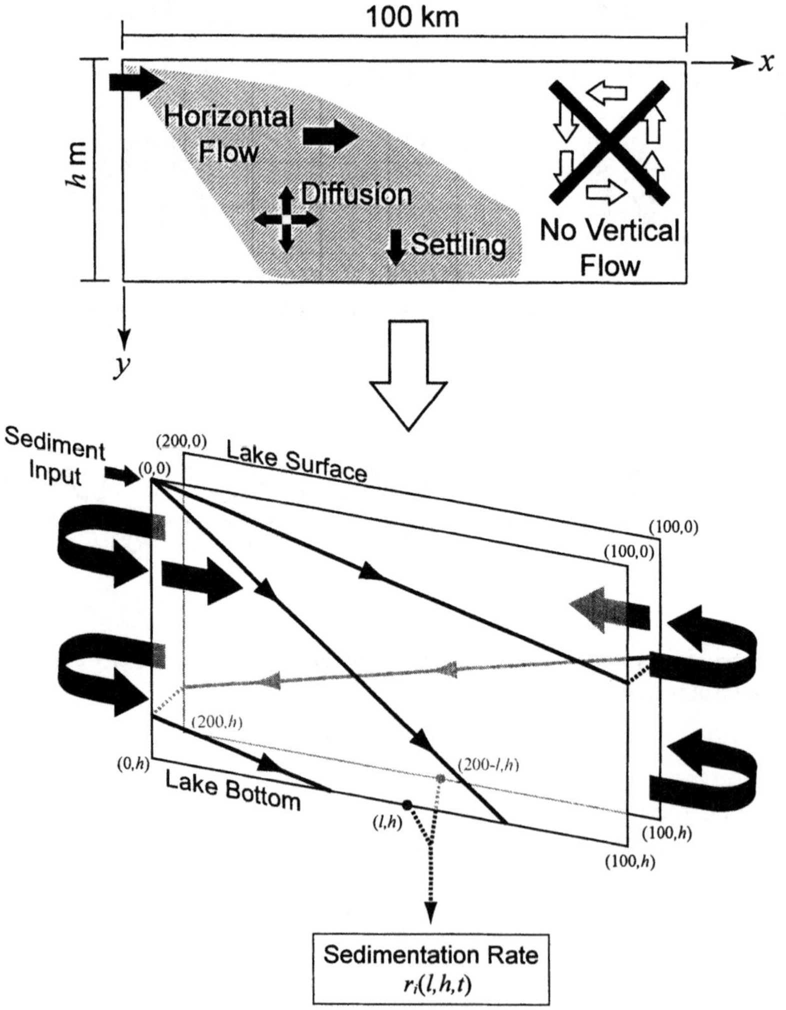

FIG. 6. Illustrations of the sedimentation model

fusion coefficient $k_x = 10\,\mathrm{m^2/s}$, and a vertical eddy diffusion coefficient $k_y = 5.0 \times 10^{-4}\,\mathrm{m^2/s}$. The horizontal flow velocity u is given on the basis of observations by Shimaraev et al. (1994). The settling velocity of grain, w_s, was calculated according to Stokes' law:

$$w_s = \frac{1}{18} g \frac{\rho_s - \rho_f}{\rho_f} \frac{1}{\nu} d^2 \tag{5}$$

where ρ_s and ρ_f are the density of grain and water, respectively, ν is the kinematic viscosity of water, g is the acceleration due to gravity, and d is the diameter of the grain. ρ_s was assumed to be $2700\,\mathrm{kg/m^3}$, by considering observed values. Values at 5°C are given for ρ_f and ν. The sediment concentration near the bottom was calculated numerically using Eq. 4.

The sedimentation rate, r_i, at the point $(x, y) = (l, h)$ for grain size d_i is defined as the downward flux calculated by the sediment concentration just above the bottom points of two planes (Fig. 6):

$$\begin{aligned} r_i(l,h,t) = &-k_y \frac{\partial c}{\partial y}\Big|_{x=l,y=h} + w_{s,i} c(l,h,t) \\ &-k_y \frac{\partial c}{\partial y}\Big|_{x=200-l,y=h} + w_{s,i} c(200-l,h,t) \end{aligned} \tag{6}$$

where l is the distance from the lake shore, and h is the depth below the lake surface, i.e., the lake water level. Then, the grain size distribution of the lake sediment is expressed as

$$q_i(l,h,t) = \frac{r_i(l,h,t)}{\sum_{j=1}^{n} r_j(l,h,t)} \tag{7}$$

where q_i is the frequency of the i-th size, and r_i is the sedimentation rate for the i-th size at the point (l, h). In this study, $n = 46$ classes (3–12 ϕ at 0.2 intervals) were used for the grain size distribution of sediment input from outside the lake, referring to observed values.

The sediment concentration for the whole lake is assumed to be zero at $t = 0$. Sediment is supplied at a constant rate ($r_{\mathrm{in},i} = 1\,\mathrm{kg/s}$) at one edge of the lake. Here, we will calculate the grain size distribution $q_i(l, h, t)$ of lake bottom sediments at $t = 260$ years, which is equivalent to the average for about the past 260 years, because settling times for 3–12 ϕ vary from 0 to 260 years based on Stokes' law. We then calculate the grain size distribution $q_i(l, h, t)$ at the point $l = 60\,\mathrm{km}$, considering the distance between Academician Ridge and the current most effective inflow river, the Barguzin River. Sediments with an ideal grain size distribution in which all grain sizes are equally contained are provided in the lake at one edge. For the sediment input, the grain size distribution for the lake bottom sediments are calculated for nine cases (flow velocity, $u = 0.02, 0.10$, and $0.18\,\mathrm{m/s}$; water depth, $h = 200, 300, 400\,\mathrm{m}$).

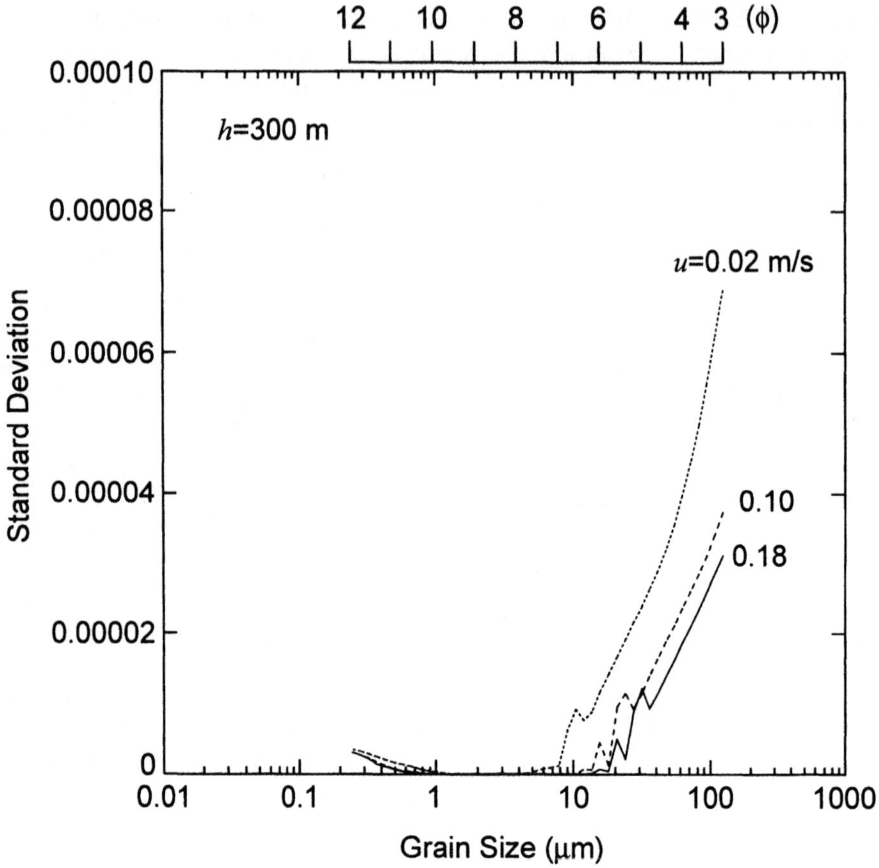

FIG. 7. Standard deviation of the sedimentation rate $r_i(x)$ from the spatial averaged sedimentation rate $r_{ave,i}$

The settling velocity for the finer grains is so slow that it takes long time for them to arrive at the lake bottom. Figure 7 shows the standard deviation of the sedimentation rate $r_i(x)$ from the spatial averaged sedimentation rate $r_{ave,i}$, which is expressed as

$$r_{ave,i} = \frac{r_{in,i}}{A} \tag{8}$$

where $r_{in,i}$ is the inflow rate of the sediment, and A is the area of the lake. Here, $r_{ave,i} = 1/10^5 = 10^{-5}\,\text{kg/m}^2\text{/s}$. The standard deviation is almost zero for grains finer than 6–8 ϕ. This indicates that grains finer than 6–8 ϕ settle uniformly in the whole lake. Large values for finer grains (10–12 ϕ) results from a lack of calculation time. To save calculation time, the sedimentation rate for grains finer than 8 ϕ is fixed at $r_{ave,i}$.

5 Results and Discussion

Figure 8 shows the calculated grain size distributions of lake sediments for three flow velocities ($u = 0.02, 0.10$, and $0.18\,\mathrm{m/s}$). The input sediment has four groups characterized by the same peak height (p) and a half-width at half-maximum (w). The mean size (m) of the four groups are given as averaged vales for all samples. The grain size distribution of lake sediment **c** was calculated as the product of the input sediment **a** and basic distribution **b** calculated as an ideal input sediment, where the basic distribution is defined as the grain size distribution calculated for ideal input sediment is which all grain size are contained equally. The basic pattern in the calculated grain size distribution is similar to that of input one, especially with finer grain. Finer grains are deposited uniformly after circulating in the lake, as shown in Fig. 7. This leads to the idea that the grain size distribution in lake sediment may reflect the grain size distribution of the input from outside the lake, i.e., the lake catchment.

Figure 8 also shows that the grain size distribution of the sediment may be related to the flow velocity in the lake, because the peak of the basic distribution of the coarse grains responds to the flow velocity. As the flow velocity becomes smaller, the peak of the coarse grains becomes smaller and the mean grain size becomes smaller.

Figure 9 shows the calculated grain size distribution for three water depths ($h = 200$, 300, and 400 m). As the lake water depth becomes small, the peak of the coarse grains and the mean grain size become smaller. However, the effect of a change in water depth seems to be smaller than the effect of flow velocity.

These results may be explained using Eq. 9. Under the assumption that grain is transported by a uniform horizontal flow and by settling, the horizontal distance l through which grains can be transported is

$$l = u\frac{h}{w_s} = \frac{18uh\rho_f\nu}{g(\rho_s - \rho_f)d^2} \tag{9}$$

where h is the depth of the lake, i.e., the lake water level. This leads to the fact that the maximum grain size, $d_{\max}$, which can be transported under the conditions flow velocity u, depth h, and distance l from the shore is given by

$$d_{\max} = \sqrt{\frac{18uh\rho_f\nu}{g(\rho_s - \rho_f)l}} \tag{10}$$

The maximum transportable grain size calculated by Eq. 10 (Figs. 8d and 9d) is equal to the peak of the coarse grains. This suggests that the grain size distribution of the lake sediment may be governed mainly by the grain size of input sediment from outside the lake, but it may also be influenced by flow velocity and water level, which are also related to external conditions.

We now discuss the actual grain size distribution of interstadial and stadial periods by considering the above-mentioned model. Figure 10 shows a summary of the results calculated for 18 cases: input sediment with or without group A,

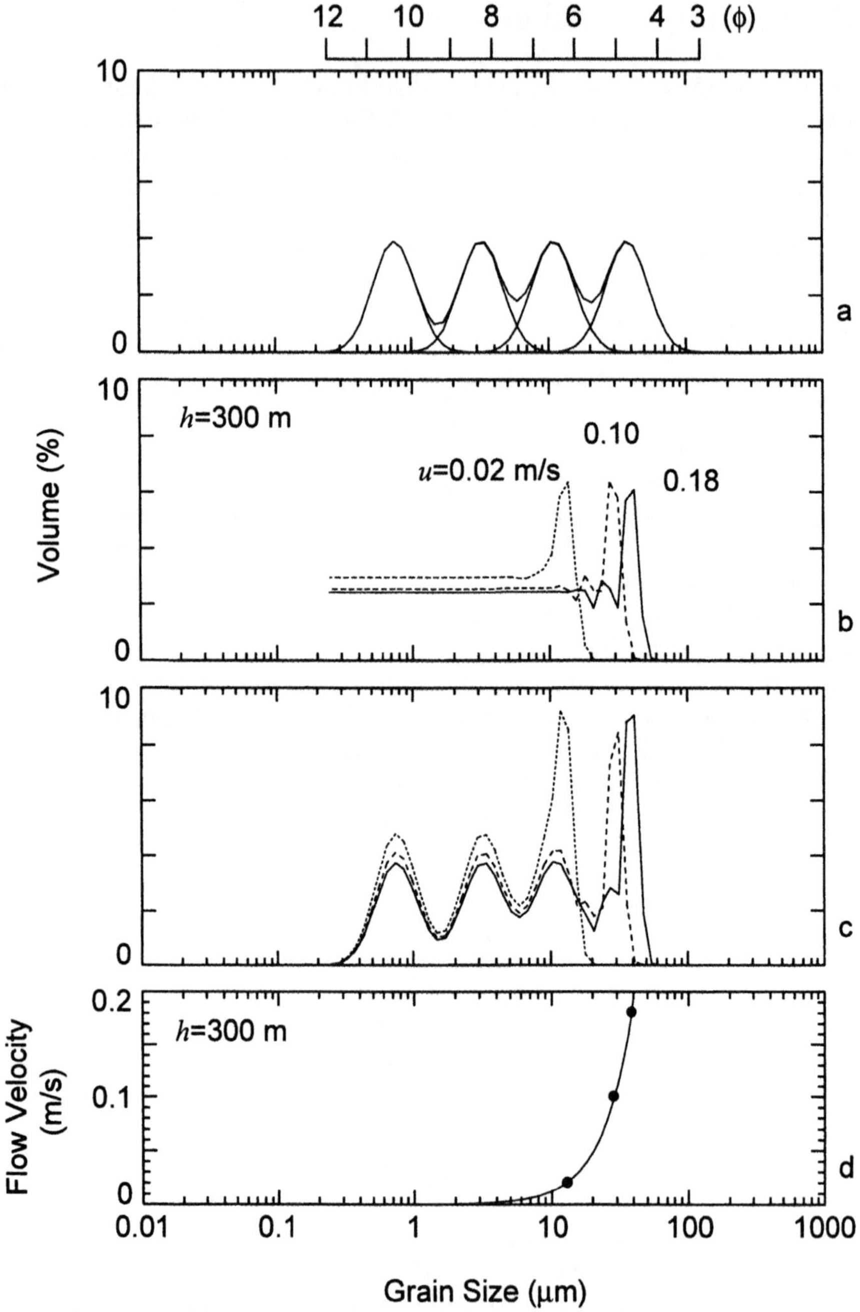

FIG. 8. Calculated grain size distributions for sediments in the four groups in three levels of flow velocity ($u = 0.02$, 0.10, and 0.18 m/s). **a** Input sediment; **b** basic distribution; **c** lake sediment calculated by the model; **d** maximum transportable grain size calculated by Eq. 10

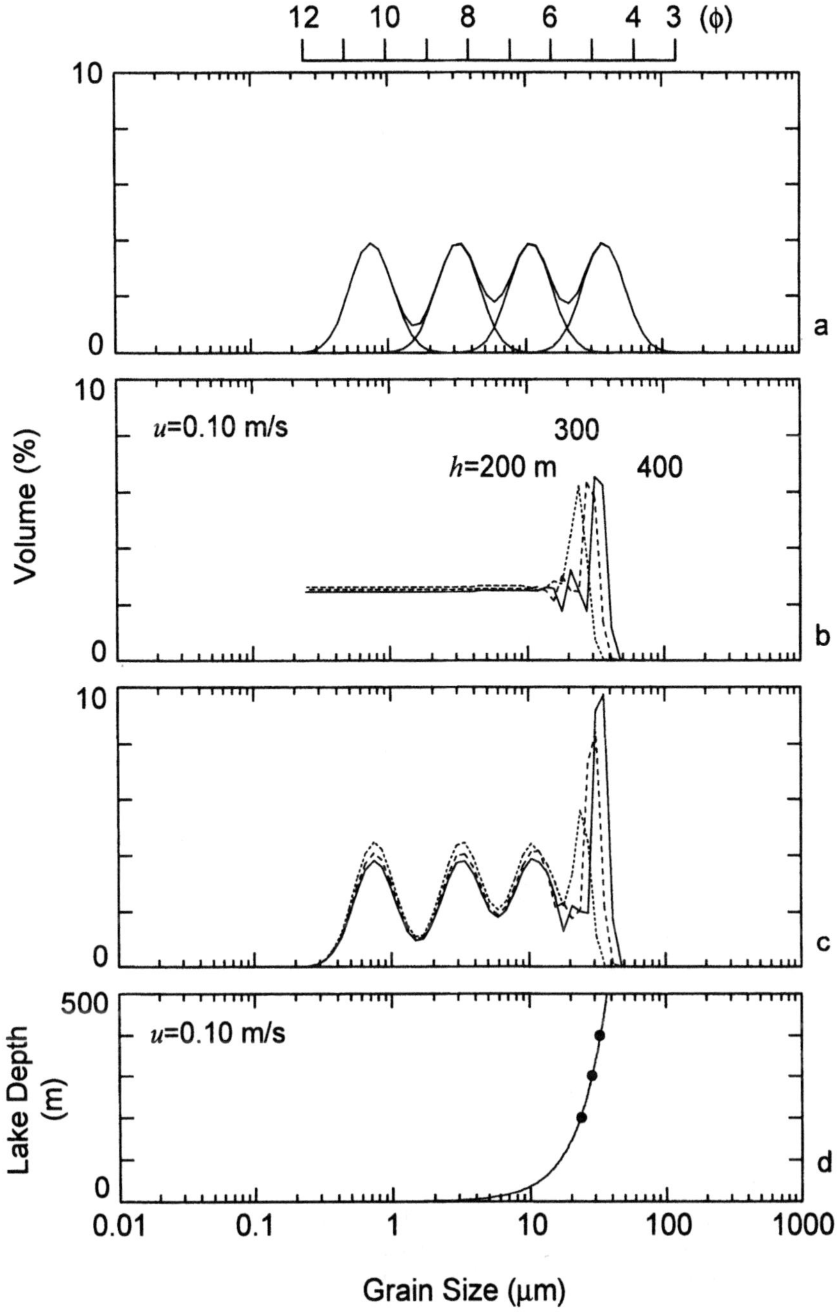

FIG. 9. Calculated grain size distributions for sediments in the four groups in three water depths ($h = 200$, 300, and 400 m). **a** Input sediment; **b** basic distribution; **c** lake sediment calculated by the model; **d** maximum transportable grain size calculated by Eq. 10

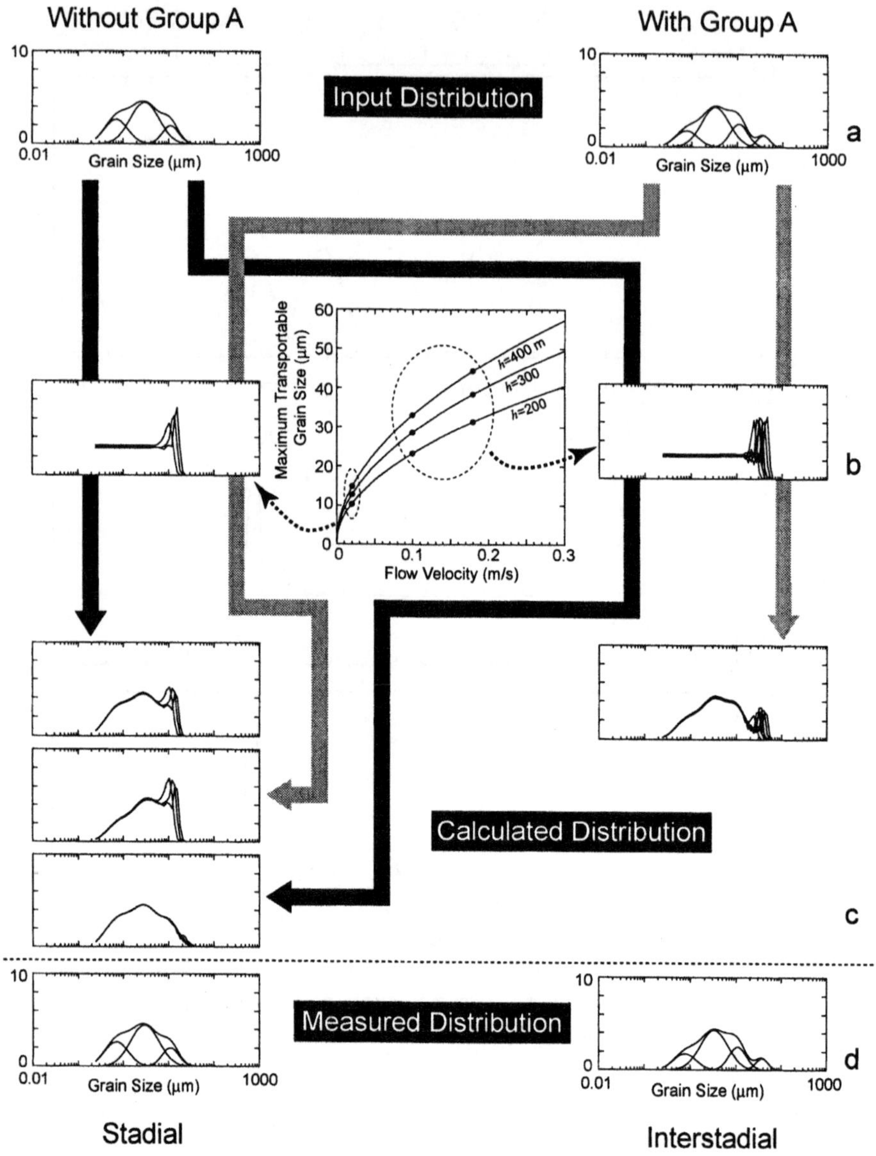

FIG. 10. Summary of the calculated results in 18 cases. **a** Grain size distribution of input sediment with or without Group A. **b** Basic distribution when the maximum transported grain size is relatively small (or large). **c** Calculated grain size distribution. **d** Measured (averaged) grain size distribution for the stadial and interstadial

three levels of flow velocity, and three water depths. When group A is contained in the input sediment and the flow velocity is large, group A is also contained in the lake sediment. When group A disappears from the input sediment and/or the flow velocity is small, group A disappears from the lake sediment. This indicates that the size of group A in the sediments could mainly be attributed to decreases in the productivity of group A in the catchment area, and also to the flow velocity and lake level. If the change in the content of group A is controlled by the input sediment, it may be possible to infer changes in the external forces in the catchment are from grain size fluctuations. It may be assumed that the external forces were large during interstadial periods and small during stadial periods.

6 Conclusions

Discussions on the grain size distribution of sediments at Academician Ridge in Lake Baikal, and a simulation based on a sedimentation model, lead to the following results. The grain size distribution of mineral grains can be classified into four groups (A, B, C, and D). The mean grain size is mainly dominated by the content of the coarsest group (A). The content of group A decreased considerably during stadials.

The sedimentation model shows that fine grains (finer than ~8ϕ) may be deposited uniformly in the lake. The grain size distribution of the lake sediment is governed by that of input sediment from outside of lake (the surrounding catchment area, etc.). The flow velocity and water level may also affect the distribution.

Acknowledgments. We thank Dr. N. Hasebe, Kanazawa University, for advice throughout this study, and H. Nakagawa, who contributed to the analysis of the physical properties of VER98ST6. We also thank Dr. T. Kawai, of the Natural Institute of Environmental Studies, for giving us a chance to obtain core samples from Lake Baikal, and colleagues in the hydrogeomorphological laboratory at Kanazawa University for their support of this study.

References

Chikita K (1990) Sedimentation by river-induced turbidity currents: field measurements and interpretation. Sedimentology 37:891–905

Chikita K, Sakata K, Hino S (1995) Transportation of suspended sediment slowly settling in a caldera lake. Jpn J Limnol 56:245–257

Colman SM, Peck JA, Karabanov EB, Carter SJ, Bradbury JP, King JW, Williams DF (1995) Continental climate response to orbital forcing from biogenic silica records in Lake Baikal. Nature 378:769–771

Ding Z, Liu T, Rutter NW, Yu Z, Guo Z, Zhu R (1995) Ice-volume forcing of east Asian winter monsoon variations in the past 800000 years. Q Res 44:149–159

Inokuchi M, Mezaki S (1974) Analysis of the grain size distribution of bed material in alluvial rivers (in Japanese, with English summary). Geogr Rev Jpn 47:545–556

Kashiwaya K (1994) A quantitative expression for external forces. In: Kirkby MJ (ed) Process models and theoretical geomorphology. Wiley, Chichester, p 85–95

Kashiwaya K, Fukuyama K, Yamamoto A (1991) Time variation in coarse materials from lake bottom sediments and secular paleoclimatic change. Geophys Res Lett 18:1245–1248

Kashiwaya K, Ryugo M, Sakai H, Kawai T (1998) Long-term climato-limnological oscillation during the past 2.5 million years printed in Lake Baikal sediments. Geophys Res Lett 25:659–662

Kashiwaya K, Ochiai S, Sakai H, Kawai T (2001) Orbit-related long-term climate cycles revealed in a 12-Myr continental record from Lake Baikal. Nature 410:71–74

Kipfer R, Hofer M, Peeters F, Imboden DM, Domysheva VM (2000) Vertical turbulent diffusion and upwelling in Lake Baikal estimated by inverse modeling of transient tracers. J Geophys Res 105:3451–3464

Laskar J, Joutel F, Boudin F (1993) Orbital, precessional, and insolation quantities for the Earth from −20 Myr to +10 Myr. Astron Astrophys 270:522–533

Minami S, Uchida T, Kawada S, Chiga Y, Okamoto T, Sasaki K (1986) Waveform data processing for scientific measurements (in Japanese). CQ, Tokyo, p 238

Nakagawa H (2000) Climato-limnological signals found in physical properties and the biogenic silica content of bottom sediments from Lake Baikal during the past several hundred thousand years (in Japanese, with English summary). Master's thesis, Kanazawa University, 65 p

Shackleton NJ, Berger A, Peltier WR (1990) An alternative astronomical calibration of the lower Pleistocene timescale based on ODP Site 677. Trans R Soc Edinburgh Earth Sci 81:251–261

Shimada S, Kashiwaya K, Hyodo M, Masuzawa T (2002) Hydro-environmental fluctuation in a lake-catchment system during the late Holocene inferred from Lake Yogo sediments (in Japanese, with English summary). Trans Jpn Geomorphol 23:415–431

Shimaraev MN, Verbolov VI, Granin NG, Sherstyankin PP (1994) Physical limnology of Lake Baikal: a review. Baikal International Center of Ecological Research, Irkutsk, p 81

Wadachi K (ed) (1987) Encyclopedia of oceanography (in Japanese). Tokyodo, Tokyo, p 589

Inorganic Characteristics of Surface Sediment from Lake Baikal: Natural Elemental Composition, Redox Condition, and Pb Contamination

Takejiro Takamatsu[1], Kenji Kashiwaya[2], and Takayoshi Kawai[3]

Summary. Fourteen short sediment cores (ca. 30 cm in length) from Lake Baikal were analyzed by inductively coupled plasma atomic emission spectrometry (ICP-AES) for Al, Ca, Co, Cr, Cu, Fe, Mg, Mn, Ni, Zn, As, Ti, V, P, S, Sr, Ba, Pb, Sc, and Y. The concentrations were comparable to those in remote unpolluted lakes in Europe and Canada. Most elements were classified into two major groups on the basis of their vertical distribution characters (Mn·Fe-oxide-philic, Fe, Mn, As, P, and Ba; terrigenous, Al, Ca, Mg, Ti, Co, Cr, Ni, Zn, V, Sr, and Sc), whereas S, Pb, Y, and Cu showed their own unique profiles. Concentrations of Cu correlated well with those of biogenic SiO_2 ($r = 0.856$), indicating the participation of diatoms in the deposition of Cu. Mn-, Fe-, and S-rich layers occurred in the same order (Mn above Fe above S) in all cores, indicating the oxidized condition of surface sediment. The oxidized layers were thin in the central basin and thick at Academician Ridge; the thickness correlated negatively with grain density ($r = -0.922$) and positively with the content of biogenic SiO_2 in the sediment. At Academician Ridge, abundant fossil diatoms may have increased the porosity of the sediment, accelerating the penetration of oxygen into the sediment. Depth profiles of Pb/Al showed distinct enrichment (pollution) of Pb in the surface sediment. The pollution began in the 1900s and became significant after ca. 1950. The pollution level (3.2–17.3, av. 13.5 $\mu g/cm^2$) was considerably lower than those in remote unpolluted lakes in Europe and Canada. It was higher in the southern basin and lower in the northern basin and at Academician Ridge. The total inventory of excess Pb in the lake was calculated at 4240 t.

Key words. Surface sediment, Elements, Redox condition, Lead, Contamination

[1] National Institute for Environmental Studies, 16-2 Onogawa, Tsukuba 305-8506, Japan
[2] Institute of Nature and Environmental Technology, Kanazawa University, Kakuma, Kanazawa 920-1192, Japan
[3] Department of Earth and Environmental Sciences, Graduate School of Environmental Studies, Nagoya University, Furo-cho, Chikusa-ku, Nagoya 464-8602, Japan

1 Introduction

Lake sediment is produced from riverine materials and atmospheric fallout via biological, chemical, and physical reactions within the lake. It records information about the origins of primary terrigenous materials and the lake environment (water temperature, redox potential, biological activity, trophic level, water depth, pollution, etc.) at the time of sedimentation as "chemical fossils." The elemental composition of sediment is also a possible indicator of lake dynamics, including currents and mixing of water, intrusion of river flow, sedimentation rate, and diagenesis. Therefore, we can decipher the local peculiarities and chronological changes of the above factors from the elemental distribution in the sediment.

Lake Baikal possesses the following remarkable features: in spite of the great water depth (maximum 1637 m), the hypolimnion always contains dissolved oxygen (Martin et al. 1993, 1998), and thus oxidized sediment covers the whole lake floor. In some places, ferromanganese crusts have formed, and grown to several centimeters thick (Takamatsu et al. 2000). The water quality and sediment composition are considerably homogeneous, because the replacement of the lake water (ca. 100 years in the southern basin, and >400 years in the northern basin; Shimaraev et al. 1994) and the residence time of dissolved materials (e.g., 20 years for ^{137}Cs, 19 years for ^{210}Pb; Edgington et al. 1991) are very long. Also, materials loaded into the lake deposit mostly within the lake (e.g., ca. 99% of Pb; Boyle et al. 1998). Such an aerobic and stable lake environment is rare not only from a limnological viewpoint, but also in sustaining the biological uniqueness.

We analyzed surface sediment cores from Lake Baikal for elements by inductively coupled plasma atomic emission spectrometry (ICP–AES) to determine the natural elemental composition (Takamatsu et al. 2000), redox condition (Martin et al. 1993, 1998), and recent pollution with Pb (Boyle et al. 1998; Mackay et al. 1998).

2 Materials and Methods

Fourteen short sediment cores (8 cm i.d. × 22–53 cm in length) were sampled with a multiple-core sampler from various sites in the lake from July 14 to 17, 1999 (Fig. 1). The cores were cut into sections 1 cm thick on a board, and brought back to the laboratory. Each section was freeze-dried and homogenized. Dried subsamples (10–20 mg) were digested with a mixture of acids (1 ml $HClO_4$, 2 ml HNO_3, 1 ml HF) in a pressurized digestion bomb (140°C for 5 h), and analyzed for elemental composition by ICP–AES (ICAP-61E Trace, Thermo Elemental, Franklin, USA). The elements investigated were Al, Ca, Co, Cr, Cu, Fe, Mg, Mn, Ni, Zn, As, Ti, V, P, S, Sr, Ba, Pb, Sc, and Y (Takamatsu et al. 2000). Matrix effects due to high concentrations of Al, Fe, Ca, Mg, and Ti were corrected for by using a *k*-factor method (Hee et al. 1985).

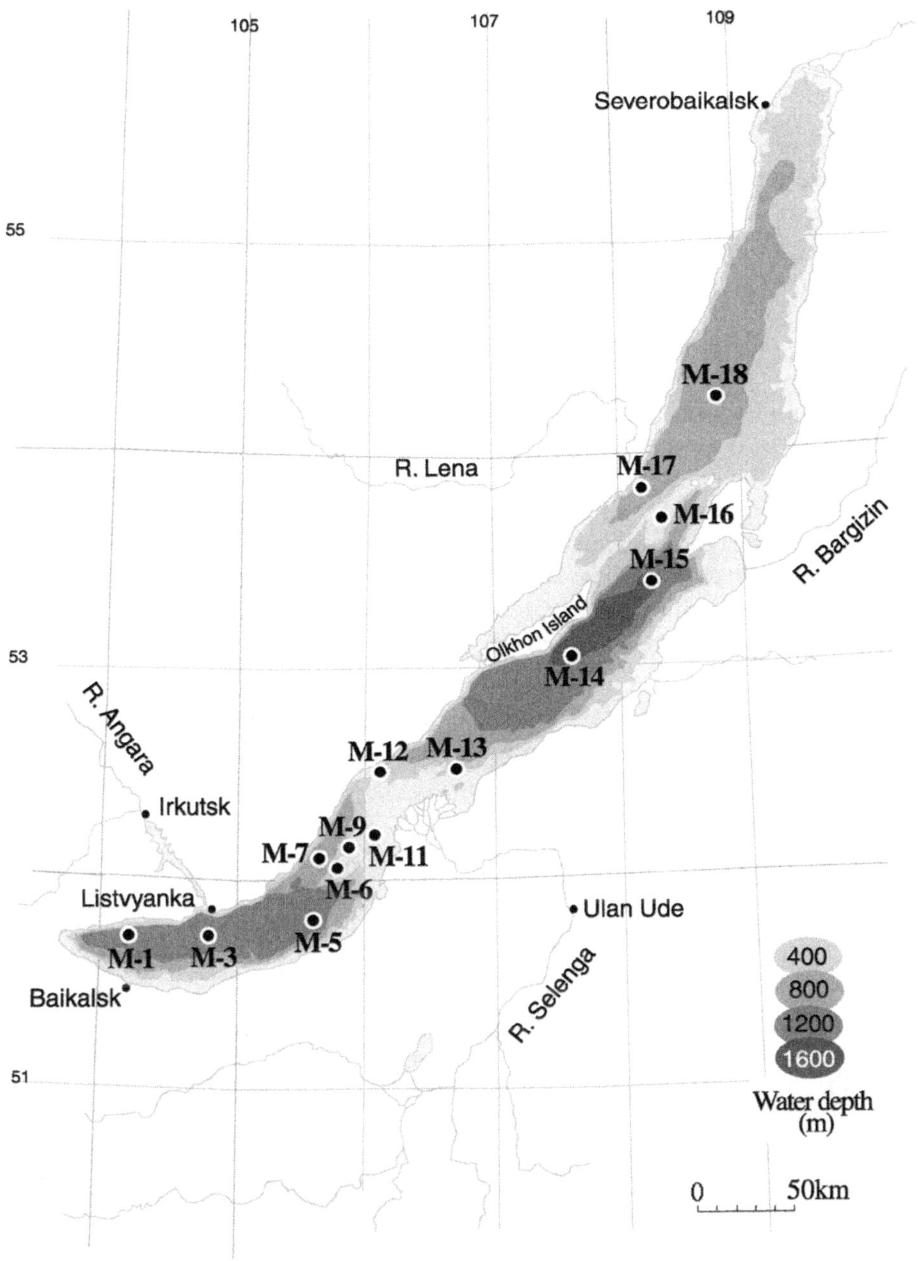

FIG. 1. Map showing the sampling sites of the short sediment cores; July 14–17, 1999

The contents of organic matter (by decomposition with H_2O_2), biogenic SiO_2 (by dissolution with Na_2CO_3 solution at 85°C), and minerals (from residue after the removal of organic matter and biogenic SiO_2), as well as some physical properties such as water content (oven-drying at 110°C for 48h), mean grain size (laser-reflection method (SALD-2000J, Shimadzu, Kyoto, Japan)), and grain density (using an autopycnometer (Accupyc 1330, Micrometrics, Norcross, USA)) were also determined (Kashiwaya et al. 2000).

The cores were dated using ^{137}Cs and ^{210}Pb. Average sedimentation rates in the upper 10cm of sediment given by ^{210}Pb were 0.015–0.058 (av. 0.027) g/cm^2/year and 0.041–0.120 (av. 0.078) cm/year, and those given by ^{137}Cs were 0.033–0.160 (av. 0.060) g/cm^2/year and 0.12–0.20 (av. 0.15) cm/year. Two ^{137}Cs fallout peaks usually appear in 1963 (main) and 1959 (supplementary) sediments (Sanada et al. 1999). However, the Lake Baikal sediment has no clear ^{137}Cs peak (Edgington et al. 1991; Mackay et al. 1998) because of the element's long residence time in the water column. Thus, we determined the sedimentation rates by assessing the date of the first detection of ^{137}Cs at 1954 (Peirson 1971). Although the rates by ^{137}Cs were about twice as high as those indicated by ^{210}Pb, and higher than those given in previous reports (e.g., 0.003–0.052 (av. 0.017) g/cm^2/year and 0.016–0.12 (av. 0.060) cm/year; Edgington et al. 1991), we used ^{137}Cs ages here, because they were determined in all cores, whereas ^{210}Pb ages were not calculated in several cores owing to unanalyzable profiles (Fig. 2).

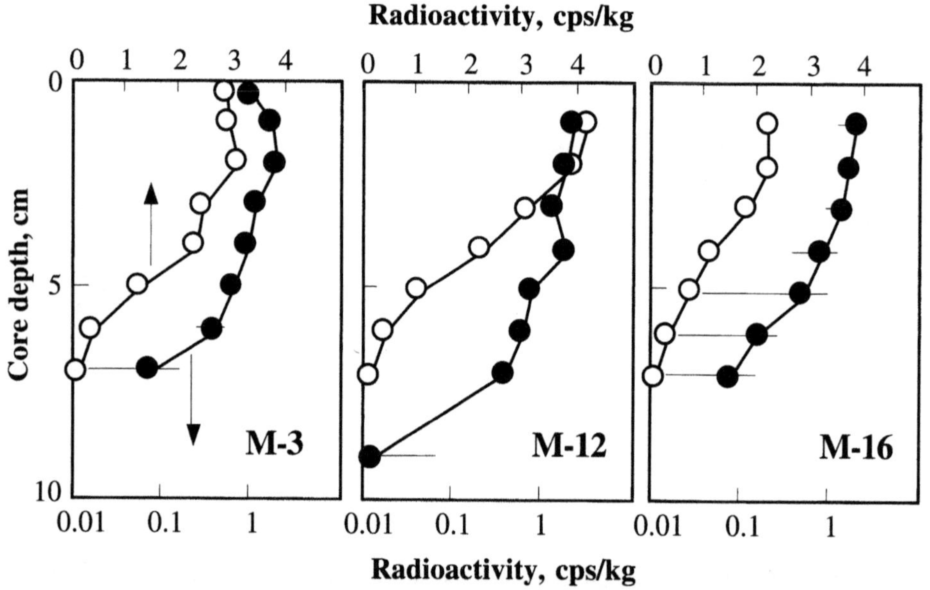

FIG. 2. Depth profiles of ^{137}Cs and ^{210}Pb in the sediment cores from the southern basin (M-3), Buguldeika Saddle (M-12), and Academician Ridge (M-16). *Open circles*, ^{137}Cs; *closed circles*, ^{210}Pb

3 Results and Discussion

3.1 Natural Composition and Distribution of Elements in Sediment

The average concentrations of elements and other components in each core are shown in Table 1, together with some physical parameters. All elements had concentrations comparable to those in remote unpolluted lakes (e.g., Hakanson and Jansson 1983; Forstner and Wittmann 1983). Although Lake Baikal is a long narrow lake with three moniliform basins, the elemental composition of the sediment (except for the layers mixed by turbidities) was generally homogeneous, as indicated by the almost constant ratios of Ti/Al and Sr/Ca (Fig. 3). This is probably because elements remain for a long time in the water column, and are thus mixed well before sedimentation.

The elemental depth profiles in the sediment were essentially similar in all cores (Fig. 4). The upper sediment showed distinct Mn- and Fe-rich layers, probably resulting from active redox cycles; Mn-rich layers always occurred above Fe-rich layers. In the lower sediment, S was enriched owing to sulfate reduction and iron sulfide formation (Takamatsu et al. 2000). Relic Mn- and Fe-rich layers were also observed in eight cores (Granina et al. 1993). Elemental concentrations were exceptionally low in the 13- to 15-cm layers of M-1 and the 23- to 26-cm layers of M-13, without any dilution with Mn·Fe-oxides, organic matter, or biogenic SiO_2, and the mean grain size was large there, indicating mixing by

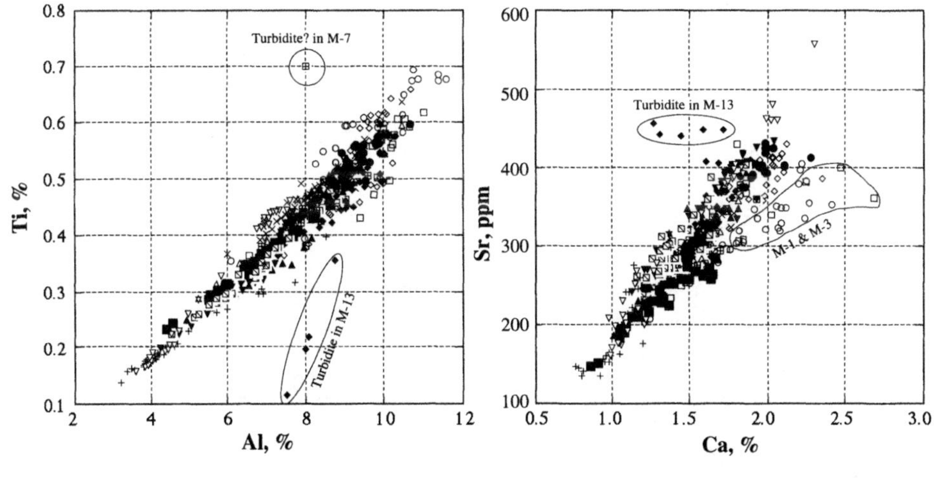

FIG. 3. Relationship between concentrations of Al and Ti and those of Ca and Sr in short sediment cores from Lake Baikal. *Open circles*, M-1; *open squares*, M-3; *open rhombuses*, M-5; *crosses*, M-6; *crossed squares*, M-7; *open triangles*, M-9; *closed circles*, M-11; *closed squares*, M-12; *closed rhombuses*, M-13; *closed triangles*, M-14; *inverted closed triangles*, M-15; *slashed squares*, M-16; *plus signs*, M-17; *inverted open triangles*, M-18

TABLE 1. Average concentrations of elements and some physical properties in short sediment cores from Lake Baikal

Element	Unit	Core No.														All cores
		M-1	M-3	M-5	M-6	M-7	M-9	M-11	M-12	M-13	M-14	M-15	M-16	M-17	M-18	
WD	m	1370	1470	1450	200	1100	400	29	365	970	1620	1650	335	930	900	914
OM	%	14.1	12.3	12.2	16.5	12.7	14.0	12.7	13.1	8.7	13.7	12.7	9.1	9.3	8.8	12.1
BS		23.2	26.8	18.8	11.8	20.4	27.4	4.0	29.6	13.9	11.4	27.3	32.3	42.9	43.7	23.8
Mi		62.8	60.9	69.0	71.8	66.9	58.6	83.4	57.3	77.4	74.9	60.0	58.5	47.8	47.6	64.1
GD	g/cm^3	2.60	2.52	2.61	2.50	2.50	2.50	2.60	2.44	2.58	2.49	2.45	2.44	2.38	2.40	2.50
Al	%	9.48	9.36	9.24	8.06	8.29	8.32	9.37	6.71	8.97	8.56	7.07	6.16	4.82	5.53	7.88
Ca		1.92	1.70	1.79	1.50	1.48	1.54	1.96	1.29	1.62	1.52	1.60	1.40	1.05	1.45	1.55
Fe		6.39	5.75	5.78	4.95	5.40	5.76	4.47	5.05	5.33	6.26	5.84	5.11	4.75	5.16	5.48
Mg		2.25	1.50	1.46	1.38	1.29	1.38	1.45	1.19	1.38	1.48	1.34	1.08	0.94	1.17	1.38
Ti		0.57	0.51	0.52	0.47	0.46	0.47	0.53	0.36	0.45	0.48	0.37	0.31	0.22	0.30	0.43
Co	p.p.m.	27	25	25	20	22	20	17	16	16	18	15	14	14	14	19
Cr		99	56	64	52	41	46	45	35	41	44	34	30	27	29	45
Cu		59	69	58	49	65	53	36	55	47	54	59	71	76	72	59
Mn		2340	3870	2380	1230	3600	4130	810	3670	1350	4530	6020	9020	9130	5370	4070
Ni		106	67	110	85	61	67	66	60	56	59	54	61	50	52	67
Zn		139	153	185	171	135	135	142	120	131	144	131	118	91	113	137
As		8.4	11.9	13	13	25	24	12	24	13	22	41	19	52	31	21.6
V		152	133	129	120	119	124	116	110	114	130	117	102	93	110	120
P		1650	1450	1780	1340	2080	2090	1080	2030	1360	2300	2430	2450	2110	2160	1910
S		1030	1470	2210	1660	1740	1970	1180	1860	2490	1450	1360	2890	618	2020	1730
Sr		323	318	352	295	291	301	403	233	357	300	350	299	194	305	308
Ba		752	833	832	740	836	918	810	788	847	777	879	1054	729	775	824
Pb		20	28	28	25	25	24	28	23	25	32	24	23	22	26	25
Sc		20	17	16	14	15	15	15	12	14	15	12	11	10	11	14
Y		24	28	29	27	29	28	30	26	27	31	26	27	26	27	28

Values are based on dried materials. WD, water depth; OM, organic matter; BS, biogenic SiO_2; Mi, minerals; GD, grain density

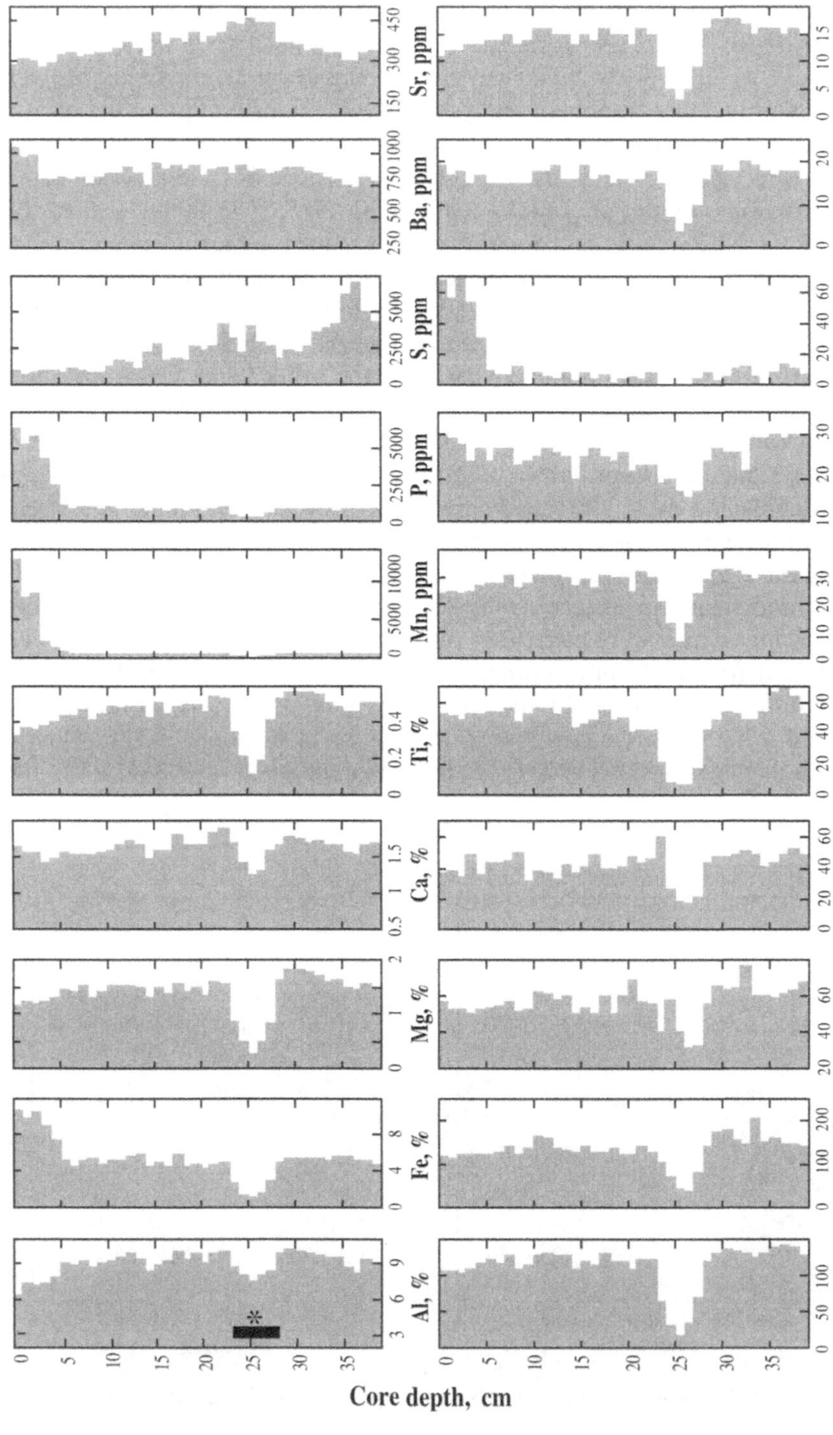

FIG. 4. Depth profiles of elements in the sediment core (M-13). *, layers contain turbidite

Si-rich turbidities. When elemental concentrations were compared between the upper oxidized layers (from the surface to tails of Fe-rich layers) and the deeper layers (including transition and reduced layers), Fe (average upper to lower ratio = 1.3), Mn (8.0), As (5.7), P (2.7), Ba (1.1), and Pb (1.2) were enriched in the upper layers, whereas S (0.68) was enriched in the lower layers. Elements were classified by principal component analysis into two major groups on the basis of their vertical distribution characters (Fig. 5): Fe, Mn, As, P, and Ba, which accumulate in Mn- and Fe-rich layers through early diagenesis (redox cycles, degradation of organic matter, etc.) (Takamatsu et al. 1985, 1993, 2000), and Al, Ca, Mg, Ti, Co, Cr, Ni, Zn, V, Sr, and Sc, which occur mainly in inert terrigenous material, and thus are hardly affected by diagenesis. The latter group was further classified into three subgroups: Ca and Sr (Ba in terrigenous fractions may be included here), Co, Cr, and Ni, and the remainder. Each set of elements may have occurred in the same mineral phase. On the other hand, Cu, S, Pb, and Y had unique depth profiles. Because the average concentrations of Pb and Y correlated well with each other ($r = 0.891$; see Table 1) (both elements showed moderately similar distributions, $r = 0.721$, also in the long BDP96-1 core; T. Takamatsu, unpublished data, 2003), and correlated positively with mineral contents (residue after the removal of organic matter and biogenic SiO_2), their major fractions may have existed in the same mineral phase. However, the Pb concentration increased near the surface owing to pollution (described later), and thus showed profiles which were somewhat different from those of Y. Sulfur showed very different profiles from the other elements owing to sulfate reduction and iron sulfide formation. In addition, Cu profiles were unique: concentrations of Cu showed a high positive correlation with those of biogenic SiO_2 ($r = 0.856$) (Fig. 6), and a negative correlation with mineral contents ($r = -0.834$), grain par-

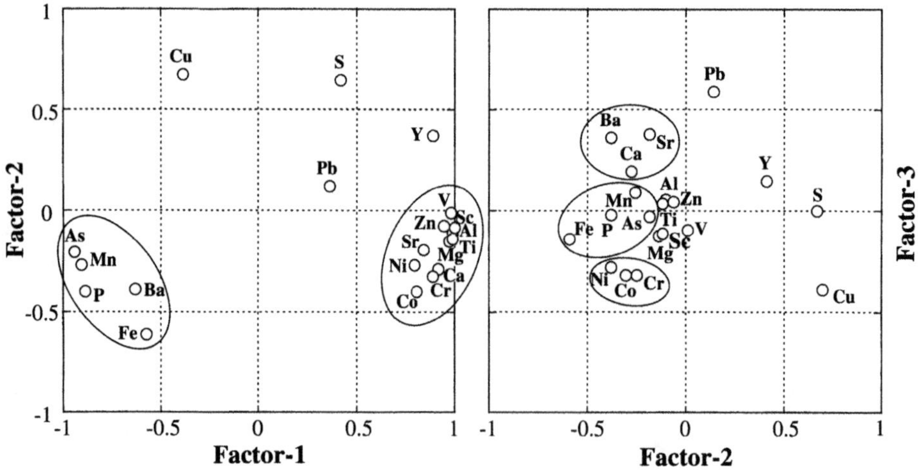

FIG. 5. Classification of elements based on principal components analysis

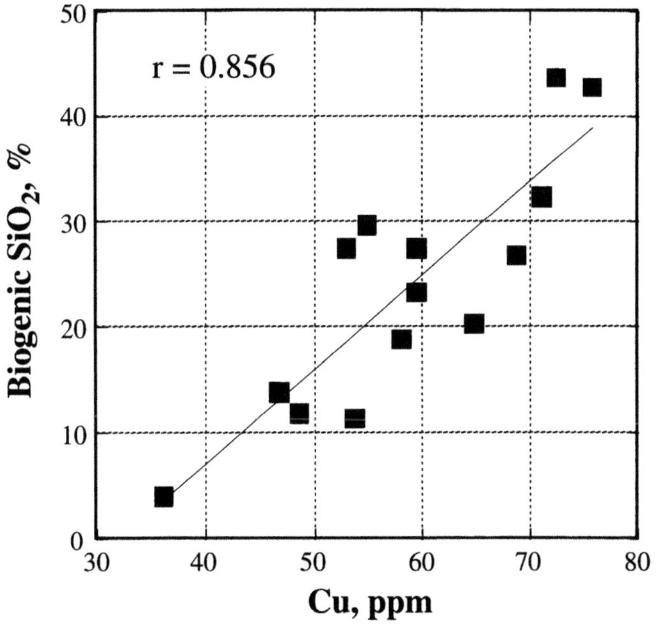

FIG. 6. Relationship between average concentrations of Cu and biogenic SiO_2 in 14 short sediment cores

ticle density ($r = -0.650$), and organic matter contents ($r = -0.483$). This may indicate the participation of diatoms in the deposition of Cu.

3.2 Redox Condition

The hypolimnion of Lake Baikal always contains oxygen, and the surface of the sediment is oxidized everywhere (Martin et al. 1993, 1998) because the lake is very oligotrophic (primary productivity 27 g-C/cm^2/year) and the water is well mixed (twice a year in the epilimnion; once every 8 years in the hypolimnion) (Weiss et al. 1991). In fact, oxidized layers of sediment (as indicated by the occurrence of Mn·Fe-rich layers) were observed in all cores analyzed. However, the thickness of the oxidized layers (defined as depth from surface to layers with the maximum enrichment of Fe) varied from 0.5 to 19.5 cm, and tended to be thin in the central basin and thick at Academician Ridge (Fig. 7). These thicknesses corresponded approximately to the values of sediment oxygen penetration depth (sediment depth at which dissolved oxygen in pore water is exhausted: 0.6 mm to >50 mm) and its regional trend, such as the relatively large depth (>38 mm) at Academician Ridge reported by Martin et al. (1993, 1998). The penetration of oxygen into sediment depends on the concentration of dissolved oxygen in the hypolimnion, the porosity of the sediment, and the amount and quality of organic matter in the sediment (Martin et al. 1998). The concentrations of dissolved

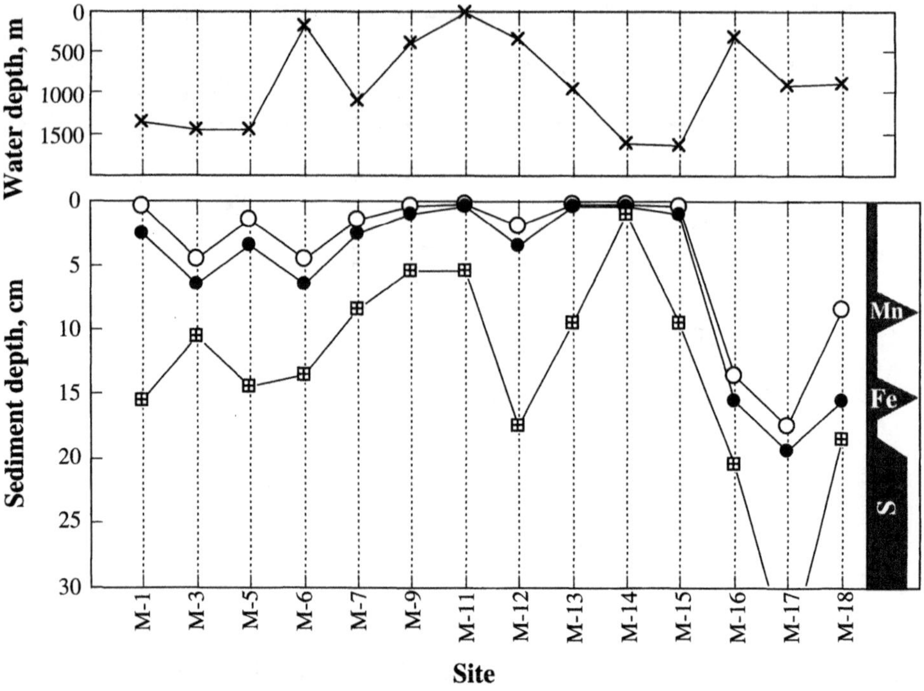

FIG. 7. Sediment depth at which the maximum accumulation of Mn and Fe was observed and the concentration of S began to increase, probably due to sulfide formation. *Open circles*, Mn; *closed circles*, Fe; *crossed squares*, S

oxygen in the hypolimnion range from 65 to 438 (av. 329) μmol/l, but are lower in the central (av. 320 μmol/l) and southern (av. 284 μmol/l) basins than at Academician Ridge (418 μmol/l) (Martin et al. 1998). Although the porosity of the sediment is not known, the thickness of the oxidized layers showed a good negative correlation with the grain density of the surface (0–5 cm) layer (Fig. 8), and a positive correlation ($r = 0.758$) with the content of biogenic SiO_2. At Academician Ridge, abundant fossil diatoms in the sediment may have increased the porosity. A limited deposition of organic matter (see Table 1), probably due to active mixing of the water (Hohmann et al. 1997), may also have contributed to the thick oxidized layers in the area.

3.3 Recent Pollution with Pb

The average concentration of Pb in each core was in the range 20–28 (av. 25) p.p.m. (see Table 1), which is comparable to the background level. However, it increased near the surface without exception (see Fig. 4; ratios of Pb concentrations in the upper 5-cm layers to those in the deeper layers: 1.06–1.54, av. 1.21), which suggests pollution. In the upper sediment, however, Mn·Fe-rich layers

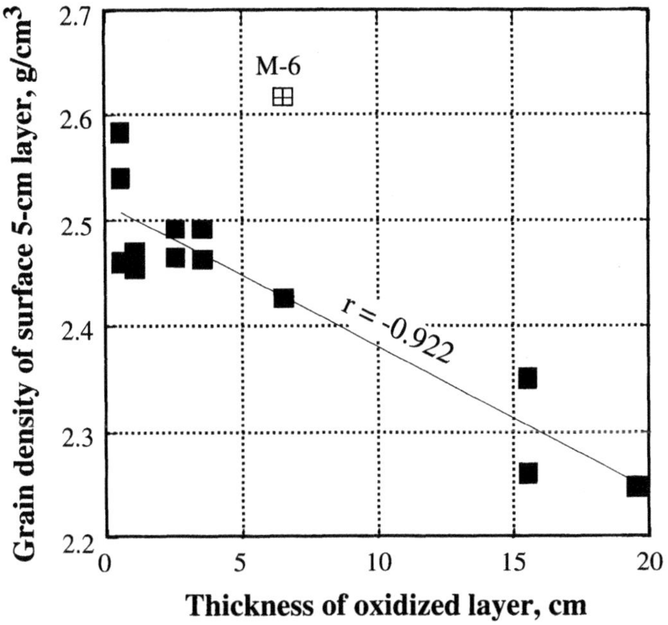

FIG. 8. Relationship between the thickness of oxidized layers (depth from surface to Fe-rich layers) and the grain density in the surface 5-cm layer. The M-6 core was excluded from the regression

often overlapped with Pb-rich layers, thereby reducing the apparent concentrations of Pb, because the Mn·Fe-oxide phase in Lake Baikal usually contains less Pb (Boyle et al. 1998; Takamatsu et al. 2000; in contrast, nodule-type oxides sometime accumulate Pb; Amirzhanov et al. 1993). Biogenic SiO_2 also reduces the apparent concentrations of Pb. Because dilution may cause the degree of pollution to be underestimated, we analyzed profiles of Pb normalized by Al (Pb/Al), and consequently identified distinguishable amounts of excess Pb in the upper parts of all sediment cores (Fig. 9). The Pb pollution began in the 1900s, and increased significantly after ca. 1950. This tendency agrees well with chronological changes in amounts of anthropogenic spherical carbonaceous particles in the sediment (Rose et al. 1998). The significant increase after ca. 1950 may have resulted from the global increase in the use of leaded gasoline after ca. 1940 (Forstner and Wittmann 1983) and a rapid increase in the output of power stations in Irkutsk after ca. 1950 (Rose et al. 1998). Lead released into the environment is loaded into the lake mainly through atmospheric processes (directly) and soil processes (indirectly), i.e., by the inflow of surface-contaminated soil due to erosion (which has become significant lately; Boyle et al. 1998; Mackay et al. 1998). Although inventories of excess Pb ranged from 3.2 to 17.3 $\mu g/cm^2$ (av. 13.5 $\mu g/cm^2$) overall, they were larger in the southern basin and smaller in the northern basin and at Academician Ridge (Fig. 10). The degree of pollution

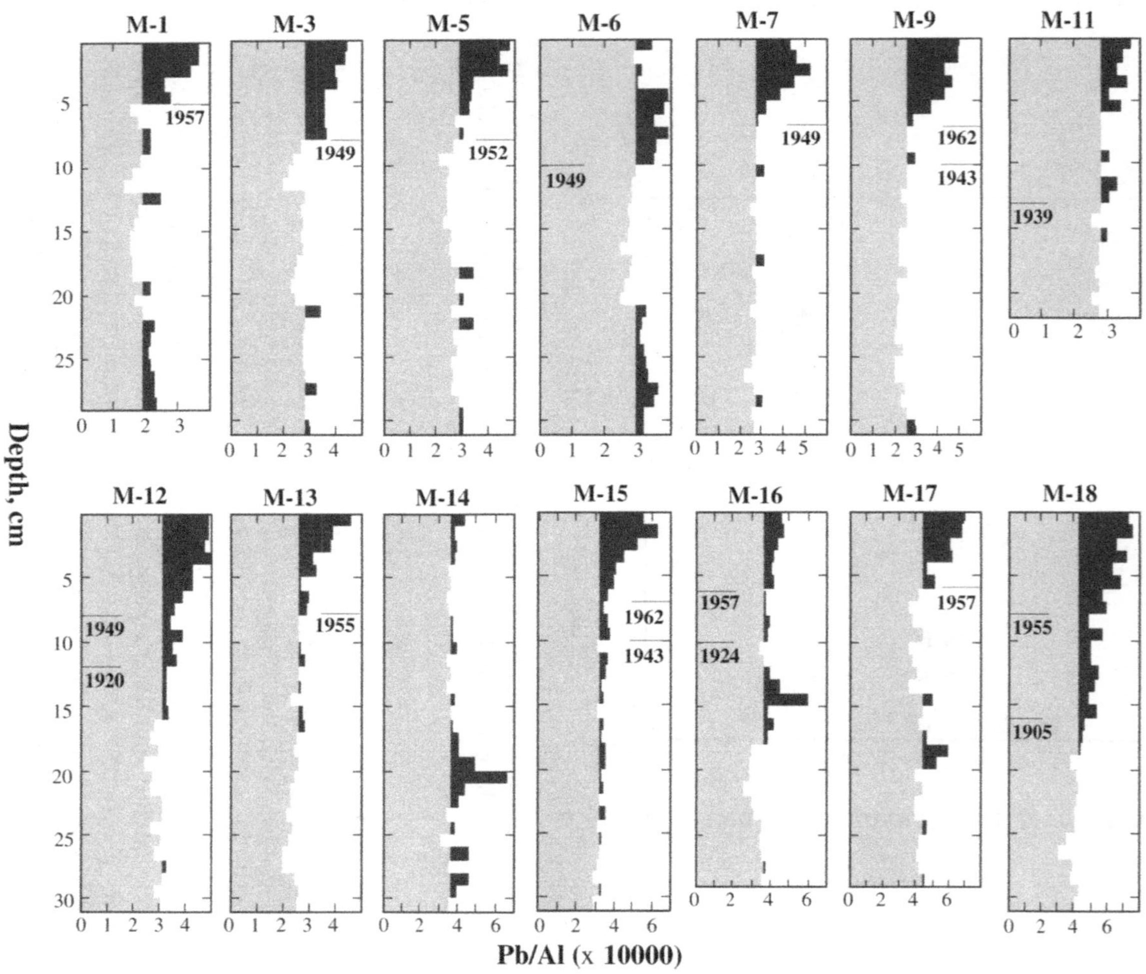

FIG. 9. Depth profiles of Pb/Al in short sediment cores dated by ^{137}Cs. *Shaded areas* show excess fractions.

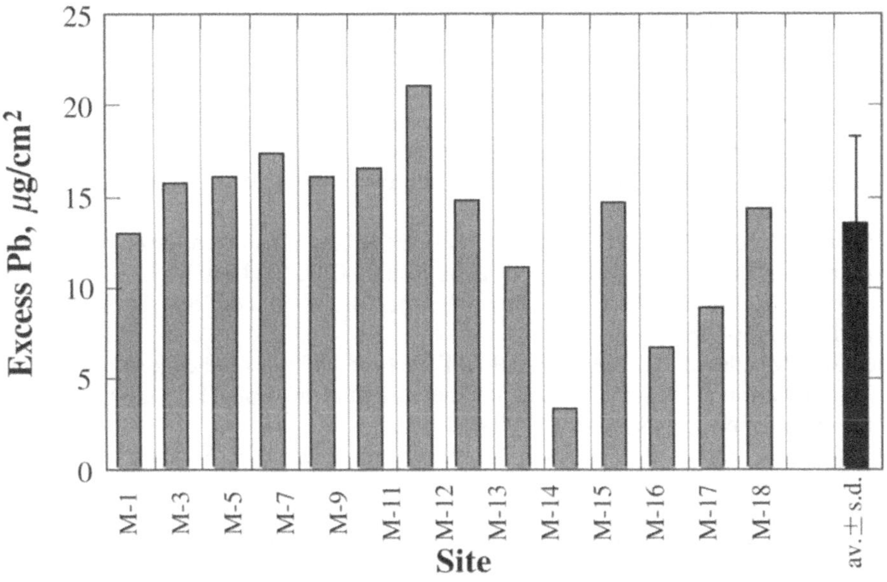

FIG. 10. Inventory in excess Pb in surface sediment in Lake Baikal

in Lake Baikal is much lower than that in remote unpolluted lakes in Europe and Canada (e.g., small lakes in Ontario, Canada: 61–77, av. 68 μg/cm^2; Dillon and Evans 1982). Multiplying the average inventory (13.5 μg/cm^2) by the lake area (31 500 km^2) gives a total Pb inventory in the lake of 4240 t. This amount seems to be almost equivalent to the total input of Pb into the lake, because ca. 99% of Pb brought into the lake deposits there (Boyle et al. 1998). The Pb pollution is not limited to the southern basin (Boyle et al. 1998; Mackay et al. 1998), but extends to the central and northern basins. This is probably because anthropogenic aerosols have been found in air masses over the northern basin (van Malderen et al. 1996), and pollution with spherical carbonaceous particles (Rose et al. 1998) and organochlorine compounds (Iwata et al. 1995) has been recognized in water and sediment from the central and northern basins.

Acknowledgment. The authors would like to thank M. Nishikawa for his help and advice in ICP–AES analysis.

References

Amirzhanov AA, Pampoura VD, Piskunova LF (1993) Rare elements in the Lake Baikal ferromanganese nodules. IPPCCE Newsl 7:25–28

Boyle JF, Mackay AW, Rose NL, Flower RJ, Appleby PG (1998) Sediment heavy metal record in Lake Baikal: natural and anthropogenic sources. J Paleolimnol 20:135–150

Dillon PJ, Evans RD (1982) Whole-lake lead burdens in sediments of lakes in southern Ontario, Canada. In: Sly PG (ed) Sediment/freshwater interaction. Junk, The Hague, p 121–130

Edgington DN, Klump JV, Robbins JA, Kusner YS, Pampura VD, Sandimirov IV (1991) Sedimentation rates, residence times and radionuclide inventories in Lake Baikal from ^{137}Cs and ^{210}Pb in sediment cores. Nature 350:601–604

Forstner U, Wittmann GTW (1983) Metal pollution in the aquatic environment. Springer, Berlin, Heidelberg, p 486

Granina LZ, Karabanov EB, Callender E (1993) Relics of oxidized ferromanganese formations in the bottom sediments of Lake Baikal. IPPCCE Newsl 7:32–39

Hakanson L, Jansson M (1983) Principles of lake sedimentology. Springer, Berlin, Heidelberg, p 316

Hee SSQ, Macdonald TJ, Boyle JR (1985) Effects of acid type and concentrations on the determination of 34 elements by simultaneous inductively coupled plasma atomic emission spectrometry. Anal Chem 57:1242–1252

Hohmann R, Kipfer R, Peeters F, Piepke G, Imboden DM, Shimaraev MN (1997) Processes of deep-water renewal in Lake Baikal. Limnol Oceanogr 42:841–855

Iwata H, Tanabe S, Ueda K, Tatsukawa R (1995) Persistent organochlorine residues in air, water, sediments, and soils from the Lake Baikal region, Russia. Environ Sci Technol 29:792–801

Kashiwaya K, Tanaka A, Sakai H, Kawai T (2000) Paleoclimatic signals printed in Lake Baikal sediments. In: Minoura K (ed) Lake Baikal: a mirror in time and space for understanding global change processes. Elsevier, Tokyo, p 53–70

Mackay AW, Flower RJ, Kuzmina AE, Granina LZ, Rose NL, Appleby PG, Boyle JF, Battarbee RW (1998) Diatom succession trends in recent sediments from Lake Baikal and their relation to atmospheric pollution and to climate change. Phil Trans R Soc Lond B 353:1011–1055

Martin P, Goddeeris B, Martens K (1993) Oxygen concentration profiles in soft sediment of Lake Baikal (Russia) near Selenga delta. Freshwater Biol 29:343–349

Martin P, Granina L, Martens K, Goddeeris B (1998) Oxygen concentration profiles in sediments of two ancient lakes: Lake Baikal (Siberia, Russia) and Lake Malawi (East Africa). Hydrobiologia 367:163–174

Peirson DH (1971) World deposition of long-lived fission products from nuclear explosions. Nature 234:79–80

Rose NL, Appleby PG, Boyle JF, Mackay AW, Flower RJ (1998) The spatial and temporal distribution of fossil-fuel-derived pollutants in the sediment record of Lake Baikal, eastern Siberia. J Paleolimnol 20:151–162

Sanada Y, Sato F, Kumata H, Takada H, Yamamoto A, Kato Y, Ueno T (1999) Estimation of sedimentation processes in Tokyo Bay using radionuclides and anthropogenic markers (in Japanese). Geochemistry 33:123–138

Shimaraev MN, Verbolov VI, Granin NG, Sherstyankin PP (1994) In: Shimaraev MN, Okuda S (eds) Physical limnology of Lake Baikal: a review. BICER, Irkutsk-Okayama, p 81

Takamatsu T, Kawashima M, Matsushita R, Koyama M (1985) General distribution profiles of thirty-six elements in sediments and manganese concretions of Lake Biwa. Jpn J Limnol 46:115–127

Takamatsu T, Kawashima M, Takada J, Matsushita R (1993) Characteristics in elemental composition of ferromanganese concretions from Lake Biwa. Jpn J Limnol 54:281–291

Takamatsu T, Kawai T, Nishikawa M (2000) Elemental composition of short sediment cores and ferromanganese concretions from Lake Baikal. In: Minoura K (ed) Lake Baikal: a mirror in time and space for understanding global change processes. Elsevier, Tokyo, p 155–164
van Malderen H, van Grieken R (1996) Chemical characterization of individual aerosol particles in central Siberia. Environ Sci Technol 30:312–321
Weiss RF, Carmack EC, Koropalov VM (1991) Deep-water renewal and biological production in Lake Baikal. Nature 349:665–669

Suspended and Dissolved Forms of Carbon, Nitrogen, and Phosphorus in Lakes Baikal and Hovsgol (Snow, Tributaries, Water, Sediments)

Eugenia N. Tarasova[1], Michael I. Kuzmin[1], Alexandre N. Gvozdkov[1], Elena A. Mamontova[1], Alexandre A. Mamontov[1], Marina Y. Khomutova[1], and Galina P. Chernyaeva[2]

Summary. This chapter gives data on the carbon, nitrogen, and phosphorus contents, and their ratios, in the suspended matter (SM), unfiltered water, and sediments of Lakes Baikal and Hovsgol, as well as in their sources (precipitation and tributaries). Both lakes have been found to show similar features (colorless water, an absence of $N\text{-}NH_4^+$ and $N\text{-}NO_2^-$, and a high content of dissolved oxygen) and distinctive factors (contents of silicon and chlorophyll-*a*, and the vertical distribution of carbon, nitrogen, and phosphorus).

The spatial distribution of carbon, nitrogen, and phosphorus in suspended organic matter (SOM) in the trophic layer of Lake Baikal is heterogeneous and depends on the season, the development of dominant plankton, and the hydrological particularities of the lake, especially currents and the descent water exchange. These contents increase during high-water periods in tributaries and during maximal phytoplankton development.

The composition of organic matter (OM) in the surface layer of sediments (1989–1990) from the south and central parts of Lake Baikal (low C:P and C:N values, and high N:P values) was similar to that of OM in water from south Baikal (1986–1989).

Key words. Organic matter, Lake Baikal, Lake Hovsgol

1 Introduction

Suspended matter (SM) is a precipitation in the initial stage of the formation of bottom sediments, which is closely associated with the physical–geographical conditions, and the biological, hydrological, and hydrochemical features of the

[1] Vinogradov Institute of Geochemistry, Siberian Branch of the Russian Academy of Sciences, Favorsky str. 1A, Irkutsk 664033, Russia
[2] Institute of the Earth's Crust, Siberian Branch of the Russian Academy of Sciences, Lermontova str. 128, Irkutsk 664033, Russia

lake environment (Lisitsyn 1974). The C:N, C:P, and N:P values are used to identify sources of organic matter (OM) (autochthonus and allochthonous) both in the water (Skopintsev 1950; Armstrong 1965; Tarasova and Mesheryakova 1992; Lukashin et al. 2002) and in the sediments (Vykhristiuk 1980; Uzaki and Ishiwatari 1989; Williams et al. 1993). These ratios in source materials such as plankton, wood, leaves of trees, and soil are all different. At present, the impact of human economic activity on the watershed is intensifying, so the determination of SM and its elemental composition (carbon, nitrogen, phosphorus) and chlorophyll-*a* is especially urgent, as all these components are stimulators of the eutrophication of water bodies.

Data on the contents of suspended and dissolved forms of carbon, nitrogen, and phosphorus in Lakes Baikal and Hovsgol are presented in this chapter. They were obtained for the whole system, i.e., precipitation (snow and rain), tributaries, lake water, sediments. The final aim of the investigations is to contribute to an understanding of changes in lakes' ecosystems, and to obtain knowledge of the supply of SM and OM from different sources.

2 Description of the Sites Studied

Lakes Baikal and Hovsgol are large freshwater, ancient, oligotrophic lakes in Asia. They belong to a single system of gigantic cavities sweeping from Lake Hovsgol in the southwest to the Muysko–Charskaya cavity in the northeast. Table 1 gives some brief characteristics of the lakes investigated.

About 40 streams flow into Lake Hovsgol. One small river (the Egiyn–Gol River) flows out of the lake. This is a tributary of the Selenga River, which is the main inflow to Lake Baikal. Thus, Lake Hovsgol has a hydrographic connection with Lake Baikal.

TABLE 1. Morphological characteristics of Lakes Baikal and Hovsgol (Galazy 1993; Filatova 1989)

	Lake Baikal	Lake Hovsgol
Site	Central Asia (Russia)	Central Asia (Mongolia)
Elevation (m)	454	1667
Length of lake (km)	636	133.4
Maximum width (km)	80	39.5
Maximum depth (m)	1620	245
Lake surface (km^2)	31500	2612
Lake basin area (km^2)	540000	6165
Water mass volume (km^3)	23000	317.5
Σ ionic (mgl^{-1})	95	248
Chlorophyll-*a* (μgl^{-1})	0.64[a]	0–0.77
Dissolved oxygen (mgl^{-1})	9.6–14.6	8.0–11.0

[a] Mean values for 1986–1989 in the 0- to 25-m layer

About 336 large and small streams flow into Lake Baikal (Galazy 1993). The Selenga River brings up to 50% of the mass of the total inflow water, the Upper Angara River provides up to 13%–14%, and the Barguzin River brings 9%. Only the Angara River flows out of Lake Baikal and flows into the Enisey River (Galazy 1993).

The basins of Lakes Baikal and Hovsgol are asymmetrical. The northwest slopes of Lakes Baikal and Hovsgol are steeper than those in the southeast (Galazy 1993; Filatova 1989).

The chemical composition of the Lake Baikal tributaries results from the leaching of Archean and Proterozoic rocks, which mainly include erupted and metamorphic rocks (Galazy 1993).

The basins of the lakes are characterized by a severe continental climate. The winter is long and cold (as low as –50°C). Lake Baikal is covered by ice from January to April, while Lake Hovsgol is covered by ice from November to May (Galazy 1993; Filatova 1989).

In the classification of Hutchinson (1969), these lakes belong to the dimictic. The temperature changes vertically twice a year. A direct thermal stratification (the water temperature decreases with depth) is typical of summer, while an inverse thermal stratification (the water temperature increases with depth) is common in winter.

The precipitation results from the humidity coming in with a cyclonic air mass. The maximum precipitation occurs in July–August, and the minimum in February. The amount of precipitation is one important factor influencing the erosion intensity in the watershed, and thus the amount of overburden brought into the lakes by the rivers. Most of this material is transported from the east coast of Lake Baikal owing to higher precipitation (Galazy 1993). The dynamics of the water mass is responsible for the further transportation of terrigenous material supplied to the lakes (Tarasova 1975). Currents (see Fig. 2) and descent water exchanges are significant factors in the water mass dynamics. The descent water exchange was described for the first time by Tolmachev (1957) using the seasonal variations of sulfate, bicarbonate, and chlorine obtained from deep sites in central Baikal (>1500 m depth). The water of the lakes investigated is weakly mineralized, and belongs to the bicarbonate class of the calcium group. The pH is about 7.1–7.2, which increases to 8.6 in the surface water during phytoplankton development.

Systematic investigations of the water chemistry in Lake Baikal, which started in 1947, indicate that the chemical composition of the water differs both horizontally and vertically. The heterogeneity in both ionic composition and trophic components (OM, SM, chlorophill-*a*, dissolved oxygen, nutrients, etc.) is due to the dynamics of the water mass, the influence of the tributaries, and the inhomogeneous development of phytoplankton. A reverse correlation between the nitrogen content in the water in winter and the chlorophyll-*a* concentration in the trophic layer of the pelagic part in spring was found in 1989 for south Baikal (Tarasova and Mesheryakova 1992; Tarasova et al. 1998). This finding provided clear evidence of the presence of toxicants in the Lake Baikal ecosystem

(Tarasova and Mesheryakova 1992). This was confirmed by the direct dependence between polychlorinated biphenyls (PCB) in the plankton and nitrogen in the water (Tarasova et al. 1997). Recent anthropogenic activity has led to a significant abundance of dioxins and related compounds in the Lake Baikal biota (Nakata et al. 1995; Kuicklick et al. 1996; Tarasova et al. 1998; Mamontov 2001; Mamontova 2001).

The concentration of SM in Lake Baikal water is extremely irregular. It amounts to 1.10–1.20 mg/l at 0- to 0.5-m depth in the pelagic part, and does not exceed 0.40–0.50 mg/l at 300 m and deeper. The SM is abundant in the mouths of large rivers. Most of the SM (up to 95%) is brought by the Selenga River, about 2% is supplied by the Upper Angara River, and 1.5% is brought by the Barguzin River (Tarasova and Mesheryakova 1992; Tarasova and Galazy 1993). About 2 million tons of SM comes into the lake annually with the main tributaries (Tarasova and Mesheryakova 1992; Tarasova and Galazy 1993). The Angara River removes 48.9 thousand tons of SM, but this is only 2.4% of the amount brought by the main tributaries. Thus, Lake Baikal is a good accumulator of SM (Tarasova and Mesheryakova 1992; Tarasova and Galazy 1993).

3 Materials and Methods

This chapter considers the concentrations of SM, suspended and dissolved forms of carbon, and mineral and organic forms of nitrogen and phosphorus in the trophic layer and in the vertical profiles of south, central, and north Baikal and its tributaries, the snow cover (1960–1989) and surface sediments (1989–1991). Water samples from Lake Hovsgol and its tributaries were collected in October 1989, May, July, and August 1990, and July and August 1991. Precipitation (rain) from Lake Hovsgol was sampled in July 1991. Sediments (0–2 cm) from Lake Baikal were sampled in 1989–1990. Lake Hovsgol sediments were collected in 1996. The sampling sites are shown in Fig. 1.

Systematic observations were carried out monthly for 16 tributaries of Lake Baikal. The area of the watersheds of the tributaries covers 92.8% of the total lake catchment. The water samples were collected each month from deep-seated water sampling sites in south (sites located in the middle of two transects: Polovinnyi Cape– Murino River, 1330 m deep; Kadil'nyi Cape– Mishikha River, 1400 m deep), central (Ukhan Cape– Turka River, 1500 m deep), and north Baikal (Elokhin Cape– Davsha River, 800 m deep). The water was sampled from depths of 0.5, 5, 10, 25, 50, and 100 m. Samples from deeper than 100 m were collected each 100 m across the bottom. The distribution of the chemical components investigated was studied for standard hydrological horizons selected by the Limnological Institute in 1960–1980. The SM was collected by filtration of the water through membrane filters (the pore size was experimentally chosen as 0.70 μm; Tarasova 1975). Chlorophyll-*a*, all forms of nitrogen (ammonium, nitrite, nitrate, and organic), phosphorus (mineral and organic) and organic carbon were determined in the water samples. Carbon, nitrogen, and phosphorus were deter-

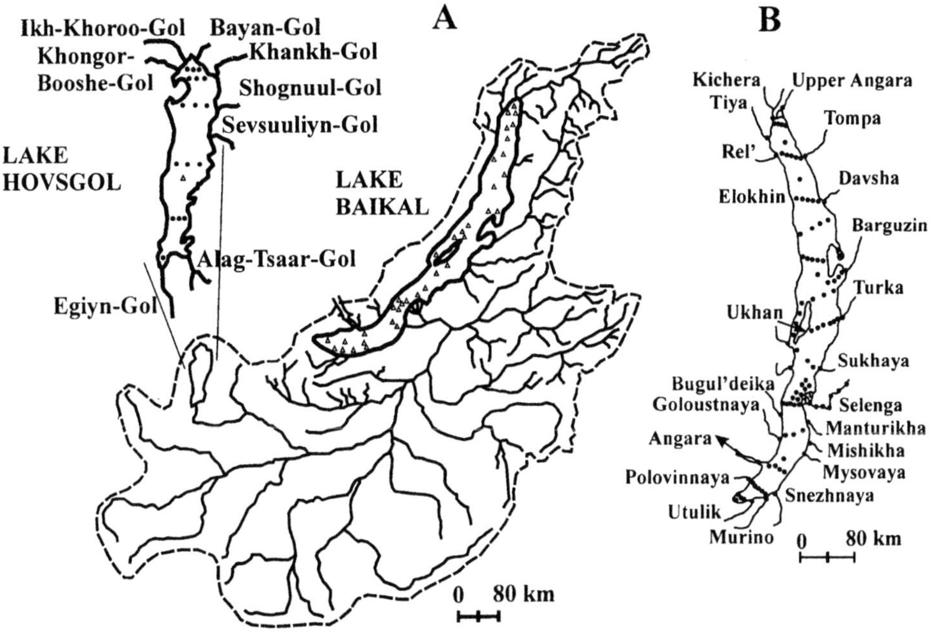

FIG. 1. Sampling sites in Lakes Baikal and Hovsgol and their tributaries. *Open triangles*, sediment; *solid circles*, water

mined in both unfiltered water and SM. Analyses of the components listed above were carried out using standard hydrochemical methods, and in particular those described by Tarasova (1975). The precision of the analyses was 3%–5%. A significantly larger error can be made by incorrect sampling. We have previously discussed errors in determining the average concentrations of elements in tributaries and in the lake proper for samples taken from deep-water sites (Tarasova and Mescheryakova 1992). Incorrect sampling could lead to significant analytical errors and thus to the wrong conclusions. The sampling period and the storage containers are two critical factors in sampling. Water was sampled from the surface through 300–500 m during the daytime, and was stored in leached, well-washed and steamed, dark glass containers.

The analyses were carried out either on the ship or at the institute laboratory within 3 days after sampling. Investigations have shown that the storage of water samples, even in dark glass containers at room temperature, resulted in a decrease in the suspended organic matter (SOM) content. The rate of OM decomposition after 266 days in water amounted to 4.4×10^{-3} per day (Tarasova 1975).

The productivity of the phytoplankton and the hydrological regime were different in the years studied. The years 1968, 1975, and 1982 had high yields of *Aulocoseira baicalensis*. Its biomass in 1968 was a maximum ($4\,g/m^3$). In 1982, the biomass of *A. baicalensis* was half that in 1968. The year 1969 showed a massive increase in *Synedra arcus*. The years 1978, 1986, and 1988 were characterized by poor development of phytoplankton. Massive increases in *Nitzschia acicularis* were observed in 1984 and 1987. The years 1971 and 1973 had heavy rain and floods.

Biogenic and abiogenic materials were determined from SM slides under a microscope. The slides were obtained by ultrafiltration of the water, and fixed with Canadian balsam. Microphotographs were taken, some of which are included here.

Sediments from Lake Baikal were sampled in 1989–1991. Water samples from Lake Hovsgol and its tributaries, as well as sediments and rain were collected during the Soviet–Mongol expedition in 1989–1996. Some samples were kindly provided by G.M. Shpeizer and V. Stal'makova from Irkutsk Univeristy.

4 Results and Discussion

The spacial and temporal variations in SOM in Lake Baikal water have been found to be significant. The surface water, which sampled was all over the lake, indicates that SOM is irregularly distributed in Lake Baikal. An increased total SM abundance occurs in the Selenga River delta, on the north and south shorelines, and in Chivyrkuy Bay. Abundant SM in June (1.10–4.60 mg/l) is due to algal development. This is confirmed by the lurid green color of the SM and slide investigations under the microscope. The SM in July is usually brown and gray, which points out the predominance of mineral substances over biogenic ones. Increases in SM contents from 1.10 to 10.0 mg/l are related to the beginning of flooding in rivers due to heavy rains. The quantity of SM in September does not decrease lower than 1.50 mg/l, and a concentration variation from 2.00 to 3.00 mg/l is found throughout the lake. The color of SM is brownish, and its abundance decreases slightly in November from 0.40 to 1.20 mg/l.

Minerals of the clay fraction, iron hydroxides, and grains of ore minerals and quartz are the main components of mineral SM. Biotite, epidote, amphibole, chlorite, plagioclase, and other minerals also occur.

Detritus is a predominant component (70%–98%) in the SOM of Lake Baikal. Only during the maximum development of *A. baicalensis* (in June 1968) was detritus scarce in the SOM (2%).

The vertical variations in SOM are insignificant. The total SM quantity in the surface layer varies from 0.70 to 2.60 mg/l. These contents decrease with depth, and were 0.38–0.50 mg/l at 1400 m deep near Cape Kadil'nyi (south Baikal), 0.30–0.50 mg/l at 1500 m deep near the Ukhan Cape (central Baikal), and 0.43–0.70 mg/l at a depth of 800 m near the Elokhin Cape (north Baikal). Increased contents (more than 1.00 mg/l) of SM are found at 350 m in south Baikal, at 200 m in central Baikal, and up to 600 m in north Baikal.

Increased SM concentrations occur episodically in deep layers of Lake Baikal. Thus, they were found at 1100 m in May 1968, and at depths of 800 and 900 m in

August 1968 in sampling sites in south Baikal (12 km from Polovinnyi Cape; Tarasova 1975). Microscopic investigations of SM slides from these horizons showed a significant abundance of live cells of *A. baicalensis* (Fig. 2). This phenomenon suggests subsidence of the surface water containing the live plankton (Tolmachev 1957).

Perennial investigations of SM concentrations from the vertical profiles of deep-water sites in south Baikal (12 km from Polovinnyi Cape) showed regular changes in the water SM content. The range of the SM content in the whole water column for 1986–1989 was significantly wider than for 1967–1969 or 1982–1985 (Table 2).

The contents of SM in deep basins in 1986–1989 were higher than in previous periods. They were found at depths of 300, 1200, and 1300 m. Although the SM content in the whole water body change insignificantly, the composition of SM for some horizons was markedly different from those observed in 1968 and 1969, which were marked by high diatomic algae development (see Fig. 2).

The distribution of suspended organic forms of carbon, nitrogen, and phosphorus in the surface water layer of Lake Baikal is analogous to the distribution of the total quantity of SM. It is inhomogeneous in the horizontal and vertical

TABLE 2. The vertical distribution of the suspended matter (SM) content in water of the sampling site in south Baikal (12 km away from Polovinnyi Cape, mg/l)

Depth (m)	1967–1969		1982–1985		1986–1989	
	Average	Min.–max.	Average	Min.–max.	Average	Min.–max.
0	1.27	0.70–2.00	1.10	0.40–2.20	1.15	0.40–3.20
5	—	—	0.86	0.65–2.00	1.14	0.28–3.60
10	1.05	0.40–2.60	1.08	0.30–2.00	1.17	0.23–1.96
25	0.91	0.40–2.20	0.86	0.60–1.25	0.96	0.26–1.60
50	0.74	0.50–1.70	0.86	0.30–1.33	0.82	0.20–1.65
100	0.68	0.40–1.80	0.46	0.30–0.80	0.65	0.30–1.60
200	0.52	0.30–0.90	0.49	0.27–0.98	0.73	0.23–1.80
300	0.54	0.20–1.10	0.48	0.20–0.85	0.78	0.33–1.50
400	0.52	0.30–1.20	0.51	0.25–0.80	0.52	0.23–0.95
500	0.46	0.30–1.10	0.49	0.22–0.80	0.50	0.20–1.37
600	0.43	0.30–0.80	0.48	0.22–1.00	0.46	0.20–1.45
700	0.42	0.30–0.70	0.46	0.20–1.14	0.30	0.15–0.60
800	0.51	0.30–1.00	0.42	0.18–1.03	0.48	0.26–1.50
900	0.49	0.30–1.00	0.41	0.20–0.90	0.44	0.10–1.50
1000	0.57	0.30–1.20	0.40	0.20–0.67	0.61	0.20–1.80
1100	0.56	0.30–1.10	0.46	0.23–0.80	0.48	0.23–4.90
1150	0.45	0.30–0.70	—	—	—	—
1200	0.46	0.30–1.00	0.48	0.26–1.00	0.70	0.48–1.75
1250	0.47	0.30–0.70	—	—	—	—
1300	0.54	0.30–0.90	0.46	0.25–0.80	0.52	0.20–1.30
Average for 0- to 1300-m layer	0.53 ± 0.10		0.46 ± 0.11		0.57 ± 0.16	

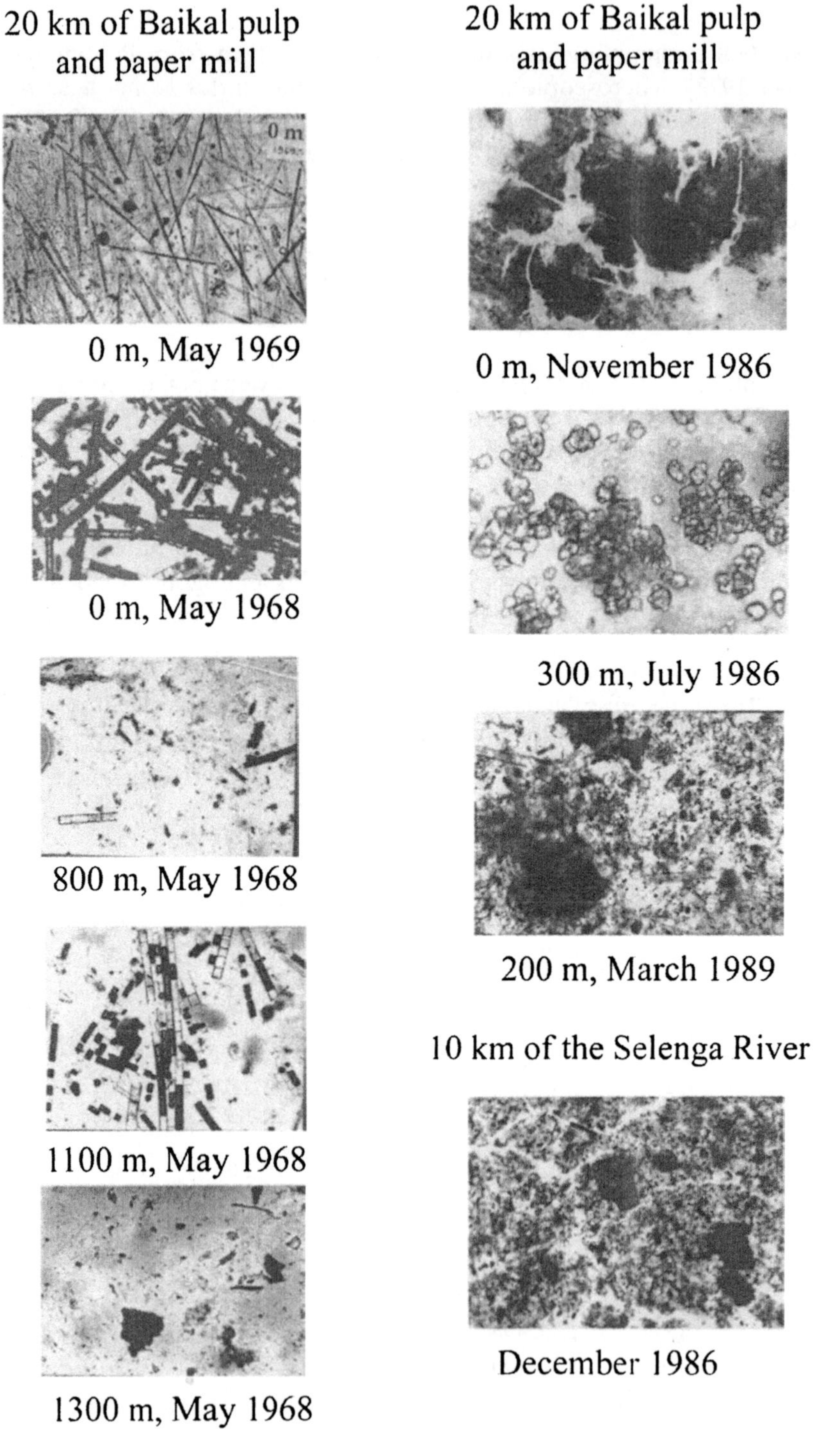

FIG. 2. Microphotographs of suspended matter (SM) in Lake Baikal (amplification 240). The volume of the filtered surface-water samples was 0.5 l. The volume of the filtered water from deep horizons was up to 10 l in 1968 and up to 1 l in 1986

profiles, and depends on the season, the development of dominant plankton organisms, and hydrological features. The quantity increases during flooding in tributaries and the maximal development of phytoplankton. Figure 3 illustrates the changes in the suspended organic carbon content of the surface water of Lake Baikal over several months. The relative content of organic carbon in the total quantity of SM is minimal in November (7%–12%) and maximal in June (18%–30%). Mineral particles constitute a large part of the SM. The proportion of mineral particles in the open lake increases in summer and winter. The carbon, nitrogen, and phosphorus contents decrease with increasing depth (Tables 3 and 4). The relative content of organic carbon in SM as well as the total organic

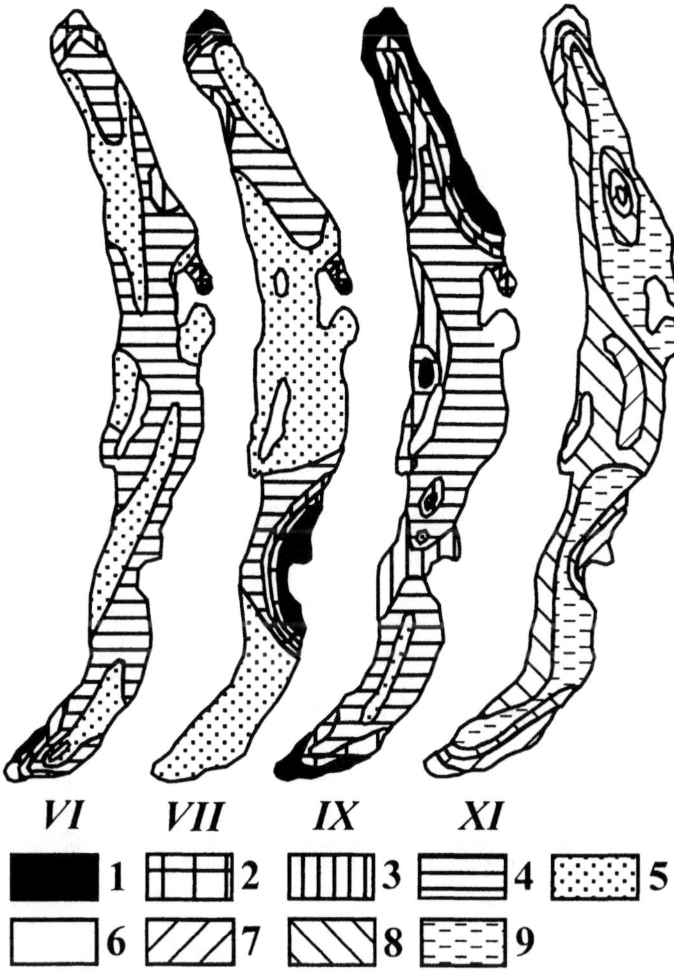

FIG. 3. The distribution of organic carbon in SM in the surface water of Lake Baikal over several months (in Roman numerals) (mg/l^{-1}). *1*, 1.00; *2*, 0.80–1.00; *3*, 0.60–0.80; *4*, 0.40–0.60; *5*, 0.20–0.40; *6*, 0.10–0.20; *7*, 0.08–0.10; *8*, 0.06–0.08; *9*, <0.06

Table 3. The vertical distribution of nitrogen and phosphorus in SM of water in south Lake Baikal (μg/l)

Depth (m)	N				P			
	1983–1984		1986–1988		1983–1984		1986–1988	
	Average	Min.–max.	Average	Min.–max.	Average	Min.–max.	Average	Min.–max.
0	12.1	3.6–22	27	5.2–61	1.70	0.23–3.96	3.22	0.54–6.5
5	—	—	36	3.2–77	—	—	4.17	1.17–6.6
10	8.1	3.6–22	26	15–37	1.00	0.21–2.62	4.33	1.47–11.
25	5.8	1.9–19	31	5.2–69	0.96	0.21–2.05	2.38	0.98–3.4
50	5.4	1.9–23	21	8.0–59	0.90	0.48–1.97	2.34	0.49–4.8
100	4.3	1.8–7.5	17	12–32	0.95	0.35–1.90	1.72	0.82–2.7
200	4.7	1.8–7.5	—	—	0.77	0.23–1.50	1.73	0.57–2.7
300	3.5	1.9–7.2	27	5.2–77	0.90	0.35–1.50	1.91	0.82–3.3
400	6.6	3.6–13	26	6.1–40	0.90	0.39–1.22	1.81	0.57–2.7
500	4.7	2.5–8.3	31	12–59	0.90	0.20–1.75	2.23	1.63–3.5
600	3.2	1.3–9.5	—	—	0.60	0.18–0.70	1.82	0.57–4.5
700	3.8	1.9–7.5	29	22–36	0.87	0.42–1.64	1.77	0.57–2.6
800	6.0	1.8–14	17	12–19	—	—	2.02	0.57–3.4
900	9.0	1.0–21	—	—	0.59	0.13–1.05	2.18	0.91–3.1
1000	3.7	1.9–5.0	20	5.2–37	0.85	0.73–1.33	1.61	0.25–4.8
1100	11.6	3.6–20	—	—	0.51	0.37–0.77	1.11	0.69–1.0
1200	—	—	11	5.2–21	—	—	2.72	1.11–5.1
1300	9.1	7.5–14	30	6.4–50	0.47	0.30–0.65	3.26	0.91–7.8
Average for 0- to 1300-m layer	5.3 ± 3.7		25 ± 24		0.75 ± 0.50		1.92 ± 0.95	

TABLE 4. The relative contents of suspended organic carbon in the total quantity of SM (1) and the total quantity of organic carbon (2) in water of south Baikal (%)

Depth (m)	1968–1969		1983–1984		1986–1988	
	1	2	1	2	1	2
0	12.0	10.3	7.6	6.2	11.9	10.5
5	—	—	—	—	9.7	9.3
10	12.5	8.9	7.2	5.5	11.9	11.1
25	11.2	7.8	11.0	7.4	11.6	10.0
50	10.5	7.8	12.4	7.6	14.0	10.3
100	8.4	5.4	16.1	6.9	16.3	9.0
200	7.8	2.8	8.2	3.9	8.6	5.5
300	6.4	—	11.0	5.1	10.8	7.5
400	4.8	—	9.4	4.7	10.2	5.1
500	4.5	—	5.7	8.6	21.8	10.9
600	1.5	2.2	12.3	5.7	15.4	6.7
700	1.11	1.4	9.8	4.3	25.7	7.6
800	1.5	3.2	8.6	3.8	20.4	10.1
900	1.3	4.4	10.5	3.4	25.0	11.1
1000	1.0	2.3	18.0	8.2	16.1	10.4
1100	2.8	—	13.7	6.4	—	14.5
1200	1.2	—	21.3	9.4	8.4	6.5
1300	3.1	1.5	20.0	7.1	27.3	15.0
Average for 0- to 1300-m layer	4	3	12	6	23	9

carbon content are given in Table 4. The relative contents of nitrogen and phosphorus in SOM are given in Table 5.

As can be seen from Tables 3–5, the composition of SOM in the water body of south Baikal has changed recently. The vertical profiles of carbon and nitrogen, carbon and phosphorus, and nitrogen and phosphorus ratios in SM is shown numerically in Table 6. The composition of SOM depends on the season. Recently, most of the annual SOM in south Baikal has been characterized by lower C:N and C:P values. The decrease in the carbon and nitrogen ratio in SM in the 0- to 1300-m layer (from 11 in 1983–1984 to 3 in 1986–1988) as well as the carbon and phosphorus ratio (from 77 to 38 in the same years) indicates changes in the composition of the SM within a short period of time.

C:N and C:P values can be used to check the genetic characteristics of the OM in natural water. Higher ratio values are typical of water flowing out of a marsh with a yellow–brown color. Such water is rich in OM. Conversely, water with low C:N and C:P values has a very slight color or none at all. The OM in such water is scarce, and has a mainly planktonic origin. Moreover, the carbon and nitrogen ratios in the initial substances are in the same ranges: for plankton organisms, C:N = 5–7; in soil humus (and modern sea dirt), C:N = 8–12; for

TABLE 5. Nitrogen and phosphorus contents in suspended organic matter (SOM) in water of south Baikal (%)

Depth (m)	N		P	
	1983–1984	1986–1989	1983–1984	1986–1989
0	7.0	9.9	1.0	1.2
5	—	1.6	—	1.9
10	5.0	9.0	0.60	1.6
25	3.0	14	0.50	1.1
50	3.0	9.0	0.50	1.0
100	3.0	8.0	0.60	0.80
200	6.0	—	0.95	1.4
300	3.3	16	0.85	1.1
400	6.9	25	0.94	1.7
500	8.4	11	1.6	1.1
600	2.7	—	0.50	1.2
700	4.0	19	0.97	1.2
800	8.0	9	—	1.0
900	10.5	—	0.69	0.99
1000	3.0	10	0.57	0.82
1100	9.0	—	0.41	0.38
1200	—	9.0	—	2.3
1300	5.0	11	0.26	1.2
Average for 0- to 1300-m layer	5	17	0.7	1.3

aquatic and terrestrial plant, C:N = 20–40 (average values). Low C:N values (10–12) correspond to low C:P values (about 150). Higher C:N (>20) values result in higher C:P values (>500).

Water in the deep open areas of Lake Baikal has very little color (0–5° on the platinum–cobalt scale; Zenin and Belousova 1988). The C:N and C:P values in a trophic water layer are characteristic of the OM of the planktonic origin. The ratios increase with depth, and this is associated with the mineralization of OM (Tarasova 1975, 1995).

The composition of SOM in the sources (tributaries, precipitation, and dominant forms of algae) was studied to explain the decrease in C:N and C:P values in the water body in recent years. As is obvious from Table 7, the tributaries of Lake Baikal import SOM with sufficiently high C:N, C:P, and N:P values. The C:N and C:P values in SOM in the snow in the Lake Baikal basin (Table 8) vary over a wide range. Low values were found near industrial emissions. Maximum C:N and C:P values in SOM in snow were found for the open water of Lake Baikal.

Table 9 shows the contents of carbon, nitrogen, and phosphorus, and their ratios in the dominant forms of phytoplankton in Lake Baikal. It is evident that

TABLE 6. Vertical profiles of C:N, C:P, and N:P weight ratios in SOM in water of south Baikal

Depth (m)	C:N		C:P		N:P	
	1983–1984	1986–1988	1983–1984	1986–1988	1983–1984	1986–1988
0	7	5	49	43	7	8
5	—	3	—	27	—	9
10	10	5	78	32	8	6
25	16	4	99	47	6	13
50	15	6	91	49	6	9
100	17	6	78	62	4	10
200	9	—	52	36	6	—
300	15	3	59	44	4	14
400	7	2	53	29	7	14
500	6	4	31	49	5	14
600	18	—	98	41	5	—
700	12	3	52	44	4	16
800	6	6	—	49	—	8
900	5	—	73	50	15	—
1000	20	5	88	61	4	12
1100	5	—	124	—	23	—
1200	—	5	—	22	—	4
1300	10	5	196	44	19	9
Average for 0- to 1300-m layer	11	3	77	38	7	13

TABLE 7. C:N, C:P, and N:P weight ratios in SM of water of the main tributaries of Lake Baikal, and the volume of flowing water

Tributaries	C:N	C:P	N:P	Flowing water (km^3/year)
Selenga	18	771	42	28.8
Upper Angara	8	462	60	9.20
Barguzin	11	562	50	4.14
Turka	8	400	53	1.57
Snezhnaya	11	564	52	1.70
Tompa	10	357	36	1.08
Tya	13	552	43	1.28
Goloustnaya	7	323	48	0.291
Bugul'deyka	8	349	44	0.138
Rel'	11	226	21	0.325
Sukhaya	7	263	36	0.137
Manturikha	11	575	50	0.317
Mysovaya	9	533	57	0.114
Utulik	11	328	29	0.547
Polovinka	7	352	25	0.097
Murino	13	253	20	1.00
Kitchera	16	449	28	—

TABLE 8. C:N and C:P weight ratios in precipitation, tributaries water, and sediments from Lakes Baikal and Hovsgol

Ratio	Lake Baikal	Lake Hovsgol
Precipitation	Snow	Rain (07.1991)
C:N	2–30	9
C:P	20–550	121
Tributaries		
C:N	7–16	6–77
C:P	226–770	8–630
Water body		
C:N	10–12(3[a])	—
C:P	83–90 (38[a])	39–136
Sediments		
C:N	7–12[b]	—
C:P	25–550[b]	47–54

[a] In the water body of south Baikal, 0–1300 m, 1986–1988
[b] In 1989–1991

TABLE 9. Carbon, nitrogen, and phosphorus contents in dominant forms of phytoplankton in Lake Baikal

Algae	Sampling season	C	N	P	C:N	C:P	N:P
		(μg/mg^{-1} dry weight)			(by weight)		
Aulocoseira baicalensis	Spring 1968	341 ± 5	32.2 ± 0.2	3.936 ± 0.002	11	87	8
Synedra arcus	Spring 1968	189 ± 4	22 ± 0.15	6.480 ± 0.004	9	29	3
Gymnodinium	Winter 1979	450 ± 5	49.7 ± 0.2	5.470 ± 0.003	9	82	9

these ratios in the algae are high enough except for C:P in *Synedra arcus*, but the biomass of this alga is not high, i.e., usually not higher than 0.2 g/m^3.

The organic carbon, nitrogen, and phosphorus contents and ratios in the surface sediments of Lake Baikal are shown in Figs. 4 and 5. The distribution of the components indicated is very irregular. The OM of the surface sediments had very low C:P and N:P values in south and central Baikal in 1989–1991. North Baikal is marked by high C:N and C:P values. It should be noted that C:N values in south and central Baikal were lower than those found in 1966–1975 by Vykhristyk (1980). A decrease in C:N and C:P values was observed in deep-water sites in the south (1968–1984 and 1986–1989). As our sediments were collected in 1989–1991, it can be assumed that the water and the sediment had a common source of OM. A previous study (Tarasova and Mescheryakova 1992) showed that anthropogenic activity could be such a source. In their study of sources of OM in surface sediments of Lake Baikal, Williams et al. (1993) showed that in terms of composition, the OM of bottom sediments in the central and south basins is different from that in the north basin.

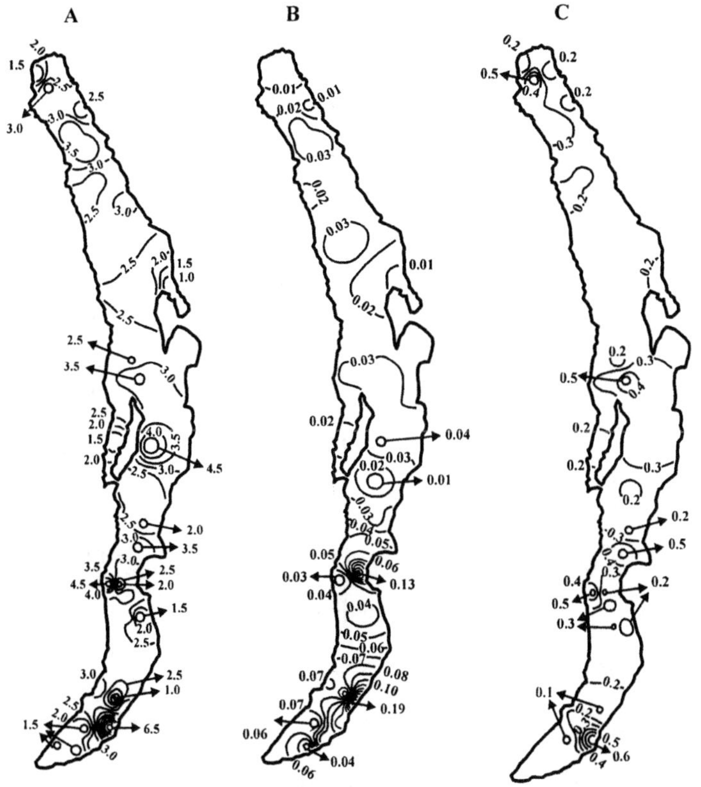

FIG. 4. Distributions of total **A** organic carbon, **B** phosphorus, and **C** nitrogen in the surface layer (0–2 cm) of Lake Baikal sediments (% of dry weight)

A comparative estimate of the SM and OM contents in the water of Lakes Baikal and Hovsgol is given in Tables 8 and 10. The water of Lake Hovsgol has little color (0–5° on the platinum–cobalt scale). As a rule, the water does not contain $N\text{-}NH_4^+$ and $N\text{-}NO_2^-$ (Tarasova et al. 1998). The total SM content is low. The maximum abundance of SM in Lake Hovsgol occurs in water near the bottom. The suspended organic carbon varies from 0.06 to 0.29 mg/l. The total organic carbon concentration in unfiltered water varies from 0.78 to 6.09 mg/l. Lake Hovsgol water has a higher silicon content and lower values of chlorophyll-*a* and $N\text{-}NO_3^-$ compared with Lake Baikal water. The ratios of mineral and organic forms of nitrogen and phosphorus in these water bodies are variable. In 1967–1969, the average mineral phosphorus content in unfiltered water of Lake Baikal in the 0–1300-m layer (11 μg/l) was slightly higher than the organic phosphorus content (4 μg/l). The same pattern was observed for nitrogen (92 and 81 μg/l) in the same period of time.

Conversely, the organic phosphorus content in the water of Lake Hovsgol (8–10 μg/l) was insignificantly higher than the mineral phosphorus content

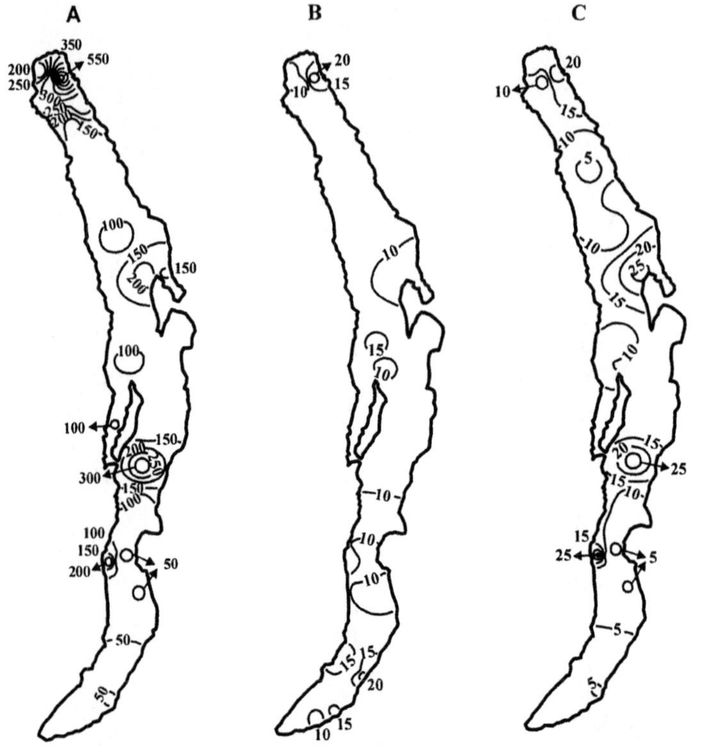

FIG. 5. Distributions of the weight ratios of **A** carbon and phosphorus, **B** carbon and nitrogen, and **C** nitrogen and phosphorus in the surface layer of Lake Baikal sediments

TABLE 10. The contents of nutrients, chlorophyll-*a*, and suspended and organic matter in water of Lake Baikal and Lake Hovsgol

Components	Lake Hovsgol	Lake Baikal
Suspended matter (mg/l)	0.30–11.7	0.18–3.20
Chlorophyll-*a* (μg/l)	0–0.77	0–3.84
C_{org} in suspended matter (μg/l)	60–290	6–370
P in suspended matter (μg/l)	0.34–7.9	0.18–6.60
N in suspended matter (μg/l)	—	1.0–77.0
C_{org} in unfiltered water (μg/l)	780–6090	600–2530
Mineral P in unfiltered water (μg/l)	2–7	1–16
P_{org} in unfiltered water (μg/l)	1–26	2–38
Mineral N unfiltered water (μg/l)	5–15	0–145
N_{org} in unfiltered water (μg/l)	35–460	11–630
Si in unfiltered water (mg/l)	2.30–4.48	0.56–2.99

(4–6 μg/l). The predominance of the organic nitrogen content (70–150 μg/l) over the mineral nitrogen content (9–10 μg/l) is of interest. This difference is reflected in the lower ratio of carbon and nitrogen, the mineral forms of nitrogen and phosphorus, the higher ratio of organic forms of nitrogen and phosphorus, and the carbon and phosphorus in Lake Hovsgol water as opposed to water of Lake Baikal. It should be noted that we do not have data for Lake Hovsgol water on the components investigated for all seasons, and the comparison was made only for the periods investigated. An essential difference between Lakes Baikal and Hovsgol was found for the vertical distribution of organic and mineral forms of nitrogen and phosphorus. The increase in mineral forms and the decrease in organic forms down the water column in Lake Baikal indicates the mineralization of OM. The vertical distribution in the water of Lake Hovsgol depends mainly on hydrological processes. In addition to a comparatively high silicon content (up to 4.48 mg/l), low values of chlorophyll-*a* (0–0.77 μg/l), and low nitrogen (3%–6%) and phosphorus (0.2%–0.5%) in OM in Lake Hovsgol showed the mainly terrigenic origin of the OM in the lake (Fig. 6).

5 Conclusions

The spatial distribution of carbon, nitrogen, and phosphorus in SOM in the trophic layer in Lake Baikal is heterogeneous and depends on the season, the development of dominant plankton, and the hydrological particularities of lakes,

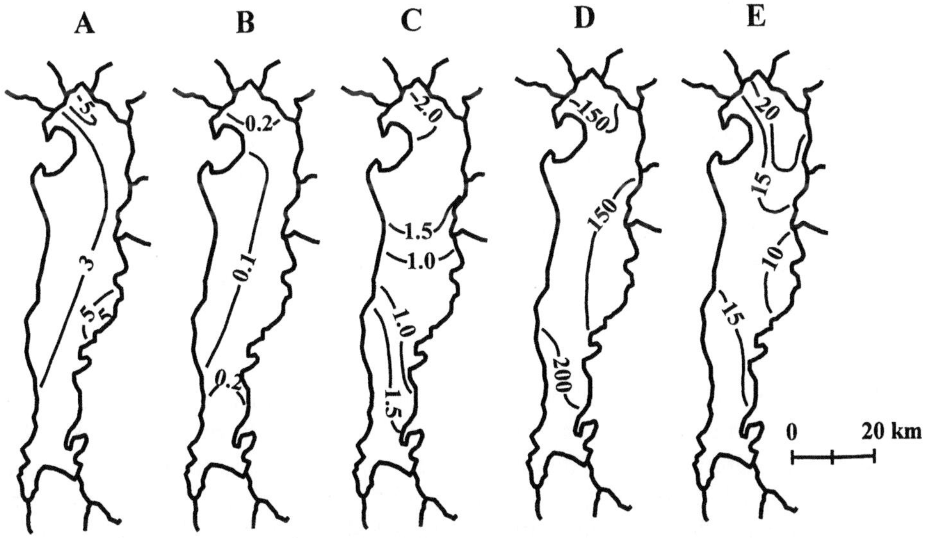

FIG. 6. Distributions of **A** SM (mg/l), **B** chlorophyll-*a* (μg/l), **C** organic carbon (mg/l), **D** nitrogen (μg/l), and **E** phosphorus (μg/l) in unfiltered water of Lake Hovsgol in July–August 1990

and especially currents and descent water exchanges. Their contents increase during high water periods in tributaries and with maximal phytoplankton development.

Carbon, nitrogen, and phosphorus contents decrease down the water column. The SOM content increases episodically in the deep basin, in particular during the homotermium. The present composition of SOM in the water of south Baikal has changed within a short period of time. The carbon and nitrogen ratios in SM in the 0- to 1300-m layer in south Baikal decreased from 11 in 1983–1984 to 3 in 1986–1988. The ratio of carbon and phosphorus decreased from 77 to 38 in the same period. Low C:P and C:N values were also found in sediments sampled in 1989–1991 from south and central Lake Baikal.

Tributaries of Lake Baikal import SOM showing high C:N, C:P, and N:P values. In snow from the Lake Baikal basin, these ratios are also high except for sites near industrial emissions. Low C:N and C:P values were found only in sewage water of the Baikalsk and Selenginsk pulp and paper mills, which are the main industrial pollution sources on Lake Baikal.

The water of Lake Hovsgol is similar to that of Lake Baikal. Neither lake contains much $N\text{-}NH_4^+$ or $N\text{-}NO_2^-$. The water in both lakes also shows little color and a low content of total SM. However, Lake Hovsgol water exhibits a higher silicon content, and a low content of chlorophyll-*a* and $N\text{-}NO_3^-$.

The proportion of mineral and organic forms of nitrogen and phosphorus in Lakes Baikal and Hovsgol is different. Lake Baikal contains more mineral nitrogen and phosphorus than organic forms, while Lake Hovsgol water shows the opposite behavior for nitrogen.

A difference between the lakes was found for in the vertical distribution of organic and mineral forms of nitrogen and phosphorus. The change in the concentration in Lake Baikal indicates the mineralization of OM. The distribution of nitrogen and phosphorus in Lake Hovsgol mainly results from hydrological processes.

The OM in Lake Hovsgol is of terrigenic origin (a comparatively high content of silicon, low values of chlorophyll-*a*, and a low proportion of nitrogen and phosphorus in OM).

Thus, this investigation contributes to an understanding of the variability of the suspended and dissolved forms of carbon, nitrogen, and phosphorus in Lakes Baikal and Hovsgol.

Acknowledgments. The authors are grateful to Academician G.I. Galazy for a deep interest and valuable suggestions in the investigation of Lake Baikal. The study was possible owing to the assistance of M.I. Grachov during sampling in Lake Baikal in 1987–1989. The authors would like to thank G.M. Shpeizer and V. Stal'makova, who provided samples from Lake Hovsgol. The investigations were made with partial financial support from the J. and K. MacArthur Foundation (individual grants for E.N. Tarasova and A.A. Mamontov) and INTAS grant (2000-00140).

References

Armstrong FAJ (1965) Phosphorus. In: Riley JP, Skirrow G (eds) Chemical oceanography. 1. Academic Press, London, New York, p 323–366

Filatova TP (ed) (1989) Atlas of Lake Khubsugul (in Russian). Federal Service of Geodesy and Cartography of Russia, Moscow, p 118

Galazy GI (ed) (1993) Atlas of Lake Baikal (in Russian). Federal Service of Geodesy and Cartography of Russia, Moscow, p 160

Hutchinson GH (1969) A treatise on limnology (in Russian). Progress, Moscow, p 592

Kuicklick JR, Harvey HR, Ostrom PH, Ostrom NE, Baker JE (1996) Organochlorine dynamics in the pelagic food web of Lake Baikal. Environ Toxicol Chem 15:1388–1400

Lisitsyn AP (1974) Formation of sediments in the ocean (in Russian). Nauka, Moscow, p 438

Lukashin VN, Isaeva AB, Serova VV, Nikolaeva GG (2002) Geochemistry of the sedimentary matter and its flows in the eastern part of the equatorial Atlantic. Geochemistry 3:306–318

Mamontov AA (2001) Polychlorinated dibenzo-para-dioxins and related compounds in the Lake Baikal ecosystem (in Russian). Academy of Earth Sciences, Moscow, p 68

Mamontova EA (2001) The hygienic estimation of dioxins and related compound pollution of the Irkutsk region environment (in Russian). SB of RAS, Novosibirsk, p 142

Nakata H, Tanabe S, Tatsukawa R, Amano M, Miyazaki (1995) Persistent organochlorine residues and their accumulation kinetics in the Baikal seal (*Phoca sibirica*) from Lake Baikal, Russia. Environ Sci Technol 29:2877–2885

Skopintsev BA (1950) Organic matter in underwater sediment (water humus) (in Russian). Proceedings of the State Oceanographic Institute, Hydrometeoizdat, L 7(29), p 142

Tarasova EN (1975) Organic matter of water of south Lake Baikal (in Russian). Nauka, Novosibirsk, p 147

Tarasova EN (1995) Components of the trophic status in water of Lakes Baikal, Hovsgol and Teletskoe (in Russian). Siberian Ecol J 5:383–390

Tarasova EN, Galazy GI (1993) About the speed of modern sedimentation in Lake Baikal. In: Galazy GI (ed) Baikal and nature. Knowledge for 100 years (in Russian). Irkutsk Regional Printing Office, Irkutsk, p 32–40

Tarasova EN, Mesheryakova AI (1992) The modern condition of the hydrochemical regime of Lake Baikal (in Russian). Nauka, Novosibirsk, p 144

Tarasova EN, Mamontov AA, Mamontova EA, Galazy GI (1997) Correlation of PCB concentration with components of the trophic status in Lake Baikal. Organohalogen Compounds 32:263–267

Tarasova EN, Mamontov AA, Mamontova EA (1998) Pollution and eutrophication in Lake Baikal. J Lake Sci 10:165–178

Tolmachev VA (1957) Some hydrochemical indexes of internal water exchange in Lake Baikal (in Russian). USSR AS Rep 113:1011–1013

Uzaki M, Ishiwatari R (1989) Carbohydrates in the sediment of Lake Yunoko: vertical distribution and implications of their origin. Jpn J Limnol 47:257–267

Vykhristiuk LA (1980) The organic matter of Lake Baikal sediments (in Russian). Nauka, Novosibirsk, p 80

Williams DF, Chin L, Karabanov EB, Gvozdkov AH (1993) Geochemical indicators of productivity and the sources of organic matter in surface sediments of Lake Baikal (in Russian). Russ Geol Geophys 34:136–148

Zenin AA, Belousova NV (1988) Hydrochemical dictionary (in Russian). Hydrometeoizdat, Leningrad, p 239

The Contribution of Eolian Material in the Composition of Suspended Matter in the Snow Cover of Ice from the Buguldeika Saddle and Academician Ridge in Lake Baikal

Galina P. Koroleva[1], Victor S. Antipin[1], Alexsandre A. Kosov[2], Michail A. Mitichkin[1], and Galina P. Sandimirova[1]

Summary. This chapter considers the results of a geochemical study of the suspended matter (SM) in the snow cover on the ice in Lake Baikal, and its probable sources. We develop the minimization of a function (geochemical distance, GD) to determine the contribution of each source to the snow SM in the lake ice. This function quantitatively expresses the degree of similarity of the mixture, using the material of potential sources and the real SM composition.

Based on the GD, the composition of SM in the snow cover was shown to be derived from the composition of the lakeshore complexes and atmospheric transfer from remote areas. The latter depends on the distance of the sampling site from the shore, and the impact of predominant and local winds.

Solving the optimization provided an estimate of the contribution of each potential source to the composition of the SM: for profiles 1 and 2 (Buguldeika Saddle), the contribution of the Goloustnaya block, the Sharyzhalgai series, Khamar–Daban, and the industrial centers of the Priangaria amounts to 10%, 10%, 15%, and 65%, respectively; for profile 3 (Academician Ridge), the contributions of the western and eastern shores, and transfer from remote areas (Bayandai) makes up 45%, 25%, and 30%, respectively.

Key words. Eolian material, Suspended matter, Baikal watershed, Academician Ridge, Buguldeika Saddle

1 Introduction

The comprehensive investigations of Lake Baikal sediments are of great importance for studies of the global changes in the environment and climate in Central

---

[1] Vinogradov Institute of Geochemistry, Siberian Branch of the Russian Academy of Sciences, Favorsky str. 1A, Irkutsk 664033, Russia
[2] Institute of Systems Dynamics and Control Theory, Siberian Branch of the Russian Academy of Sciences, Irkutsk, Russia

Asia. The Buguldeika Saddle, separating the central basin from the southern basin, and Academician Ridge, separating the north and south basins, are of particular significance. As earlier investigations indicated, these two sites are marked by the lowest sedimentation rate. The sediment sequence from Baikal Drilling Project data contains alternating sandy–silty mud, which contains diatom and clays in addition to the significant share of sand and silt (BDP Members 1998; Grachev et al. 1997). The sediment is deposited from material in the water column, and a supply of matter from water flows, ice, and wind.

The share of silty suspension brought by rivers and shore flows to the Buguldeika Saddle and Academician Ridge is not significant, as these sites are separated from the shores by basins.

The ice and eolian transfer should make a greater contribution. The possibility of a transfer of clastogene material is considered both for the lake depression as a whole (Agaphonov 1990; Taisaev 1982), and for Academician Ridge only (Vologina and Potemkin 2001; Koroleva et al. 2001). The chemical composition of aerosols over the Baikal water area in different seasons, and in particular in the southern and central parts of the lake, was studied without considering the sources or transportation mechanisms (Potemkin and Khodzher 1987; Obolkin et al. 1994; Khodzher et al. 1997; Vetrov et al. 1983; Vetrov and Kuznetsova 1997). These data cannot be used to study the geochemical composition of the mountains surrounding Lake Baikal, the suspensions in the snow cover, or the bottom sediments.

This chapter describes the results of a geochemical study of the suspended mater (SM) in the snow cover on the ice of Lake Baikal, and its probable sources.

2 Geological–Geographic Peculiarities and the Composition of the Bedrocks of Lake Baikal

The processes of continental sedimentation are defined mainly by the geological–geomorphic peculiarities of the region, the composition of the bedrocks, and climatic changes. The eolian transfer of the suspended matter to the ice or snow cover during winter is one important mechanism of the transportation of sediments to the water area of Lake Baikal. The mineralogical and geochemical characteristics of the sediments are largely inherited from the bedrocks of the watershed, which varies greatly in different petrographic provinces of the Baikal area. In western Pribaikalia, the prevailing acid magmatic occurrences consist of associations of the Proterozoic Primorsky Rapakivi granite complex, potassium volcanics of the Akitkan series, and subalkaline products of the granitization of the Chuya uplift among metamorphosed rocks of the Siberian platform basement. The main petrochemical characteristics of all these rocks are their high alkali contents, with a predominance of K over Na, and a high value for the Fe/Fe + Mg ratio in the rocks and minerals. The spatial closeness of the complexes studied and their petrogeochemical similarities suggest that they were all produced by a single magmatic process at different depths (Petrova et al. 1997). The recent geochemical data on the Goloustnaya migmatite–granitic block

(Petrova 2001) indicated that the leucogranites of this region differ greatly from the Primorsky rapakivi granites of the second stage. The geochemical similarity of the leucogranites of the Goloustnaya block to those of the Ol'khon Island area is a very important factor in analyses of the probable sources of substances supplied during the formation of sediments.

In southern Pribaikalia, the metamorphic rocks of the crystalline blocks of the Precambrian basement are widespread. They have the composition features of the basic rocks. The rocks of the Khamar–Daban Range belong to the Khamardaban group, which includes flyshoid formations, i.e., psammite, aleuropelite, metapelite, aleurolite, marble, coaly shale, sandstone, and gritstone. The plutons of calc-alkaline and subalkaline granitoids, and Riphean migmatites and pegmatites, as well as dikes and intrusions of the monzonitoids and rare-metal granites occur within these rocks. The acid magmatism of the eastern Pribaikalia produced huge areas of outcrops. The largest in central Asia is the late Paleozoic Angara–Vitim batholith, which spreads over the Barguzin, Ulan-Burgasy, and Ikat ranges. The main stage of the batholith is represented by calc-alkaline granitoids. The massifs of monzonite–granosyenite composition which crossed the ancient rocks of the Precambrian basement (gneisses, crystalline schists, amphibolites, marbles, and quartzites) were formed in the western and southern parts of the Angara–Vitim batholith. The granitoids of the Angara–Vitim batholith is crossed by large rivers (Selenga, Barguzin, Turka, and Verkhnya Angara) flowing into Lake Baikal (Fig. 1).

This means that the individual provinces of the Baikal area differ in their geological–geomorphic features. The eastern and southern parts of Pribaikalia are part of the Sayan–Baikal–Stanovoe highlands, with characteristic geomorphic contrast forms and the development of glacial morphostructures. Eastern Pribaikalia is part of the Trans Baikalian middle-height relief, with subdued forms and abundant Cenozoic depressions of the "Baikal" type related to the main rivers of the region. Investigations of the bedrocks of the Baikal watershed (Antipin et al. 1997; Petrova et al. 1997; Makrygina et al. 2000) indicate essential distinctions in the mineral and chemical compositions of different parts of these rocks. These peculiarities must be taken into account when examining the probable sources and transfer mechanisms of the sedimentary material to the water area of Lake Baikal.

3 Methods of Studying the Composition of the Suspended Matter in the Snow Cover

Snow from the ice of Lake Baikal was sampled in March 2001 using the following profiles: 1, Goloustnoe–Mishikha (7 samples); 2, Goloustnoe–Posolskoe (9 samples); 3, Rytyi Cape–Turka (17 samples). The samples were collected at 6-km intervals. The location of the profiles is across the direction of the total transfer of air masses at the sites under investigations (from long-term observations, the winds are from the west and northwest; see Fig. 1). The samples were

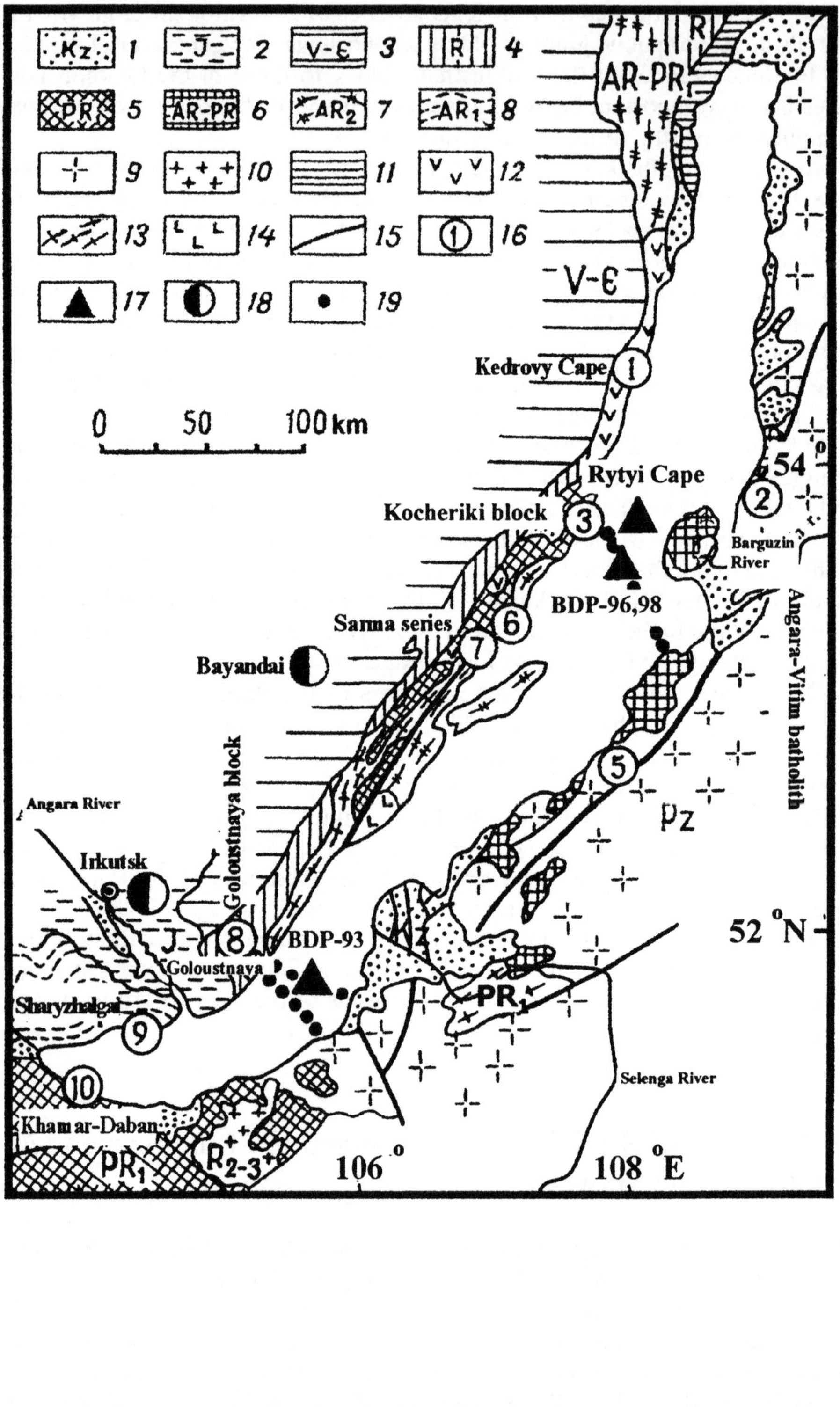
Kz 1
J 2
V-Є 3
R 4
PR 5
AR-PR 6
AR2 7
AR1 8
9
10
11
12
13
14
15
16
17
18
19
0 50 100km
AR-PR
V-Є
Kedrovy Cape
Rytyi Cape
Kocheriki block
Barguzin River
BDP-96,98
Sarma series
Bayandai
Angara-Vitim batholith
Goloustnaya block
Angara River
Irkutsk
Pz
52 °N
BDP-93
Goloustnaya
J
PR1
Sharyzhalgai
Selenga River
Khamar-Daban
PR1
R2-3
106 °
108 °E
54 °

collected and analyzed using standard methods (Koroleva et al. 1998). The SM in the snow cover was analyzed in 50% of the samples collected. The chemical composition of the SM and the water phase were determined by Inductively Coupled Plasma–Mass Spectrometer. The chemical compositions of the rocks of the mountains framing the western and eastern Baikal shores Kedrovyi Cape (acid volcanoes, 1a, 1b); Kocheriki block (gabbro and granites, 3a, 3b); Sarma series (plagioschists and quartzite schists, 6, 7a, 7b, 7c); Goloustanaya block (8); Sharyzhalgai series (9); Angara–Vitim batholith (2a, 2b, 5); Khamar–Daban (granitized metarocks, 10) were taken from literature sources (Petrova et al. 1997; Petrova 2001; Antipin et al. 1997; Makrygina et al. 2000, 2001).

A mathematical theory for recognizing images based on a distance function was used to differentiate the elements in the SM in the snow on the Baikal ice, and to compare them with the suggested sources (the bedrocks of the western and eastern Baikal shores) (Gonzales 1978).

4 Results

The total air transfer at the sites under investigation is from the west and northwest (Galazy 1993). The maximum velocity of the winds during autumn–winter and spring is over 10 m/s. A comparison with earlier results (Vologina and Potemkin 2001) on the eolian transfer in summer from the coast to the lake water indicates that in winter the eolian transfer is stronger. These stronger winter winds are able to raise particles and move them both in the air and over ice with the snow. During the glaciation, the ice cover on Lake Baikal was preserved for a very long time, and the power of the winds was even greater. Therefore, it can

FIG. 1. Scheme showing the geological structure of the Baikal rift zone. *1*, Cenozoic deposits of sands, pebbles, clays, brown coal; *2*, Jurassic sandstones, conglomerates, argillites (Irkutsk Jurassic basin); *3*, Vendian–Cambrian sandstones, gritstones, siltstones, limestones, siliceous rocks; *4*, Riphean sandstones, dolomites, limestones, schists, gritstones, conglomerates, carbonate–siliceous rocks (Goloustnaya, Uluntuy, Cochergat suites); *5*, lower Proterozoic metamorphic complex (Khamar–Daban); *6*, Archean–Proterozoic migmatites (Chuisky series) and crystalline schists, plagiogneisses, metaultrabasic rocks, migmatites, quartzites, carbonate rocks (Olkhon series); *7*, Archean gneisses, crystalline schists, migmatites, marbles, quartzites, amphibolites; *8*, Archean crystalline schists, gneisses, quartzites, marbles, metaultrabasic rocks, migmatites (Sharyzhalgai suites); *9*, Paleozoic granitoids (Angara–Vitim batholith); *10*, Riphean granitoids (Sayan or Khamar–Daban complex); *11*, lower Proterozoic amphibolites and gabbro (Njurundukan series); *12*, lower Proterozoic basalt porphyrites, dacites, rhyodacites, tuffs, rhyolite-porphyries (Akitkan series); *13*, lower Proterozoic rapakivi and rapakivi-like granites (Primorsky complex); *14*, lower Proterozoic gabbro, gabbro–diorites, gabbro-pyroxenites; *15*, deep fault zones; *16*, bedrocks of the Baikal watershed obtained for comparison with SM in the snow cover; *17*, the site of BDP-93, -96, -98; *18*, site of snow sampled in the Bayandai area; *19*, sites of snow sampled on the ice of Lake Baikal

be suggested that in the glaciation, the eolian component in bottom sediments might be more significant (Vologina and Potemkin 2001; Shimaraev et al. 1995). Tables 1–4 give the concentrations of chemical elements in the SM, in the snow, and in the rocks framing the lake (western and eastern shores).

The water phase of the snow contains few minerals, varying from 7.2 to 73.3 mg/l. The greatest mineralization is found in the snow samples from profiles 1 and 2, while the smallest is observed in samples from profile 3. This corresponds to literature data concerning the south and central basins, as aerosol pollution from industrial centers in the Baikal area is brought to the lake basin predominantly from the south. The concentration of most of the elements in the central and northern areas is 1.5–5 times lower. This is because of their remoteness from anthropogenous sources (Obolkin et al. 1994; Khodzher et al. 1997; Vetrov and Kuznetsova 1997). In terms of Sr^{87}/Sr^{86}, the snow waters of the profiles are typically meteoric, and normal in Lake Baikal water. In all samples, the value was 0.708. However, the river waters of the Lake Baikal catchment area (Solzan River, Goloustnaya River, south basin) showed a ratio of 0.711–0.712, i.e., they provide information about the surrounding rocks (e.g., for the Primosrkyi complex, the value of Sr^{87}/Sr^{86} is 0.750–1.626, while for the Khamar–Daban it is 0.708–1.083).

The SM in the snow contains alumosilicate matter, which includes the main rock-forming elements (Si, Al, Fe, Ca, Na, etc.). As compared with the Clarke number of the earth's crust (Voitkevich et al. 1990), the suspension is enriched with Ba, Co, Pb, Cu, Zn and Sn. The contents of the rest of the elements are similar to the Clarke number of the earth's crust, and that is also typical of the Baikal sediments (Wei et al. 1997). The minerals of the bedrocks include quartz and feldspar, and in some cases mica or other Fe–Mg minerals. The size of the grains rarely reaches 1 mm, and frequently these are dust-like particles or agglutinated aggregates. The main SM component is soot-like aggregates of industrial origin. Wool and other fibers are also found. The laminated, sponge-like, columnar emanations of the authigene minerals, which require special diagnostics, were found. When comparing the composition of SM in the snow samples from the southern and northern parts of Lake Baikal, there is clearly greater pollution by soot-like aggregates in the snow cover in the south.

Profiles 1 and 2 are located in the southern basin of Lake Baikal, and thus the SM came from the west (rocks of the Goloustnaya block, 8), the south-east (Khamar–Daban rocks, 10), and the southwest (Sharyzhalgai series rocks, 9) (see Fig. 1). In addition, the industrial centers located in the Angara River valley have a great impact on the composition of the precipitation over the southern basin. The composition of SM in the snow collected in the vicinity of Irkutsk was used in the calculations as one of sources of the eolian substance transfer.

The location of profile 3 suggests that if the wind is from the west or northwest, matter is brought into the profile from the west, i.e., the Kocheriki block rocks (gabbro and granites, 3a, 3b) and the Sarma series (plagioschists and quartz-schists, 6, 7) which occupy the major part of this area. Another direction which supplies eolian material to the Academician Ridge site is the east, which includes the rocks from the Angara–Vitim batholith (granitoids, 2, 5). We did not

TABLE 1. Contents of chemical elements in the suspended matter in the snow collected from profiles 1 and 2 near the Buguldeika Saddle drill site (south basin, Lake Baikal) and Irkutsk (weight %)

Element	Profile 1			Profile 2					Irkutsk
	601 (1 km)[a]	604 (19 km)	607 (37 km)	609 (10 km)	611 (20 km)	613 (32 km)	615 (42 km)	616 (48 km)	
Al	7.1	11.3	11.2	8.2	9.3	9.2	9.5	9.0	15.0
Mg	1.1	1.2	1.1	1.3	1.2	0.97	1.1	1.1	2.0
Ca	1.8	2.2	1.6	2.3	1.7	2.0	2.2	1.7	3.0
Fe	4.2	4.2	4.6	4.4	3.9	4.0	4.0	3.6	6.0
Na	0.74	0.65	0.53	0.69	0.72	0.48	0.57	2.1	2.0
Ba	0.248	0.063	0.058	0.057	0.048	0.046	0.061	0.420	0.15
Sr	0.054	0.042	0.038	0.032	0.027	0.042	0.046	0.052	0.05
Mn	0.053	0.055	0.048	0.073	0.044	0.051	0.051	0.051	0.15
Ti	0.38	0.34	0.38	0.34	0.37	0.4	0.37	0.47	0.4
Co	0.028	0.014	0.015	0.018	0.011	0.012	0.014	0.006	0.006
V	0.0082	0.0134	0.0123	0.0067	0.0098	0.0120	0.0108	0.0049	0.02
Sn	0.0064	0.0042	0.0036	0.0021	0.0029	0.0030	0.0034	0.0014	0.003
Be	0.00058	0.00079	0.00062	0.00052	0.00056	0.00057	0.00061	0.00033	0.0008
Sc	0.0026	0.0034	0.0033	0.0022	0.0024	0.0029	0.0028	0.0029	0.005
Ga	0.0054	0.0076	0.0078	0.0044	0.0056	0.0063	0.0056	0.0033	0.005
Pb	0.132	0.043	0.029	0.041	0.043	0.044	0.040	0.018	0.03
Cu	0.050	0.068	0.041	0.035	0.028	0.036	0.038	0.016	0.03
Zn	0.45	0.06	0.08	0.16	0.07	0.20	0.25	0.06	0.1

[a] Sample number and distance from the mouth of the Goloustnaya River

TABLE 2. Contents of chemical elements in the suspended matter in the snow collected from profile 3 near Academician Ridge drill site (central basin) and the Bayandai settlement (weight %)

Element	630 (2 km)[a]	632 (10 km)	634 (22 km)	629 (36 km)	627 (46 km)	625 (56 km)	623 (68 km)	621 (79 km)	619 (89 km)	618 (95 km)	Bayandai
Al	4.3	2.6	3.5	9.0	7.0	7.5	6.9	8.0	8.0	8.8	15
Mg	0.82	0.55	0.7	1.1	1.2	1.3	1.1	1.6	1.1	1.4	1.5
Ca	1.4	0.79	1.4	2.0	1.9	3.0	2.6	2.2	1.4	1.8	2.0
Fe	2.3	1.2	1.8	3.3	3.7	3.9	3.7	4.5	4.1	4.0	6.0
Na	0.85	0.62	0.46	1.8	1.0	0.79	0.81	0.81	0.68	1.1	2.0
Ba	0.812	0.141	0.106	0.096		0.072	0.151	0.136	0.114	0.314	0.100
Sr	0.046	0.019	0.029	0.035	0.045	0.041	0.035	0.038	0.035	0.040	0.035
Mn	0.03	0.017	0.026	0.046	0.051	0.06	0.05	0.055	0.046	0.05	0.08
Ti	0.22	0.22	0.022	0.3	0.42	0.37	0.4	0.45	0.42	0.41	0.5
Co	0.0122	0.0061	0.0205	0.0096	0.0120	0.0141	0.0119	0.0140	0.0159	0.0115	0.003
V	0.0059	0.0032	0.0049	0.0055	0.0056	0.0060	0.0056	0.0088	0.0053	0.0056	0.015
Sn	0.0037	0.0034	0.0035	0.0025	0.0044	0.0041	0.0029	0.0039	0.0051	0.0023	0.003
Be	0.00032	0.00013	0.00026	0.00037	0.00027	0.00045	0.00032	0.00033	0.00045	0.0030	0.001
Sc	0.0018	0.0014	0.0027	0.0030	0.0019	0.0019	0.0017	0.0019	0.0022	0.0017	0.003
Ga	0.0036	0.0017	0.0028	0.0034	0.0031	0.0041	0.0033	0.0038	0.0063	0.0035	0.003
Pb	0.062	0.034	0.058	0.026	0.079	0.057	0.037	0.049	0.045	0.038	0.030
Cu	0.040	0.027	0.050	0.041	0.046	0.044	0.029	0.025	0.042	0.028	0.020
Zn	0.54	0.13	0.23	0.19	1.44	0.14	0.16	0.15	0.28	0.18	0.10

[a] Sample number and distance from Rytyi Cape

TABLE 3. Concentrations of chemical elements in the shoreline rocks of Lake Baikal near the Buguldeika drill site (weight %)

Element	8[a]	9	10
Si	21.85	30.78	33.92
Al	9.22	8.68	7.84
Mg	5.39	1.44	0.27
Ca	7.42	2.87	0.96
Fe	9.26	4.11	1.51
Na	2.17	2.24	2.05
Ba	0.032	0.091	0.048
Sr	0.030	0.017	0.024
P	0.066	0.052	0.044
B	0.001	0.0007	0.0038
Mn	0.15	0.06	0.03
Ti	0.40	0.28	0.21
Ni	0.020	0.005	0.014
Co	0.005	0.002	0.001
V	0.023	0.010	0.011
Cr	0.051	0.0078	0.033
Sn	0.00012	0.00027	0.00038
Be	0.000095		
Sc	0.0025		
Pb	0.00022	0.0017	0.0012
Cu	0.00012	0.0036	
Zn	0.017	0.0074	0.018

[a] Numbers correspond to the designations of rocks in Fig. 1

exclude the influence of the more remote transport of eolian material from the submontane valley areas of the Ust-Orda and Ol'khon regions. To check this, we analyzed the chemical composition of the SM in snow collected near the Bayandai settlement, which is remote from the industrial centers.

A comparison of the chemical compositions of the SM in the snow and of the rocks did not show any simple pattern to reveal the sources of the winter eolian transfer. Therefore, we used mathematical processing for all geochemical data, which includes a theory for recognizing images based on a distance function, as well as a solution of the optimization problem, when assessing the contribution of potential sources (Gonzales 1978).

4.1 Anomalous Elements

The rocks from the shore line and particles deposited from the atmosphere and brought by the wind, possibly from remote primary sources such as Irkutsk, Bayandai (atmospheric transfer), were considered as potential sources of solid particles arriving at the ice of Lake Baikal.

The main hypothesis about the preservation of elements in the SM all the way from the source to the Baikal ice was applied. In this, the concentrations of all

TABLE 4. Contents of elements in the shoreline rocks of Lake Baikal near Academician Ridge drill site (weight %)

Element	Western shore									Eastern shore			
	1a[a]	1b	3a	3b	7a	7b	7c	6	Average	2a	2b	5	Average
Si	32.81	34.34	23.26	36.13	33.33	38.52	28.36	36.24	32.87	29.84	33.48	33.75	32.36
Al	7.78	7.43	8.68	8.56	6.85	4.8	10.98	6.67	7.72	10.07	8.68	8.67	9.14
Mg	0.29	0.11	3.82	0.65	0.65	0.46	0.9	0.07	0.87	0.89	0.27	0.33	0.5
Ca	1.28	0.76	6.45	1.13	0.86	0.08	0.13	0.43	1.39	2.42	1.05	1.08	1.52
Fe	3.54	2.79	9.049	2.07	4.44	2.61	1.38	1.58	3.43	2.96	1.64	1.51	2.03
Na	2.46	2.63	2.02	3.39	1.6	0.06	0.06	1.98	1.77	3.17	2.76	2.82	2.92
Ba	0.19	0.14	0.046	0.09	0.062	0.033	0.15	0.16	0.11	0.18	0.12	0.12	0.14
Sr	0.015	0.015	0.033	0.023	0.005	0.0042	0.0045	0.0076	0.013	0.1	0.04	0.061	0.07
P	0.065	0.026	0.052	0.026	0.043	0.026	0.017	0.013	0.034	0.109	0.043		0.08
B	0.00056	0.00074	0.00088	0.0021	0.0088	0.00086	0.015	0.0018	0.0038				
Mn	0.046	0.031	0.145	0.031	0.054	0.015	0.015	0.023	0.045	0.09	0.046		0.068
Ti	0.31	0.21	0.58	0.15	0.35	0.18	0.41	0.11	0.29	0.32	0.15	0.13	0.2
Ni	0.0006	0.00064	0.0011	0.0025	0.0058	0.0021	0.0013	0.00032	0.0018	0.00006	0.00048	0.0011	0.00073
Co	0.00075	0.00034	0.004	0.0026	0.0014	0.00078	0.00037	0.00024	0.0013	0.00092	0.00044	0.0004	0.00059
V	0.0033	0.00049	0.048	0.0058	0.018	0.045	0.017	0.00023	0.017	0.0074	0.0025	0.0015	0.0038
Cr	0.0014	0.00098	0.016	0.0037	0.013	0.0044	0.017	0.00091	0.007	0.00089	0.00075		0.00082
Sn	0.003	0.00063			0.00048	0.00034	0.00066	0.00055	0.00094	0.00096	0.00048	0.00017	0.00054
Be	0.00016	0.00022			0.00017	0.00013	0.00054	0.00036	0.00026	0.0003	0.00018	0.00029	0.00026
Sc	0.0015	0.00089	0.007	0.0007	0.0024	0.00053	0.0022	0.0012	0.0021	0.00092	0.00035		0.00064
Pb	0.0033	0.0046	0.00086	0.0026	0.002	0.0028	0.003	0.0034	0.0028	0.0032	0.0065	0.0022	0.004
Cu	0.0036	0.0048	0.013	0.0028	0.005	0.0023	0.0023	0.001	0.0044	0.0012	0.0017	0.0018	0.0016
Zn	0.0073	0.017	0.013	0.0052	0.0086	0.007	0.0065	0.0046	0.0086	0.0098	0.0058		0.0078
Ag					0.000011	0.000003	0.000023		0.000012				

[a] Numbers correspond to the designations of rocks in Fig. 1

the elements in the SM lie within a range of variations (from a_i^{min} to a_i^{max}) in all sources. The minimum and maximum values for element i were found to lie within $d_i = a_i^{max} - a_i^{min}$.

A comparison of the ranges of all possible variations in all potential sources with data from a determination of the real element concentrations in the SM in snow collected from profiles 1–3 on the Baikal ice showed anomalies for 9 elements: for three elements (Al, Fe, Ga), the samples had a lower concentration than the minimum in the source (Academician Ridge); for seven elements (Ba, Co, Sn, Ga, Pb, Cu, Zn), the samples had a higher content than the maximum in the source. The data on these anomalous elements are given in Table 5. The number of samples with anomalies out of the total number of samples is given for each element. For example, the entry 5/8 in the 3rd column indicates that 5 out of 8 samples collected from the Buguldeika Saddle have a concentration higher than the maximum in the source, while the entry 3/10 in the 4th column means that an Al content lower than the minimum in the source was observed in 3 out of 10 samples taken from Academician Ridge.

Anomalies with concentrations lower than the minimum in the source were found for the Academician Ridge site only (profile 3). This could have resulted from features of the mineral composition of the SM in the snow which was collected close to the western shore of the lake (Rytyi Cape), where anomaly minima were observed. The absence of similar anomalies near the Buguldeika Saddle could be related to the strong influence of anthropogenic pollution in the southern basin.

Some anomalous elements significantly exceeded the maxima in the sources (e.g., the Zn content can exceed the average content in the earth's crust at times). The distribution of such samples is similar in the Buguldeika Saddle and Academician Ridge (Table 5), which shows that there is a single reason for these anomalies. However, finding an explanation of all the anomalies is a separate task. We note only that the increased content of some elements in the solid phase of the Baikal snow most likely results from anthropogenic pollution, and it is possible to estimate this contribution to the composition of the SM in the snow.

TABLE 5. Anomalous elements

Element	Buguldeika Saddle		Academician Ridge	
	<min	>max	<min	>max
Al			3/10	
Fe			1/10	
Ba		2/8		2/10
Co		8/8		10/10
Sn		5/8		7/10
Ga		6/8	2/10	1/10
Pb		6/8		9/10
Cu		6/8		6/10
Zn		4/8		10/10

As the anomalous elements do not correspond with the main hypothesis of element composition preservation, they were not used to determine the geochemical distance between samples (see below).

4.2 Geochemical Distance (GD) between Samples

The content (x_i) of element i was determined in sample (x). In sample (y), the content of the same element is (y_i), and the range of variations of element concentrations from all potential sources in the mixtures used to prepare both samples was d_i. Thus, the value $\rho_i(x, y) = |x_i - y_i|/d_i$ between samples x and y is called GD $\rho_i(x, y)$ from element i. It is obvious that the inequality $0 \leq \rho_i(x, y) \leq 1$ will always be true for elements which are not anomalous. Moreover, the GD between samples for any given element is symmetrical. It is equal to zero only when the concentrations in both samples coincide, and it reaches the maximum value (equal to 1) only in the case when the element concentrations in different samples lie on the limits of the range of possible variation for all sources.

The value $\rho(x, y)$ is

$$GD\,\rho\,(x, y) = \frac{1}{n} \cdot \sum_{i=1}^{n} \rho_i\ (x, y)$$

for all n elements between samples x and y. It is presumed that only GDs from anomalous elements are summed. From these values, the GD shows how two samples differ from each other in terms of element composition. It should be noted that only items within the significance limits, as selected by numerical experiments, were considered for the calculations. This was done in order to reduce the influence of incidental factors (possible errors in calculation, misprints, etc.).

The GD values for the composition of the mountains framing the southern basin and the SM in the snow from Irkutsk were calculated for eight samples from profiles 1 and 2 (the Buguldeika Saddle). The results are given in Table 6 and Fig. 2. These show that the rocks of Sharyzhalgai and Khamar–Daban, and those which are more remote (Irkutsk), have a greater impact on the composition of the SM in the snow than rocks from the Goloustnaya block. The insig-

TABLE 6. Geochemical distance (GD) to sources (Buguldeika Saddle)

GD	Sampling sites							
	Profile 1			Profile 2				
	601	604	607	609	611	613	615	616
To Goloustnaya	0.516	0.534	0.484	0.454	0.357	0.407	0.489	0.470
To Sharyzhalgai	0.182	0.272	0.320	0.233	0.295	0.257	0.147	0.304
To Khamar–Daban	0.196	0.337	0.281	0.315	0.260	0.196	0.302	0.293
To Irkutsk	0.339	0.371	0.349	0.354	0.326	0.316	0.349	0.414

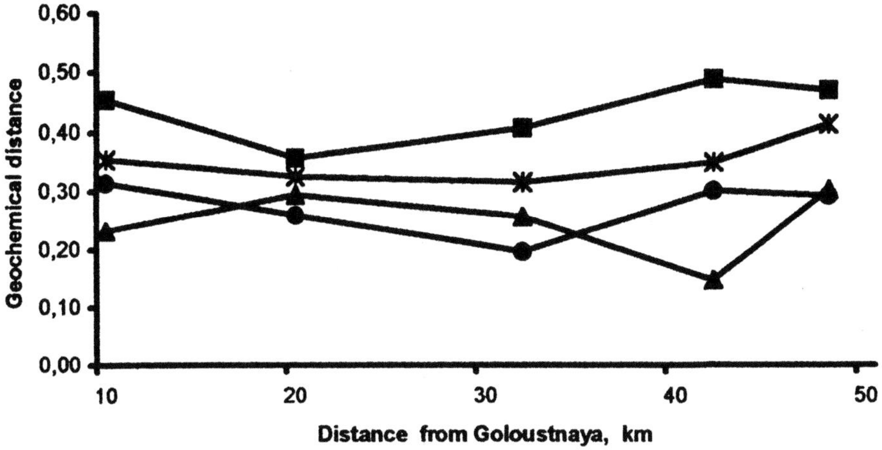

FIG. 2. Geochemical distance (GD) along profile 2 (Goloustnoe–Posolskoe). *Squares*, Goloustnaya; *triangles*, Sharyzhalgai; *circles*, Khamar–Daban; *stars*, Irkutsk

nificant range of GD values along the profiles results from the local winds, e.g., the southeastern "Kultuk" (Galazy 1993). When the western and northwestern winds are predominant, the location of profile 1 from north to south results in a smaller supply of rock particles from the Goloustnaya block to the samples from profile 1 than to those from profile 2, which runs from west to east. The industrial centers of the Priangaria influence the composition of the SM in snow from both profiles almost equally.

The GD values of the average composition of the mountains framing the west and east, and the SM in the snow from Bayandai, were calculated for ten samples from profile 3 (Academician Ridge). The GD of rocks of the Kocheriki block and the Sarma series were calculated after considering the location of profile 3 and the direction of local winds. The results are given in Table 7 and Figs. 3 and 4. Figures 3 and 4 show a similar GD to sources in the middle of the profile (37 km) in the eastern direction. It is most likely that the results indicate intensive mixing of the material in the SM in the snow, owing to the influence of the eastern wind "Barguzin", which becomes stronger in winter. The GD of the western shore is less than that of the eastern shore, which shows the stronger transport of substances by westerly winds. Closer to Rytyi Cape (on the Western shore), remote transfer from Bayandai is screened by the Primorsky Range (up to 2000 m above sea level). This is confirmed by the corresponding increase in the GD to this source.

The location of profile 3, from north to south, leads to a different influence from rocks of the Kocheriki block and Sarma series, as substance transfer by westerly and northwesterly winds is predominant (Fig. 4). In the whole profile, the influence of rocks from the Sarma series is much greater than that of rocks from the Kocheriki block. Therefore, the GD of SM from the Sarma rock series is lower than that of SM from rocks of the Kocheriki block. The impact of the

TABLE 7. GD to sources (Academician Ridge)

GD	Sampling sites on profile 3									
	630	632	634	629	627	625	623	621	619	618
To western shore	0.160	0.209	0.172	0.085	0.119	0.148	0.126	0.140	0.133	0.122
To eastern shore	0.192	0.263	0.236	0.166	0.199	0.224	0.220	0.236	0.237	0.188
To Bayandai	0.385	0.476	0.406	0.240	0.279	0.270	0.287	0.238	0.256	0.254

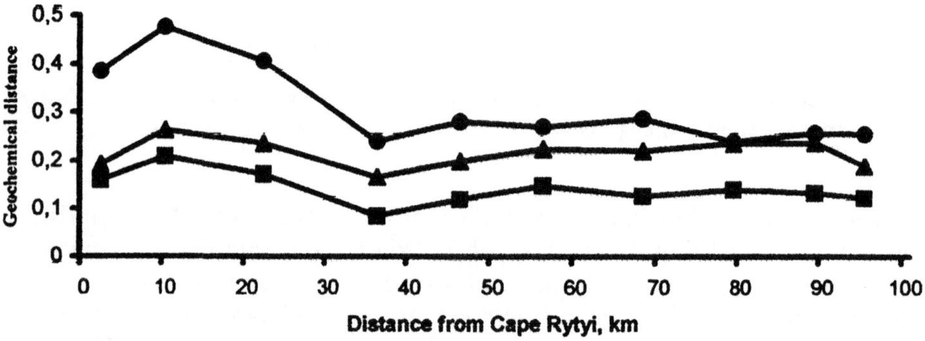

FIG. 3. GD along profile 3 (Rytyi Cape–Turka). *Squares*, to the average composition on the western coast; *triangles*, to the average composition on the eastern coast; *circles*, to the composition on Bayandai

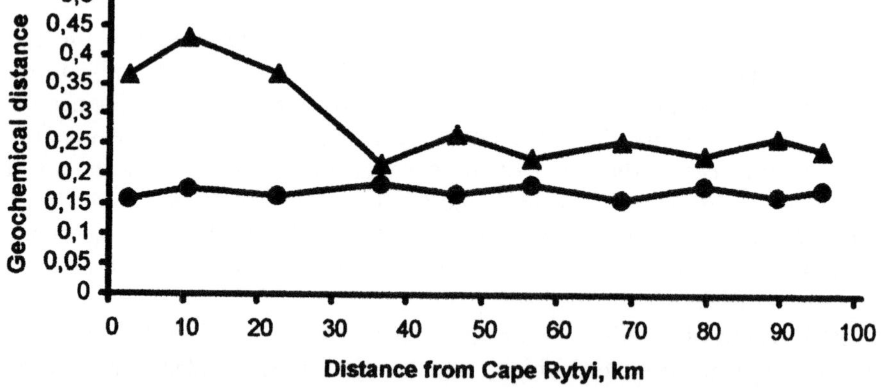

FIG. 4. GD along profile 3 (from Kocheriky; from Sarma). *Triangles*, to the average composition on the Kocheriky; *circles*, to the average composition on the Sarma

eastern wind “Barguzin” starts from 37 km (st. 627). The values of the GD do not differ significantly from each other.

5 Share of Shore and Anthropogenic Sources

We now consider the minimization of a function (GD or its analog) to determine the contribution of each source into the SM in the snow on the lake ice. This function quantitatively expresses the degree of similarity of mixture, and is compiled from potential source material and the real composition of the SM. Let us assume a source m for the profile, along which k samples of SM were collected. The composition of n elements was determined, and the composition of all these elements is known from all sources. x_i^j is the concentration of element i in sample j of the profile, and y_i^s is the concentration of the same element in source s. The minimum, maximum, and average values of the content each element in the profile are to be determined.

$$X_i^{\min} = \min_{j=1,k} x_i^j, \qquad X_i^{\max} = \max_{j=1,k} x_i^j, \qquad X_i^0 = \frac{1}{k} \cdot \sum_{j=1}^{k} x_i^j$$

The sum of the positive numbers $c_1, \ldots, c_m$, is 1 (contributions from each source in the mixture), while the content of element i in the admixture will be

$$Y_i = \sum_{s=1}^{m} c_s y_s$$

We consider the function

$$F(c) = \frac{1}{n} \cdot \sum_{i=1}^{n} \frac{|Y_i(c) - X_i^0|}{X_i^0}$$

as a measure of the similarity of the elemental composition of the synthesized mixture and all samples of SM from the given profile. We determine the contribution of each source as follows: select nonnegative numbers $c_1, \ldots, c_m$ which sum to 1 in such a way that the composition of the mixture obtained from all elements will lie within the variation along the profile, i.e., $X_i^{\min} \leq Y_i(c) \leq X_i^{\max}$ and function $F(c)$ will acquire its lowest values.

This task was solved for the Buguldeika Saddle (profiles 1 and 2) and Academician Ridge (profile 3), and the results are given in Figs. 5 and 6. The figures for each element on the ordinate axis are calculated as $(Y_i(c) - X_i^0)/X_i^0$. These are the relative deviations of the element concentration in the synthesized mixture from the average value of the concentration in all samples from the Baikal area.

The contribution from the Golustanaya block, the Sharyzhalgai series, and Khamar–Daban, and remote transfers to the Buguldeika Saddle make up 10%, 10%, 15%, and 65%, respectively (Fig. 5). For such values, the synthesized

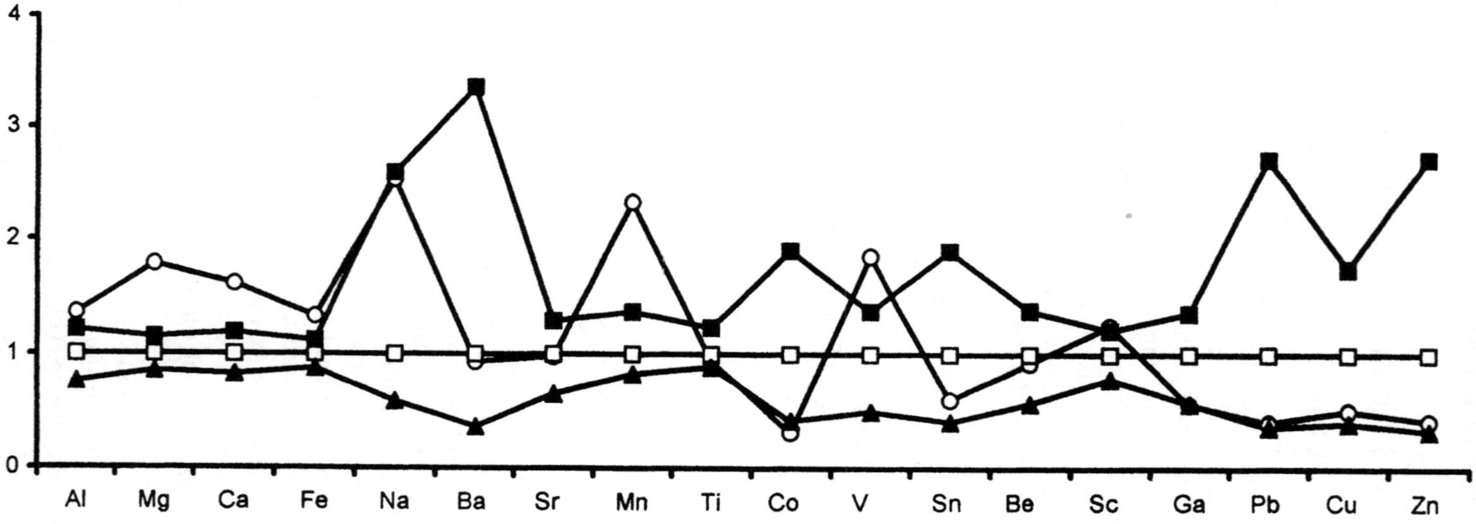

FIG. 5. Variations of suspended matter composition in the snow of profiles 1 and 2 (Buguldeika Saddle), and composition of the mixture obtained from the potential sources of supply (Goloustnaya : Sharyzhalgai : Khamar–Daban : Irkutsk = 0.1 : 0.1 : 0.15 : 0.65). *Open circles*, mixture; *open squares*, average; *solid triangles*, minimum; *solid squares*, maximum

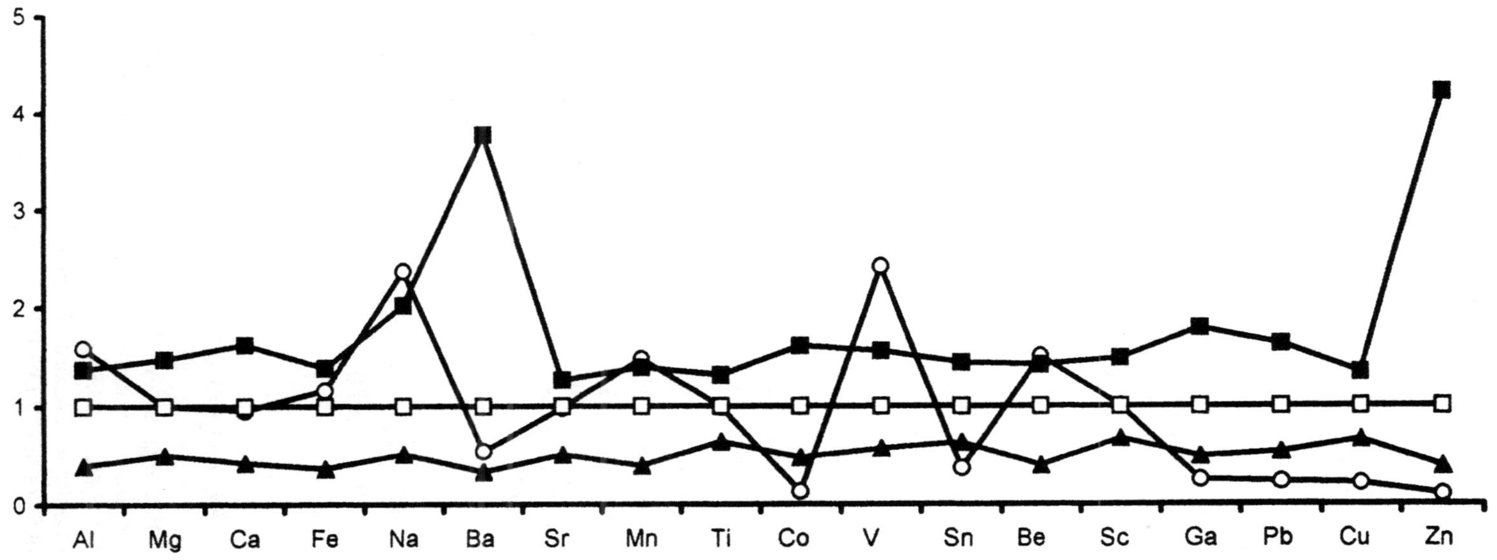

FIG. 6. Variations of the suspended matter composition in the snow of profile 3 (Academician Ridge), and composition of the mixture obtained from the potential sources of supply (west : east : Bayandai = 0.45 : 0.25 : 0.30). *Open circles*, mixture; *open squares*, average; *solid triangles*, minimum; *solid squares*, maximum

mixtures from the rock-forming elements Al, Mg, Ca, Fe, and Mn lie outside the range of variations along profiles 1 and 2. This is because of the marked anthropogenous pollution from anomalous elements (Co, Sn, Pb, Cu, and Zn) in the region. When solving the minimization task, the *F* criterion depends more on these than on the rock-forming elements.

The contribution of rocks from the western and eastern shores, and remote transfer to Academician Ridge amounts to 45%, 25%, and 30%, respectively (Fig. 6). The figure shows that the anomalous elements (Co, Sn, Pb, Cu, and Zn) lie outside the variation range along profile 3. The lack of agreement from these elements could result from the presence of anthropogenic pollution.

The slight increase in the contribution of rocks from the eastern shore (unlike the results of Koroleva et al. 2001) from 15% to 25% is most likely due to the different locations of the profiles studied and the influence of local winds (Barguzin).

6 Conclusions

1. The analysis of "anomalous" elements showed that those in excess of the maxima for these elements indicate the impact of industrial outbursts on the composition of SM in the snow on the Baikal ice.
2. The composition of SM in the snow showed that it was derived from the shore complexes and by atmospheric transfer from remote areas, depending on the distance of the sampling site from the shore and the impact of predominant and local winds.
3. Solving the optimization allowed an estimate of the contribution of each potential source: for profiles 1 and 2 (Buguldeika Saddle), the contribution of the Goloustanya block, the Sharyzhalgai series, Khamar–Daban, and the industrial centers of the Priangaria amounted to 10%, 10%, 15%, and 65%, respectively; for profile 3 (Academician Ridge), the contribution of the western and eastern shores and remote transfer (Bayandai) made up 45%, 25%, and 30%, respectively.

Acknowledgments. This research was supported through grants from RFBR-Baikal (projects 01-05-97251 and 01-01-97201).

References

Agaphonov BP (1990) Exolitodynamics of the Baikal rift zone. Nauka, Novosibirsk, p 176

Antipin VS, Tomilov BV, Goreglyad AV, Kovalenko VI, Budnikov SV (1997) Chemical composition of the Angara–Vitim batholith as a source of sediments in the eastern part of Lake Baikal. IPPCCE Newsl 10:5–10

BDP Members (1998) Continuous record of climatic changes in Baikal sediments for the last 5 Ma. Russ Geol Geophys 39:139–156

Galazy GI (ed) (1993) Baikal atlas. Federal Service of the Geodesy and Cartography, Moscow, p 160

Gonzales P (1978) Principles of image recognition. Mir, Moscow, p 411

Grachev MA, Likhoshwai EB, Vorob'eva SS, Khlystov OM, Bezrukova EV, Veinberg EV, Goldberg EL, Granina LZ, Kornakova EG, Lazo FI, Levina OV, Letunova PP, Otinov PV, Pirog VV, Fedotov AP, Yaskevich SA, Bobrov VA, Sukhorukov FV, Rezchikov VI, Fedorin MA, Zolotarev KV, Kravchinskii VA (1997) Signals of paleoclimates of the upper Pleistocene in Baikal sediments. Russ Geol Geophys 38:957–980

Khodzher TV, Obolkin VA, PotemkinVI (1997) Seasonal variation of the element composition of the atmospheric aerosols over Lake Baikal. Chem Interests Sustainable Dev 5:547–551

Koroleva GP, Gorshkov AG, Vinogradova TP (1998) Studies of the pollution of the snow cover as the deponent media (south Pribaikalia). Chem Interests Sustainable Dev 6:327–337

Koroleva GP, Kosov AA, Geletii VF, Vologina EG (2001) Geochemical characteristics of the eolian material on Academician Ridge, Lake Baikal. Russ Geol Geophys 42:254–257

Makrygina VA, Petrova ZI, Gantimurova TP (2000) Comparative geochemistry of rocks of the Sarma series and metasomatites of the zone of the Primorskyi fault (western Pribaikalia). Geochemistry 2:131–146

Makrygina VA, Gantimurova TP, Petrova ZI (2001) Correlation of the main magmatism in the lithosphere blocks of the western and north Pribaikalia (from geochemical data). Russ Geol Geophys 41:414–427

Obolkin VA, Potemkin VL, Khodzher TV (1994) Elementary composition and main sources of the atmosphere aerosol of the south Baikal. Geogr Nat resource 4:75–81

Petrova ZI (2001) Geochemistry of rocks of the Goloustnaya in the south framing the Siberian platform. Geochemistry 6:593–606

Petrova ZI, Makrygina VA, Antipin VS (1997) Petrologic geochemical correlation of the rapakivi granites and acid volcanics in the south framing the Siberian platform. Petrology 5:291–311

Potemkin VL, Khodzher TV (1987) Correlation analysis and spatial changeability of the supply of substances from the atmosphere into the Lake Baikal region. Regional monitoring of Lake Baikal. Gogidrometeoizdat, Leningrad, p 63–71

Shimaraev MN, Granin NG, Kuimova LN (1995) Experience of reconstructing the geophysical conditions in Lake Baikal in the late Pleistocene and Holocene. Russ Geol Geophys 36:97–102

Taisaev TT (1982) Eolian processes in the Priolkhonie and Ol'khon Island (western Pribaikalia). Dokl AN SSSR 265:948–951

Vetrov VA, Kuznetsova AI (1997) Microelements in natural environments of the Lake Baikal region. Publishing House of SB RAS, OIGGM SB RAS, Novosibirsk, 236 p

Vetrov VA, Mel'nikova RD, Kabirova GM (1983) Neutron-activation analysis of the snow cover in the background regions of Lake Baikal. Semiconductive spectrometry in geology and geochemistry. Nauka, Novosibirsk, p 110–119

Voitkevich GV, Kokin AV, Miroshnichenko AE, Prokhorov VG (1990) Reference book on geochemistry. Nedra, Moscow, p 480

Vologina EG, Potemkin VL (2001) Characteristics of the eolian transfer in winter at Academician Ridge and its sources (from the data of the snow cover survey). Russ Geol Geophys 42:258–266

Wei R, Ikeda K, Sawatari H, Haraguchi H (1997) Concentrations of the 100-m sediment core samples (BDP-93-2) from Lake Baikal as determined by ICP-AES and ICP-MS. IPPCCE Newsl 10:134–144

Key-word Index

GPSR Compliance

The European Union's (EU) General Product Safety Regulation (GPSR) is a set of rules that requires consumer products to be safe and our obligations to ensure this.

If you have any concerns about our products, you can contact us on ProductSafety@springernature.com

In case Publisher is established outside the EU, the EU authorized representative is:

Springer Nature Customer Service Center GmbH
Europaplatz 3
69115 Heidelberg, Germany

Batch number: 10682207

Printed by Printforce, the Netherlands